Statistics for Biology and Health

Series
M. Gail
K. Krickeberg
J. Samet
A. Tsiatis
W. Wong

For other titles published in this series, go to
http://www.springer.com/series/2848

Karl W. Broman · Śaunak Sen

A Guide to QTL Mapping with R/qtl

Springer

Karl W. Broman
Department of Biostatistics
& Medical Informatics
University of Wisconsin–Madison
1300 University Ave.
Madison, WI 53706-1510
USA
kbroman@biostat.wisc.edu

Śaunak Sen
Department of Epidemiology & Biostatistics
University of California, San Francisco
185 Berry St., Suite 5700
San Francisco, CA 94107-1762
USA
sen@biostat.ucsf.edu

Portions of the authors' articles published in *Genetics* are reprinted with permission of the Genetics Society of America.

Linux® is a registered trademark of Linus Torvalds.

Google™ and Google Groups™ are trademarks of Google Inc.

Microsoft®, Windows®, and Excel® are registered trademarks of Microsoft Corporation.

Mac®, Mac OS®, and Macintosh® are registered trademarks of Apple Computer, Inc.

UNIX® is a registered trademark of The Open Group.

ISSN 1431-8776
ISBN 978-0-387-92124-2 e-ISBN 978-0-387-92125-9
DOI 10.1007/978-0-387-92125-9
Springer Dordrecht Heidelberg London New York

Library of Congress Control Number: 2009929238

Printed on acid-free paper

Springer is part of Springer Science+Business Media (www.springer.com)

To Aimee and Suheeta

Preface

QTL are quantitative trait loci: genetic loci that contribute to variation in a quantitative trait. QTL mapping is the effort to identify QTL through an experimental cross.

In this book, we give an overview of the practical aspects of the analysis of QTL mapping experiments based on inbred line crosses, with explicit instructions on the use of the R/qtl software (an add-on package for the general statistical software, R). We give some of the details of the statistical methods, but we mostly focus on how to get and make sense of results. Real data examples are included throughout.

The intended audience includes scientists who are performing QTL mapping experiments and participating directly in the analysis. We expect the reader to have a general understanding of statistical methods, including maximum likelihood estimation and linear regression. Some readers will be statisticians analyzing data from QTL experiments with a basic understanding of genetics. We provide limited introduction to either statistics or genetics. Readers with a limited understanding of statistics may wish to first study Rice (2006). Readers with a limited understanding of genetics may wish to first study Brown (2006). Alternatively, one might consider *The Cartoon Guide to Statistics* (Gonick and Smith, 1993) and *The Cartoon Guide to Genetics* (Gonick and Wheelis, 1991), which are more gentle and entertaining (but less complete) introductions to the subjects.

In line with our aim to describe the practical aspects of QTL mapping, the book contains extensive discussion of the R/qtl software. We have attempted to separate the discussion of R/qtl into subsections, so that readers who wish to focus on the basic ideas and skip over the software considerations may do so. In some places (e.g., Chap. 3, on data diagnostics), this was not feasible.

While much can be accomplished with R/qtl (and much of this book may be read) with a limited understanding of R, efficient use of the software (and an understanding of more complex R/qtl code) requires a more detailed understanding of R. We provide very little discussion of R itself, and refer the reader to Dalgaard (2002), for a gentle introduction to R, and Venables and Ripley (2002), for a more comprehensive discussion of R.

The content of the book is ordered according to the way in which QTL analyses might proceed. (There is one exception: we postpone the discussion of experimental design to Chap. 6, as it requires a reasonably complete understanding of QTL mapping.) We begin with an introduction (Chap.1), including an overview of the structure of data from a QTL mapping experiment and the basic statistical problems. In Chap. 2, we explain how to import QTL mapping data into R/qtl, we describe some of the example data sets that will be considered further in later chapters, and we demonstrate how one may simulate QTL mapping data in R/qtl. At the end of the chapter, we describe the internal structure of QTL mapping data within R/qtl; this section should probably be skipped at first reading. In Chap. 3, we describe the various diagnostic procedures for assessing the quality and integrity of QTL mapping data.

Chapter 4 is the heart of the book. There, we discuss the basic approach to QTL mapping (interval mapping), the assessment of statistical significance in a genome scan, and the calculation of confidence intervals for QTL location. We focus on the case that residual variation in the phenotype follows a normal distribution. In Chap. 5, we consider several extensions of standard interval mapping for non-normal phenotypes.

In Chap. 6, we describe various experimental design issues, including the choice of cross, marker density, and sample size, and selective genotyping strategies. We consider both the power to detect a QTL and the precision of localization of QTL. We focus on the use of the R/qtlDesign software (another add-on package for R), but also describe how one may estimate power and precision through computer simulation with R/qtl.

In Chap. 7, we describe the use of covariates in QTL mapping. We initially consider the inclusion of additive covariates (in which the effect of the QTL is constant, independent of the value of the covariate), but we also discuss the investigation of QTL × covariate interactions. We conclude the chapter with a discussion of composite interval mapping (CIM), in which genetic markers are included as covariates.

The first seven chapters focus almost exclusively on single-QTL models. In Chap. 8, we take the first step towards multiple-QTL models by considering two-dimensional, two-QTL genome scans. Such two-dimensional scans offer the opportunity to assess evidence for linked or interacting QTL. In Chap. 9, we provide a more comprehensive discussion of the identification and exploration of multiple-QTL models. The problem is viewed as one of model selection in multiple linear regression, though with a number of special features.

We conclude the book with two case studies (Chap. 10 and 11), in order to illustrate the entirety of the process of mapping QTL. We bring together all of the tools discussed in the previous chapters to demonstrate their combined use in order to solve two moderately difficult problems.

The book has been written with a variety of possible readers in mind, including experienced QTL mappers interested in adopting the R/qtl software, postdoctoral researchers new to QTL mapping, and statistics graduate students interested in exploring applications of statistics. We do not expect that the book will be often read front-to-back in a linear fashion, and different readers will likely wish to approach the book differently.

The experienced QTL mapper might start with Chap. 2, on importing QTL mapping data sets, but would then likely skip about, making liberal use of the Contents and Index to identify sections of particular interest. The reader new to QTL mapping should start with the Introduction (Chap. 1), but might skip Chap. 2 and 3 at first reading and jump right into Chap. 4, in which the essentials of QTL mapping are described.

We have created a web site with on-line complements for the book (see `http://www.rqtl.org/book`). Included on that site are files with all of the R code used in the book, including the detailed code used to create the figures. We have also created an R package, R/qtlbook, containing all of our example data sets (except those already included in R/qtl).

We thank Victor Boyartchuk, Bill Dietrich, Mehmet Guler, Krista Nichols, Virginie Orgogozo, Sarah Owens, Bev Paigen, Karlyne Reilly, Noel Rose, Andy Smith, Michelle Southard-Smith, and Gary Thorgaard for providing data and for allowing its distribution. The public distribution of data is invaluable for statistical genetic methods development, and for learning. We further thank Aimee Teo Broman, Ken Manly, Krista Nichols, Virginie Orgogozo, Abraham Palmer, and several anonymous reviewers for suggestions to improve the book, and Sungjin Kim for identifying a number of typographical errors. Our ideas on QTL mapping were greatly influenced by Gary Churchill, Mark Neff, and Terry Speed; we thank them for many years of stimulating discussions. Our efforts were supported, in part, by NIH grants R01-GM074244 and R01-GM078338.

The book was created using R version 2.8.1, R/qtl version 1.11-12, R/qtlDesign version 0.92, and R/qtlbook version 0.16-3. Later versions of these software may have some minor differences; important changes will be described in the on-line complements (`http://www.rqtl.org/book`). The book was constructed with LaTeX and Sweave; we don't know how we could have done it otherwise. We thank the developers of R, LaTeX, and Sweave for making this work possible.

Madison, Wisconsin; San Francisco, California
June, 2009

Karl W. Broman
Śaunak Sen

Contents

1

Introduction

Many phenotypes (traits) of biomedical, agricultural, or evolutionary importance are quantitative in nature. Examples include blood pressure (to study hypertension), milk output (in dairy breeding), and number of seeds produced per plant (to study evolutionary fitness). Many phenotypes such as coat color of mice, or cancer tumor aggressiveness, may not be strictly quantitative, but may be studied by a derived quantitative measure. We may classify mice by whether or not they have an agouti coat color, a 0/1 measure, or grade tumors by aggressiveness on a scale of 1 to 4 by examining tumor biopsies.

Variation in such quantitative traits is often due to the effects of multiple genetic loci as well as environmental factors. Knowledge of the number, locations, effects, and identities of such genetic loci (called quantitative trait loci, QTL) can lead to new biological insights. The information from QTL can be used to develop new therapeutic drugs, assist the selection for improved agricultural crops or breeds, and improve our understanding of natural selection.

Studies to identify, or "map," QTL may be undertaken in humans or in nonhuman species including model organisms such as mice or Drosophila (fruit flies). In these studies, we assemble or create a genetically diverse population. Then we associate genetic variation with phenotypic variation in the study population. Regions of the genome that show convincing evidence of association are flagged as QTL.

In this book, we consider the problem of mapping QTL in an experimental cross formed from two inbred lines. Such crosses are the simplest populations in which we can perform QTL mapping. They are the easiest to understand biologically, as well as mathematically. For this reason, they are the staples of experimental geneticists, and the most common QTL study population. QTL mapping in more complex populations, including in humans, may be viewed as generalizations. Thus, for both biologists and quantitative methodologists, understanding QTL mapping in experimental crosses is an excellent launching point for more ambitious investigations.

We focus on QTL mapping with the R/qtl software, an add-on package for the R statistical software, for the analysis of QTL experiments. In the following

K.W. Broman, Ś. Sen, *A Guide to QTL Mapping with R/qtl*,
Statistics for Biology and Health, DOI 10.1007/978-0-387-92125-9_1,

three sections, we elaborate on the idea of a QTL experiment, describe the basic crosses and data, and introduce the central statistical problems in QTL mapping. Subsequently, we describe R and R/qtl, as well as some of the alternative programs for QTL mapping. We conclude the chapter with a brief description of the general work flow of QTL data analysis.

1.1 Why perform a QTL experiment?

As mentioned above, the fundamental idea underlying QTL mapping is to associate genotype and phenotype in a population exhibiting genetic variation. Conceptually, the most straightforward study would be in a natural population of the organism of interest. For example, to understand hypertension, we may study genetic associations with hypertension in a large cohort such as the Nurses Health Study (London *et al.*, 1989). While extremely useful, this approach presents a few problems. First, human studies are expensive; second, the phenotypic characterization may be noisy because we cannot control the subjects' environment and life history; and finally, associations do not necessarily imply causation because of possible confounding due to population structure.

QTL mapping in experimental crosses provides an excellent alternative. By suitably choosing a model organism we can home in a particular aspect of the phenotype of interest. For example, we may study hypertension in mice, or alcohol addition in Drosophila, by appealing to the evolutionary connection between humans, mice, and Drosophila. We have greater control over the phenotyping accuracy, environment, and life history than in natural populations. For example, we can feed all mice the same diet, and keep the mouse rooms climate-controlled, and phenotype the mice at the same day in identical experimental conditions. We can perform phenotyping that may be invasive, impractical, or unethical in humans (such as examining the liver in mice on a high-fat diet). We also have greater control over the genetic composition of the population in experimental crosses. This allows us to magnify the genetic effect of a putative QTL by judiciously choosing the strains to cross. By crossing two (or more) strains, we also effectively randomize genetic variation in the progeny population. This allows us to conclude that genetic associations detected in a cross are causal. Because experimental crosses segregate genetic variation that is simpler relative to a natural population, we can perform more complex statistical modeling to identify epistasis (QTL $\times$ QTL interactions), and QTL $\times$ environment interactions.

Experimental crosses do have some disadvantages. If the cross is in a model organism, the degree to which the conclusions apply to the target organism depends on how good the model is. For example, if we use Arabidopsis to study drought tolerance, the conclusions may not be directly applicable to grapes, although there is a good chance that the pathways implicated in Arabidopsis are also relevant to grapes. Identification of a QTL does not necessarily help us identify a gene, since the region spanned by a QTL may contain tens, and sometimes thousands, of genes. A QTL may not be successfully replicated in a congenic strain (identical to one strain at all but the QTL region derived from a different strain) because of epistatic interactions with the genetic background. Developing statistical and experimental solutions to these shortcomings is an active research area.

QTL mapping in experimental crosses is an excellent first step towards more expensive investigations. The simple genetic structure of experimental crosses provides a relatively tractable framework to study the conceptual and statistical principles of genetic mapping.

1.2 Crosses and data

We focus on experimental crosses between inbred lines. An inbred line is formed by repeated sibling mating (or, in many plants, selfing) to obtain individuals who are completely homozygous: both chromosomes are identical at all positions. Inbred lines are essentially "immortal" since all individuals in an inbred line are genetically identical to one another, and to their progeny (apart from sex).

If two inbred lines show a consistent phenotypic difference, despite being raised in a common environment, one may be confident that the strain difference has a genetic basis. The identity of QTL underlying such phenotypic differences may be revealed analyzing crosses between the strains. By crossing the strains, we obtain progeny whose genomes are shuffled versions of the parental genomes.

The simplest cross to describe is the backcross (Fig. 1.1). Two inbred strains, denoted A and B, are crossed to obtain the first filial (F_1) generation. F_1 individuals receive a copy of each chromosome from each of the two parental strains; wherever the parental strains differ, the F_1 generation is heterozygous. The F_1 individuals are crossed to one of the two parental strains. For example, if an F_1 individual is crossed with its A strain parent, the backcross progeny receive one chromosome from the A strain and one from the F_1. Thus, at each autosomal locus, they have genotype AA or AB. The chromosome received from the F_1 parent may be one of the original parental chromosomes intact but is generally a mosaic of the two parental chromosomes, as a result of recombination at meiosis (the process of cell division that gives rise to sex cells). The points of exchange are called crossovers.

While the backcross is the simplest possible experimental cross, the intercross (Fig. 1.2) is also commonly used. In an intercross, one crosses F_1 siblings (or, in many plants, one may self an F_1 individual) to obtain the F_2 generation, who receive a recombinant chromosome from each parent and so, at any autosomal locus, have genotype either AA, AB, or BB. The intercross allows the detection of QTL for which one allele is dominant, while in a backcross to the A strain, one may detect a QTL only if the A allele is not dominant.

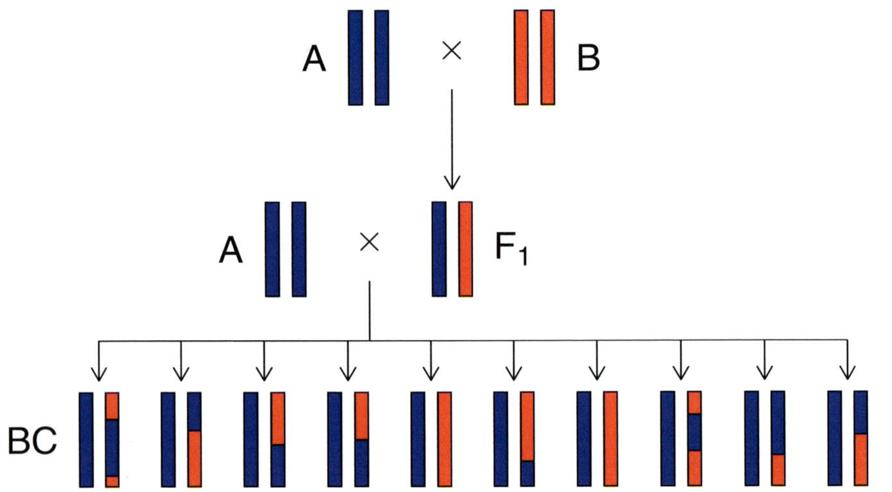

Figure 1.1. Schematic representation of the autosomes in a backcross experiment. The two inbred strains, A and B, are represented by blue and pink chromosomes, respectively. The F_1 generation, obtained by crossing the two strains, receives a single chromosome from each parent, and all individuals are genetically identical. If we cross an individual from the F_1 generation "back" to one of the parental strains (the A strain, in this example), we obtain a population exhibiting genetic variation. The backcross individuals receive an intact A chromosome from their A parent. The chromosome received from their F_1 parent may be an intact A or B chromosome, but is generally a mosaic of the A and B chromosomes as a result of recombination at meiosis. Any given locus has a 50% chance of being heterozygous and a 50% chance of being homozygous. This figure represents the autosomes only. When considering the X chromosome, four backcross populations ("directions") are possible (see Fig. 4.22 on page 110).

Moreover, the intercross allows one to estimate the degree of dominance at a QTL.

Another strategy would be to use recombinant inbred lines (Fig. 1.3). These are constructed by beginning with an intercross, and then mating pairs of F_2 siblings, followed by a parallel series of repeated sibling mating to construct a new panel of inbred lines whose genomes are a mosaic of the two initial lines. In organisms that allow selfing, one may following the initial cross by repeated selfing, multiple times in parallel; the progress to inbreeding in recombinant inbred lines by selfing is more rapid.

Recombinant inbred lines (RILs) have a number of advantages. Since they are immortal, we need genotype each line only once; we can phenotype multiple individuals from each line to reduce individual, environmental and measurement variability; we can obtain multiple invasive phenotypes on the same set of genomes; and, as the breakpoints in RILs are more dense than those

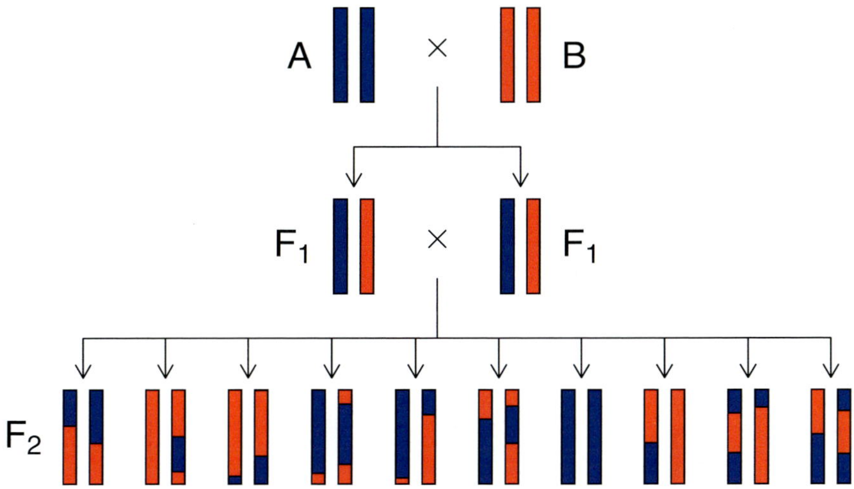

Figure 1.2. Schematic representation of the autosomes in an intercross experiment. The two inbred strains, A and B, are represented by blue and pink chromosomes, respectively. As with the backcross strategy, the F_1 generation, obtained by crossing the two strains, has a chromosome from each parent, and all F_1 individuals are genetically identical. By crossing two individuals in the F_1 generation (or by selfing when possible), we create genetic variation in the resulting F_2 population. The three possible genotypes AA, AB, and BB appear in a 1:2:1 ratio. This figure represents the autosomes only; the behavior of the X chromosome is displayed in Fig. 4.23 on page 111.

that occur in any one generation, we can achieve better mapping resolution. However, constructing and maintaining RILs is expensive. For this reason, they are more commonly used by plant biologists whose costs are lower relative to animal biologists. Most of the examples in this book will focus on the backcross and intercross. However, we will further consider the RIL design in the chapter on experimental design (Chap. 6).

Before embarking on a QTL experiment, the experimenter will usually phenotype several individuals from the two parental strains, and often some F_1 hybrid individuals. Hypothetical phenotype data for two inbred strains, the F_1 hybrid, and a backcross population are shown in Fig 1.4. Individuals within each of the parental strains are genetically identical, and so any variation in the phenotype within the strains is nongenetic (due to a combination of measurement error, environmental variation, and individual developmental noise). One generally chooses parental strains that show systematic phenotypic differences. Note that the within-strain variation is not necessarily the same in the two parental strains. The F_1 individuals are also genetically identical, and so any phenotypic variation in the F_1 is again nongenetic. The phenotype distribution of the F_1 individuals is often intermediate between the parental

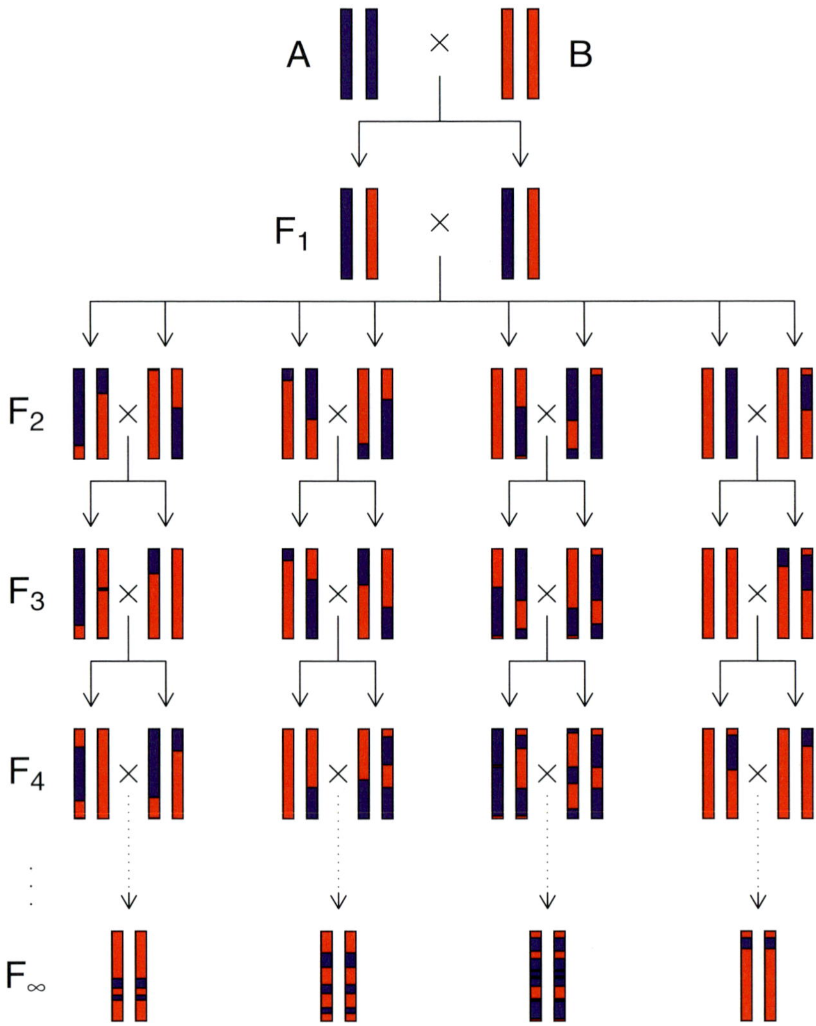

Figure 1.3. Schematic representation of the autosomes during the breeding of recombinant inbred lines by sibling mating. The first two generations are identical to an F_2 intercross. In subsequent generations, siblings are mated producing progeny that are less and less heterozygous. If continued indefinitely, this process will produce individuals that are completely homozygous at every locus, but with chromosomes that are a mosaic of the parental chromosomes. The frequency of the breakpoints between the AA and BB genotypes is determined by the breeding scheme (sib-mating or selfing, or some other scheme). In practice, 10–20 generations of inbreeding are actually performed.

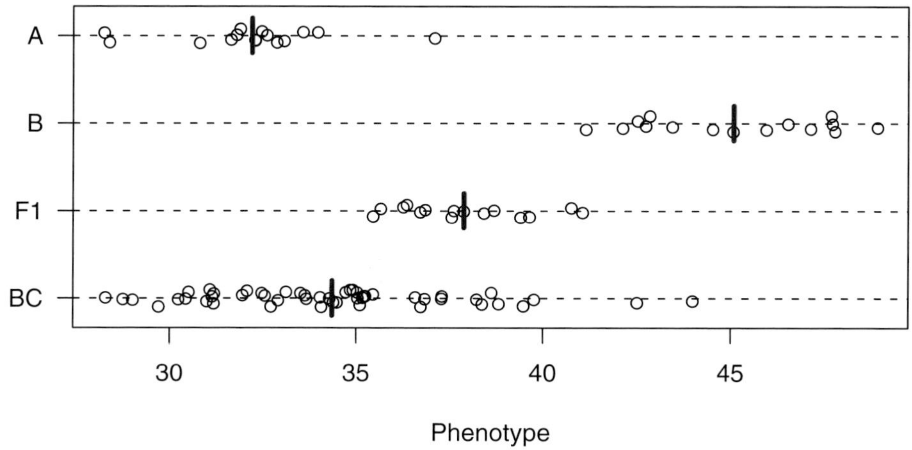

Figure 1.4. (Hypothetical) phenotype data for the two parental strains, the F_1 hybrid, and a backcross population. Vertical line segments are plotted at the within-group averages. In this simulated example, the difference in the phenotype means between strains is quite large relative to the within-strain variation. The mean phenotype in the F_1 generation is intermediate between the two strains, while the backcross progeny exhibit a wider spectrum of phenotypic variation and resemble the A strain more than the B strain.

lines' distributions, but this is not necessarily the case. Individuals in the backcross population are not genetically identical, and so they should exhibit greater phenotypic variation.

The key idea in QTL mapping is to obtain phenotype data on a number of backcross or intercross progeny and then identify regions in the genome where genotype is associated with the phenotype. However, the genotype is not observed at every possible position along the chromosomes, but only at a set of discrete landmarks called genetic markers. These are biochemical assays that reveal the identity of a defined genomic region. Microsatellite and SNP markers are the most common types of markers used for QTL mapping.

QTL mapping data has three interrelated data structures: the phenotypes, the genotypes, and the marker map. The phenotype data consists of observable characteristics of each individual in the population. These would include traits of interest such as blood pressure, or body weight, but also covariates such as sex, cross direction, and environmental conditions such as diet. Typically one has data on 100–1000 individuals. The genotype data consists of a set of genetic markers spanning the genome. In a typical mouse experiment, one may start with about 100 markers approximately evenly spaced along the genome. The genetic map specifies the markers locations on the chromosomes in terms of genetic distance. If a genetic map is not available, one would infer marker order and position with the available data.

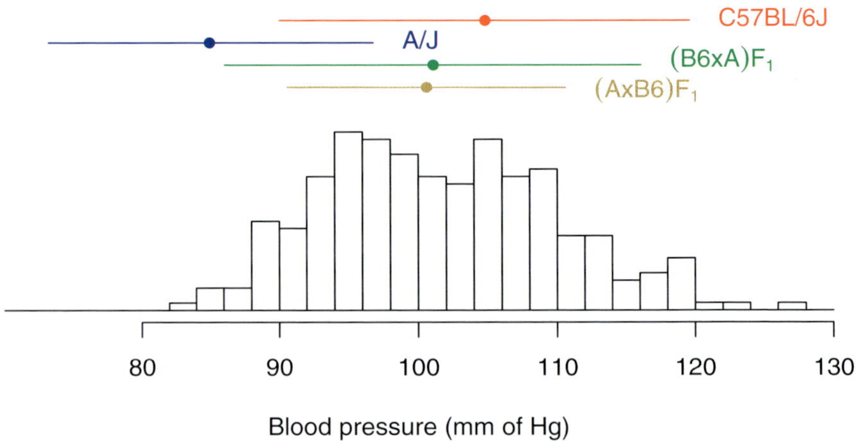

Figure 1.5. Histogram of systolic blood pressure for 250 backcross mice from Sugiyama *et al.* (2001). Also shown are the phenotypic ranges of parental and F_1 hybrid strains (mean $\pm$ 2 SD). The A/J (or A) strain is normotensive, while the C57BL/6J (or B6) strain is hypertensive. The F_1 hybrids and the backcross resemble the hypertensive B6 strain. Notice that the range of the phenotype is not unlike humans.

While a physical map specifies the physical position of markers on the chromosomes, in a genetic map distance is measured by the rate of crossover events at meiosis. Two markers are d centiMorgans (cM) apart if there is there is an average of d crossovers in the intervening interval in every 100 products of meiosis.

1.2.1 Mouse hypertension data as an example

As an example, we consider the data of Sugiyama *et al.* (2001) on salt-induced hypertension in the mouse. They measured systolic blood pressure in 250 male mice from a backcross between the hypertensive C57BL/6J (B6) and normotensive A/J (A) strains. (B6xA)F_1 mice were mated to B6 mice to produce a total of 250 male mice. The data are included with R/qtl, and will be referred to as the `hyper` data. (Further details will be provided in Sec. 2.3.) A histogram of the blood pressures of these backcross mice is shown in Fig. 1.5.

A total of 173 markers were genotyped; the genetic map of the markers is shown in Fig. 1.6. For most regions of the genome, markers were placed at a spacing of 10–20 cM. In regions for which an initial analysis indicated some evidence for a QTL (for example, chromosomes 1 and 4), additional markers were added.

The actual genotype data are shown in Fig. 1.7; individuals have been sorted by their phenotype: mice with lower blood pressure are at the bottom, and mice with higher blood pressure are at the top. By careful study of Fig. 1.7,

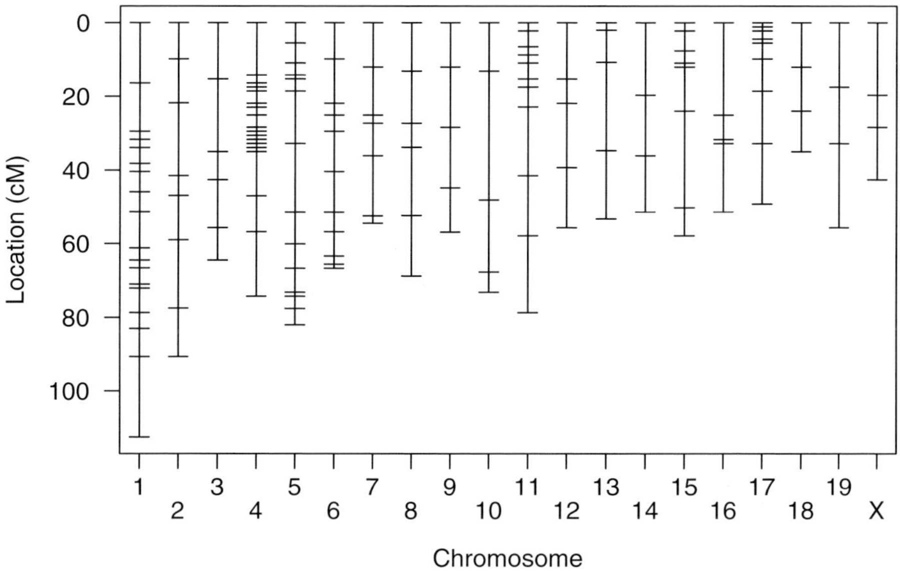

Figure 1.6. The genetic map of markers typed in the data from Sugiyama *et al.* (2001). Almost all marker intervals are less than 20cM. Some regions, most notably on chromosomes 1 and 4, have a higher density of markers.

one can guess the identity of a few putative QTL. For example, on each of chromosomes 1 and 4, there is mostly blue at the bottom of the figure and mostly red at the top. Homozygotes (red) tend to have higher blood pressure than heterozygotes (blue). On chromosome 6, the opposite pattern is seen: the bottom is mostly red and the top is mostly blue. This may indicate a QTL with an effect of the opposite sign. Visual examination of the raw data has an important role in detecting data quality issues, and can also provide informal evidence for QTL. However, for objective evidence for QTL formal statistical methods are required.

1.3 Central statistical problems

Investigators perform QTL experiments with a variety of goals in mind, and so the details of the appropriate statistical methods to meet those goals will also vary. For example, an evolutionary biologist may be particularly interested in the number and effects of QTL, while for a biomedical researcher the goal is to identify the gene (or genes) underlying at least one QTL; the number and effects of QTL may be of minor interest.

The principal goals of QTL mapping are nevertheless well defined. First, we seek to detect QTL (and, potentially, interactions among QTL). Second, we seek confidence regions for the locations of the QTL. Finally, we seek to

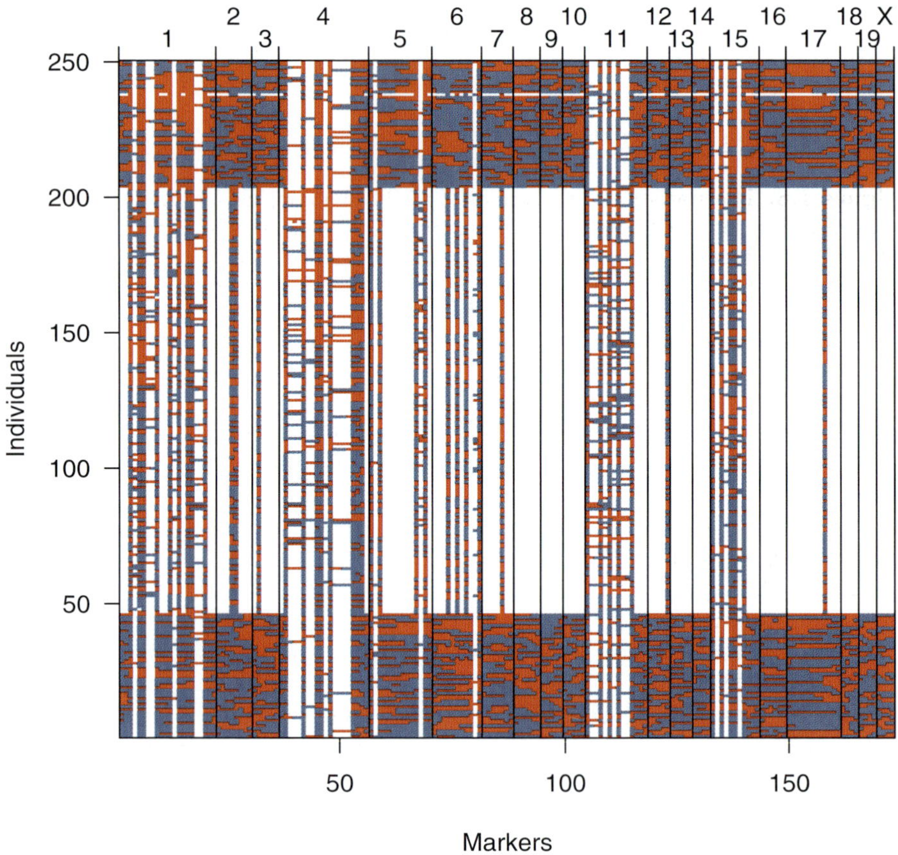

Figure 1.7. Genotype data for the backcross from Sugiyama *et al.* (2001). Red and blue pixels correspond to homozygous and heterozygous genotypes, respectively. White pixels indicate missing genotype data. Black vertical lines indicate the boundaries between chromosomes. Individuals are sorted by their phenotype (low blood pressure at the bottom, high blood pressure at the top). A three-step selective genotyping strategy was used for this cross. Individuals at the extremes of the phenotype distribution were genotyped for a set of framework markers spanning the genome. On selected chromosomes all individuals were typed for framework markers; dense genotyping was performed on recombinant marker intervals to narrow down the locations of recombination events.

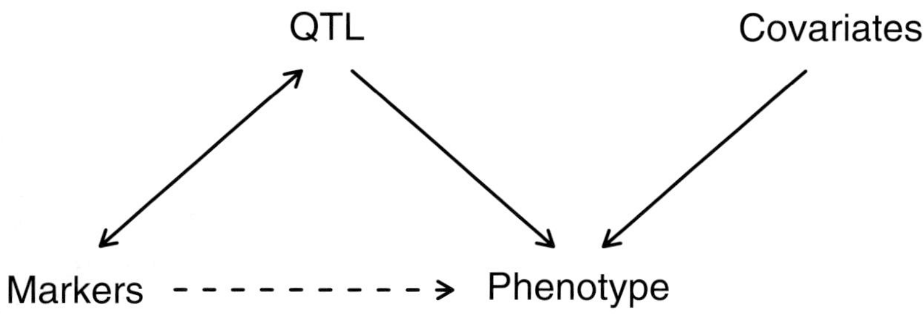

Figure 1.8. The statistical structure of the QTL mapping problem. The QTL and covariates are responsible for phenotypic variation (indicated by the directed solid arrows). The markers and the QTL are correlated with each other due to linkage (indicated by the bidirectional solid arrow). The markers do not directly cause the phenotype; some markers may be associated with the phenotype via linkage to the QTL (indicated by the directed dashed arrow).

estimate the effects of the QTL (that is, the effect, on the phenotype, of substituting one allele for another). These are generally viewed to be of decreasing importance. (For example, there is little need for a confidence interval for a QTL that has not been clearly detected). In this book, we will focus primarily on detecting QTL and their interactions.

The QTL mapping problem is best split into two distinct parts: the *missing data problem* and the *model selection problem.*

As illustrated schematically in Fig. 1.8, the phenotype is influenced by the genotype at QTL plus possible covariates such as sex, treatment, or environmental effects. However, we generally do not observe the genotype at the QTL, but only at marker loci. The genotypes at the markers and the QTL are associated due to linkage, which results in an association between the phenotype and the marker genotypes.

If one knew the genotype of each individual at each position in the genome, QTL mapping would reduce to the identification of the set of sites in the genome that matter in producing the phenotype and of how these sites combine together (and with other covariates) to produce the phenotype. This is the *model selection problem.* However, since we observe individuals' genotypes only at a discrete set of genetic markers and wish to consider positions between markers as the possible locations of QTL, we must use the marker genotype data to infer the genotypes at intervening locations. This is the *missing data problem.*

There are several solutions to the missing data problem, which are all satisfactory when the markers are reasonably dense. Although solutions to the model selection problem have been proposed, it remains challenging. The general problem of variable selection in regression is an active area of research for which no generally acceptable solution is known.

We complete this section with a more detailed discussion of probability models for the processes that give rise to QTL data. For the missing data problem, in which one uses marker genotype data to infer the genotype at intervening positions, one requires a model for the recombination process. More important, however, are models that connect the genotype and phenotype.

1.3.1 Models for recombination

Solutions to the missing data problem mentioned in the previous subsection rely on models for recombination. These models help us probabilistically connect unobserved genotypes to observed genotypes. Genotypes are missing for all individuals at locations between typed markers; they may be missing for untyped individuals at typed markers. All approaches for dealing with missing data rely on the calculation of the genotype probabilities at a putative QTL on the basis of the available multipoint marker data. To do so, one must have a model for the recombination process.

The most convenient model is that of no crossover interference: that the locations of crossovers on a meiotic product are according to a Poisson process. Informally, this means that crossover locations are random. Under the no interference model, recombination events in disjoint intervals are independent. As a result, the genotypes at markers along a chromosome form a Markov chain. In other words, conditional on the genotype at a particular locus, the genotypes at positions to the left are independent of the genotypes at positions to the right. We should emphasize that we also assume that there is no segregation distortion (i.e., that the frequencies of genotypes at an autosomal locus are in the ratios 1:1 in a backcross and 1:2:1 in an intercross).

The convenience of the no crossover interference assumption is best illustrated by an example. In Fig. 1.9, we display hypothetical genotype data for three different backcross individuals at a set of six markers along a chromosome; each row corresponds to a different individual. (The dashes are meant to indicate missing data.) We seek the probability that an individual is AA or AB at the locus (perhaps a putative QTL) indicated by the triangle, given its available marker data.

If no crossover interference is assumed to hold, one need only consider the genotypes at the nearest flanking typed markers. For the first individual, we need only consider the genotypes at markers M_3 and M_4; we can ignore the genotypes at the other markers. If r_{iQ} is the recombination fraction between marker M_i and the putative QTL, and if r_{ij} is the recombination fraction between markers M_i and M_j, then the probability that the individual has genotype AA at the putative QTL is $(1-r_{3Q})(1-r_{4Q})/(1-r_{34})$; the probability that it is AB is $r_{3Q}r_{4Q}/(1-r_{34})$.

Note that the fact that the individual showed a recombination event between markers M_4 and M_5 is irrelevant here, as under the no interference model, recombination events in disjoint intervals are independent. However, most organisms exhibit positive crossover interference (crossovers tend not to

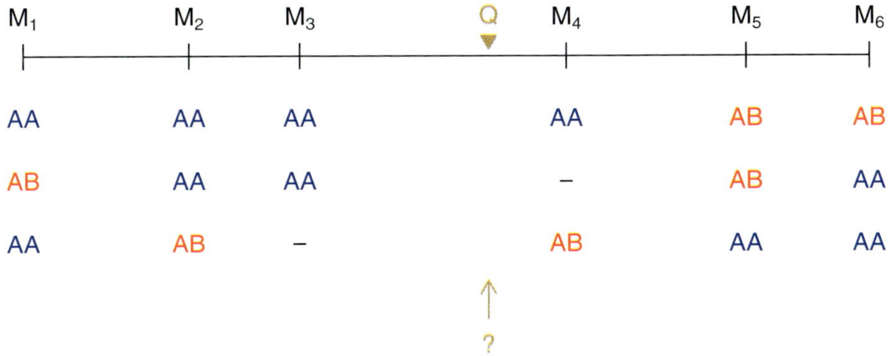

Figure 1.9. Illustration of the problem of inferring missing genotype data. Each row is the marker genotype data for a different backcross individual. Dashes indicate missing data. We seek the probability that each individual has genotype AA or AB at the putative QTL indicated by the triangle.

occur too close together). In particular, the mouse exhibits extremely strong crossover interference, with crossovers seldom being separated by less than 20 cM. In the presence of positive crossover interference, the recombination event between markers M_4 and M_5 would result in a decreased chance for a double recombinant in the interval between M_3 and M_4, and so there would be a greater probability that the individual is AA at the putative QTL. However, calculations are much simpler under the no interference model, and the difference has been seen to have little influence on QTL mapping results.

For the second individual, we consider the genotypes at markers M_3 and M_5, as the genotype at marker M_4 is missing. The probability that the individual is AA at the putative QTL is $(1 - r_{3Q})r_{5Q}/r_{35}$. The probability that it is AB is $r_{3Q}(1 - r_{5Q})/r_{35}$.

Finally, for the third individual, we consider the genotypes at markers M_2 and M_4. The probability that it is AA at the putative QTL is $r_{2Q}r_{4Q}/(1-r_{24})$. The probability that it is AB at the putative QTL is $(1-r_{2Q})(1-r_{4Q})/(1-r_{24})$.

In summary, an essential task in QTL mapping concerns the reconstruction of missing genotype data conditional on the observed marker genotypes. This requires a model for the recombination process. While meiosis generally exhibits positive crossover interference (with crossovers not occurring close together), we generally assume no crossover interference, as calculations are then greatly simplified. We have illustrated the value of no crossover interference with some simple examples. In general, one may use algorithms for hidden Markov models (HMMs) for these sorts of calculations, as one can then allow for the presence of genotyping errors and more simply deal with partially informative genotypes (such as the case of dominant markers in an intercross). HMM algorithms form the core of R/qtl, and are described in detail in Appendix D.

One last point: it may be worthwhile to discuss the role of map functions. A map function relates the genetic length of an interval (which is generally not estimable) to its recombination fraction (which generally is estimable); that is, it relates the expected number of crossovers in an interval to the probability of an odd number of crossovers (as a recombination event implies an odd number of crossovers). A model for crossover interference may imply a map function (though not necessarily, as if there is variation in the level of interference along a chromosome, an interval of a given genetic length would not correspond to a fixed recombination fraction), though the converse is not true. Common map functions include the Haldane map function (corresponding to no interference), the Kosambi map function (corresponding approximately to the level of interference in humans) and the Carter–Falconer map function (corresponding approximately to the level of interference in mice).

In calculating the genotype probabilities at a putative QTL given the available marker data, and using the no interference assumption, a map function is used to convert genetic lengths to recombination fractions. But if one's own data is used to estimate the genetic map, and if that is done (as is typical) under the no interference assumption, it is the recombination fractions that are actually estimated; the estimated genetic distances are derived via a map function. In this case, it will not matter what map function is used, provided that the same map function used to derive genetic distances is also used to convert back to recombination fractions. The only role of the map function concerns the scale on which results are plotted, as the results are generally plotted as a function of genetic distance. One should not treat estimated genetic distances with much reverence; most important are the markers themselves, which tie one's results back to the DNA sequence.

1.3.2 Models connecting genotype and phenotype

Most important are models connecting genotype and phenotype. Consider a single backcross individual, and let y denote its phenotype and $\boldsymbol{g}$ its whole-genome genotype (that is, its genotype at all polymorphisms between the parental strains).

Imagine that there are just p sites that matter in producing the phenotype, where p is some small proportion of the total number of polymorphisms, and let $g_1, \ldots, g_p$ denote the individual's genotype at these p QTL. We then have $\mathrm{E}(y|\boldsymbol{g}) = \mu_{g_1 \cdots g_p}$ and $\mathrm{var}(y|\boldsymbol{g}) = \sigma^2_{g_1 \cdots g_p}$. That is, the expected value (i.e., mean or average) and variance of an individual's phenotype, given its whole-genome genotype, depends only on its genotype at the p QTL; all other polymorphisms are inconsequential.

Thus we may split individuals into 2^p groups (3^p groups in an intercross) that are genetically identical at all polymorphisms contributing to the phenotype. The residual variation in each group, $\sigma^2_{g_1 \cdots g_p}$, will be entirely nongenetic (measurement error, environmental variation, and individual developmental noise).

We often make a number of simplifying assumptions. First, we may assume constant variance (homoscedasticity): that $\sigma^2_{g_1 \cdots g_p} \equiv \sigma^2$. In other words, the individual, residual variation (which includes measurement error and environmental variation) is constant within each of the 2^p groups. To the contrary, we often see that such variation increases with the average phenotype, but the constant variance assumption is convenient.

Second, we may assume that the residual variation follows a normal distribution: $y|\boldsymbol{g} \sim N(\mu_{g_1 \cdots g_p}, \sigma^2)$. That is, the phenotype distribution within each of the 2^p groups follows a normal curve, though with different averages. Note that this is not the same as to say that the marginal phenotype distribution follows a normal distribution; rather, the marginal distribution follows a *mixture* of normal distributions, the components of the mixture being the 2^p groups with distinct genotypes at the p QTL.

Finally, we often assume that the QTL act *additively*, so that

$$\mathrm{E}(y|\boldsymbol{g}) = \mu_{g_1 \cdots g_p} = \mu + \textstyle\sum_j \Delta_j z_j$$

where $z_j = 0$ or 1, according to whether g_j is AA or AB. That is, the effect of QTL j is Δ_j, no matter the genotype at the other loci.

Any deviation from additivity is called epistasis. The term *epistasis* is often reserved for a particular type of interaction, in which the effect of a mutation at a locus may be masked by the presence of a mutation at a second locus. However, statistical geneticists have come to use the term more generally.

As an illustration of epistasis, consider the hypothetical data plotted in Fig. 1.10. Focus first on Fig. 1.10A; we split backcross individuals into four groups according to their joint genotypes at two QTL. Dots are plotted at the average phenotype for each of the two-locus genotypes; line segments are drawn between the averages for a fixed genotype at QTL 2.

The average phenotype for individuals with genotype AA at both QTL is 10, while the average phenotype for individuals with genotype AB at QTL 1 and AA at QTL 2 is 40, and so the effect of QTL 1 is 30 in individuals with genotype AA at QTL 2. The average phenotype for the individuals with genotype AA at QTL 1 and AB at QTL 2 is 60, while the average phenotype for individuals with genotype AB at both QTL is 90, and so the effect of QTL 1 is 30 in individuals with genotype AB at QTL 2. Thus the effect of QTL 1 is the same, no matter the genotype at QTL 2; similarly, the effect of QTL 2 is the same, no matter the genotype at QTL 1. Thus the QTL are said to be additive. (Sometimes, in this case, the QTL are said to be *independent*, but this can be confused with whether or not the QTL are linked on a chromosome, and so we prefer the term *additive*.)

In Fig. 1.10B, on the other hand, the effect of QTL 1 is 30 when the genotype at QTL 2 is AA, but is 65 when the genotype at QTL 2 is AB; similarly the effect of QTL 2 depends on the genotype at QTL 1. Thus the QTL are said to interact (or to be *epistatic*): the effect of one QTL depends on the genotype at the other QTL.

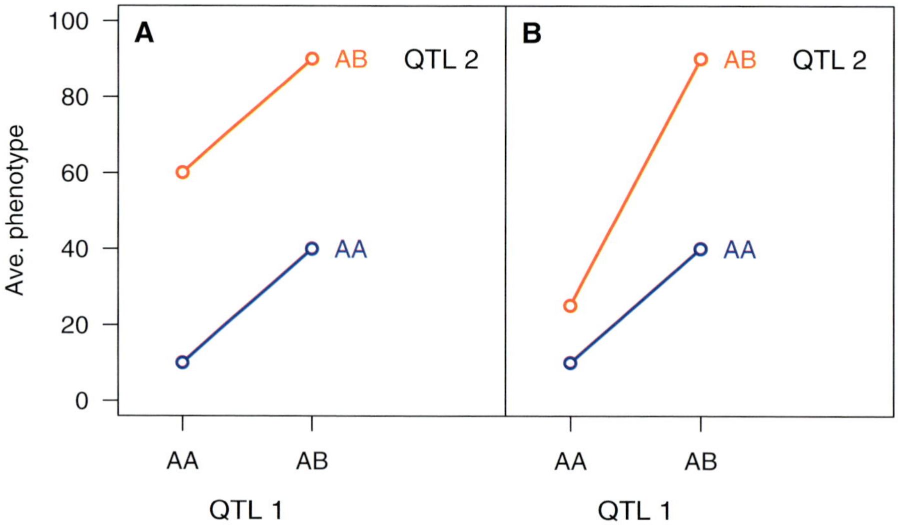

Figure 1.10. Illustration of possible effects of two QTL in a backcross. **A.** Additive QTL. **B.** Interacting QTL. Points are located at the average phenotype for a given two-locus genotype. Line segments connect the averages for a given genotype at the second QTL.

Figure 1.11 provides the analogous illustration for an intercross. The position of the average phenotype for the AB group between the averages for the AA and BB groups concerns additivity at a locus (versus dominance or recessivity). Two loci are additive (as in panel A) if the pattern of effect at one QTL is the same no matter the genotype at the other QTL: that the three curves in Fig. 1.11A are parallel. In Fig. 1.11B, the pattern of effect of QTL 1 is different for different genotypes at QTL 2, and vice versa, and so the two QTL are said to interact.

Sometimes a distinction is made between epistasis that concerns simply differences in the sizes of effects (as in Fig. 1.10B) versus changes in the sign of effect (as in Fig. 1.11B), as the former might be eliminated by a change in the scale on which the phenotype is measured, while the latter cannot be. We should emphasize further: strict additivity of QTL is rare and would be lost with a transformation of the phenotype. Thus, the question is not whether two loci interact but by how much. Further, deviation from additivity does not necessarily imply physical interaction or even a shared pathway. Polymorphisms in multiple genes may underly each QTL, and so conclusions regarding potential biological interactions are quite tenuous.

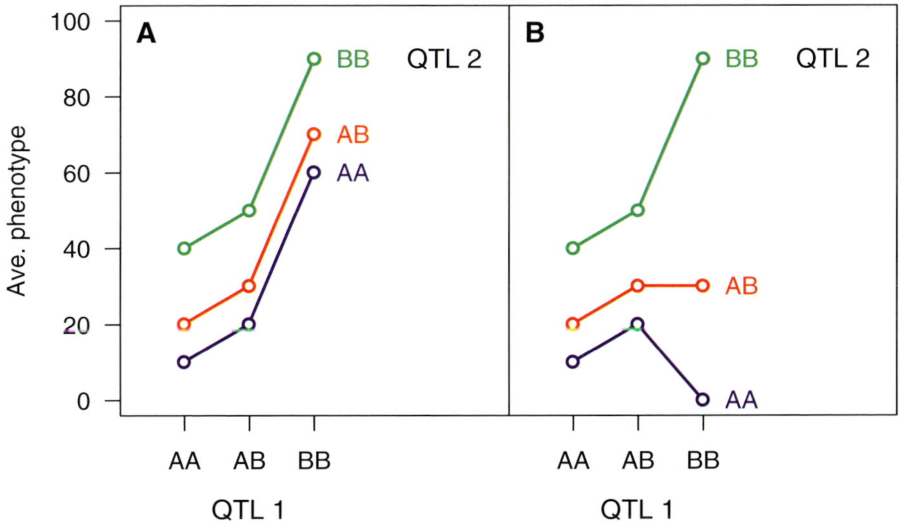

Figure 1.11. Illustration of possible effects of two QTL in an intercross. **A.** Additive QTL. **B.** Interacting QTL. Points are located at the average phenotype for a given two-locus genotype. Line segments connect the averages for a given genotype at the second QTL.

1.4 About R and R/qtl

The development of the R/qtl software was begun at the suggestion of Gary Churchill. Our primary goal was to make complex QTL mapping methods widely accessible and allow users to focus on modeling rather than computing. We further sought to develop an extensible platform for QTL mapping: to have a fast implementation of the hidden Markov model technology for dealing with the problem of missing genotype information, which forms the core of all QTL mapping methods, and to make these intermediate calculations readily accessible to the sophisticated user, so that specially tailored mapping methods can be more easily implemented.

R/qtl has been implemented as an add-on package to the general statistical software, R. R is an open source implementation of the S language, is widely used by academic statisticians, and is extensively used for microarray analyses (see the Bioconductor Project, `http://www.bioconductor.org`). As described on the R project homepage (`http://www.r-project.org`):

> R is a system for statistical computation and graphics. It consists of a language plus a run-time environment with graphics, a debugger, access to certain system functions, and the ability to run programs stored in script files.
>
> The core of R is an interpreted computer language which allows branching and looping as well as modular programming using

> functions. Most of the user-visible functions in R are written in R. It is possible for the user to interface to procedures written in the C, C++, or FORTRAN languages for efficiency. The R distribution contains functionality for a large number of statistical procedures. Among these are: linear and generalized linear models, nonlinear regression models, time series analysis, classical parametric and nonparametric tests, clustering and smoothing. There is also a large set of functions which provide a flexible graphical environment for creating various kinds of data presentations. Additional modules are available for a variety of specific purposes.

The development of R/qtl as an add-on to R allows us to take advantage of the basic mathematical and statistical functions, and powerful graphics capabilities, that are provided with R. Further, the user benefits by the seamless integration of the QTL mapping software into a general statistical analysis program.

Much of the source code for R/qtl is written in R (particularly the portion that concerns data manipulation and graphics). However, most functions that require fast computation (such as those concerning hidden Markov models) were written in C.

R and R/qtl are freely available for Windows, Unix and Mac OS X, and may be downloaded from the Comprehensive R Archive Network (CRAN, `http://cran.r-project.org`). Also see the R/qtl web site, `http://www.rqtl.org`. For instructions regarding downloading and installing R and R/qtl, see Appendix A.

Much can be accomplished with R/qtl (and much of this book may be understood) with little detailed knowledge of R. Learning R may require a formidable investment of time, but it will definitely be worth the effort, both for increased facility in the use of R/qtl and for more general statistical analysis. Numerous free documents on getting started with R are available at CRAN. In addition, a growing list of books on R are available [for example, see Dalgaard (2002) or Venables and Ripley (2002)].

In learning R and R/qtl, as with any computer language or program, it is important to fiddle about: try out the example code, and explore what happens when the code is modified. In addition, one should refer to the extensive documentation included with both R and R/qtl. See Sec. A.5 for details on accessing the documentation.

1.5 Other software

There are numerous other computer programs for QTL mapping. We do not attempt to cover these exhaustively.

MapMaker/QTL was the first computer program for QTL mapping, but it has not been updated since 1994, and only allows the fit of single-QTL models. QTL Cartographer provides more extensive facilities for the fit of multiple-QTL models, and has a graphical user interface (GUI) for Windows. Map Manager QTX is a GUI-based program that some users find most intuitive, but QTL mapping is performed exclusively with Haley–Knott regression, which can be inefficient and prone to artifacts. Commercial software programs include MapQTL and MultiQTL.

R/qtl is one of the few open source QTL mapping programs; its extensible structure, which simplifies the implementation of specially tailored mapping methods, is unique. R/qtl includes many of the important diagnostics present in MapMaker, but also allows the fit of multiple-QTL models, as in QTL Cartographer.

1.6 Work flow

In this section, we briefly describe the general work flow in QTL mapping. One must start with the design of the experiment: the choice of strains, of phenotypes to measure, of whether to perform a backcross or an intercross, what markers should be genotyped and which individuals should be genotyped. Aspects of design are discussed in Chap. 6.

After the data have been obtained, the first task is to assemble them into a computer file or files and import them into software. The import of QTL mapping data into R/qtl is discussed in Chap. 2. Second, one performs a variety of diagnostic checks on the data to identify possible errors, such as mistakes in data entry, errors in marker order, and genotyping errors. This task is discussed in Chap. 3, and is particularly important, as our ability to map QTL will be eroded by low-quality data. Aspects of the phenotype distribution may lead one to consider phenotype transformations or the use of special phenotype models, such as the two-part model discussed in Sec. 5.3.

Next, one uses interval mapping (or one of its variants), performing a genome scan with a single-QTL model, to detect loci with important marginal effects. Interval mapping is discussed in Chap. 4 and 5. The statistical significance of putative QTL is established, taking into account the genome-wide scan, generally by a permutation test (Sec. 4.3). Interval estimates for the locations of QTL may also be obtained (Sec. 4.5).

One next turns to two-dimensional, two-QTL scans of the genome (discussed in Chap. 8). Such two-dimensional scans provide the first opportunity to identify interactions between QTL, including the possibility of detecting QTL with limited marginal effects, whose importance may be seen only by considering their interaction with other loci. In addition, evidence for two linked QTL (versus a single QTL on a chromosome) is best obtained by considering an explicit two-QTL model.

Finally, one will bring all of the putative QTL and QTL $\times$ QTL interactions together into an overall multiple-QTL model (Chap. 9). In the context of a global model, some QTL may then be omitted, while the exploration of additional QTL or interactions may lead to the addition of further terms to the model. The fit of the global model may allow some refinement in the location of QTL, and provides the most reliable estimates of the QTL effects.

The result of the analysis is some set of inferred QTL, with some understanding of their effects, locations, and possible interactions. These results will be used to guide further experiments, perhaps with the aim of fine-mapping the QTL and ultimately identifying the underlying gene or genes.

1.7 Further reading

There are numerous review articles on QTL mapping (e.g., Doerge *et al.*, 1997; Broman and Speed, 1999; Jansen, 2007; Broman, 2001). We particularly like the review of Jansen (2007). Broman (2001) is one of the few reviews written for nonstatisticians.

A number of books have some discussion of QTL mapping: Falconer and Mackay (1996) has a chapter on QTL mapping, and Lynch and Walsh (1998) and Liu (1998) each have several. Silver (1995) is a very nice book on mouse genetics, and it is freely available online at `http://www.informatics.jax.org/silver`. Also see the recent book by Wu *et al.* (2007).

McPeek (1996) provides a useful introduction to recombination and crossover interference. See also McPeek and Speed (1995). For a discussion of map functions, see Speed (1996) and Zhao and Speed (1996). Broman *et al.* (2002) studied crossover interference in the mouse. Strickberger (1985) contains an excellent chapter on epistasis.

Ihaka and Gentleman (1996) is the paper introducing R. Broman *et al.* (2003) is the original article reporting R/qtl. The most important book on R is Venables and Ripley (2002); every user of R should have a copy. Dalgaard (2002) provides a more gentle introduction.

Lander *et al.* (1987) is the original paper on MapMaker; Manly *et al.* (2001) described Map Manager. The only clear reference on QTL Cartographer is the manual (Basten *et al.*, 2002), available online at `http://statgen.ncsu.edu/qtlcart/manual`.

2

Importing and simulating data

One of the more frustrating tasks associated with the use of any data analysis software concerns the importation of data. Data can be imported into R/qtl in a variety of formats, but users often have trouble with this step. In this chapter, we describe how to import QTL mapping data into R for use with R/qtl. We further discuss the simulation of QTL mapping data. In an optional section, we describe the internal format that R/qtl uses for QTL mapping data.

As this may be the reader's first exposure to R, we will introduce some of the basic aspects of R as we go along. We should again emphasize that the novice user will benefit by spending a couple of days reading Dalgaard (2002) and playing with R.

Before you do anything, you must install R and the R/qtl package; this is described in Appendix A. After invoking R, you must type `library(qtl)` to load the R/qtl package. (In R, R/qtl is known as the `qtl` package or library.) It is best to create a `.Rprofile` file containing this command, so that the package will automatically be loaded whenever you invoke R. (See Sec. A.3.)

Essentially all tasks in R are performed via *functions*, such as the `library` function mentioned above. Appendix B contains partial list of the functions in R/qtl. A complete list may be viewed by typing the following.

```
> library(help=qtl)
```

The `>` symbol is the R prompt, which you will observe when R is ready to accept input commands. R commands may be spread over several lines, in which case the R prompt turns into the `+` symbol, indicating a continuation line. (Appearance of the `+` prompt when one believes one's command is complete may indicate imbalance in parentheses. Press the escape key to cancel the command.) R input will be shown in *`a slanted typewriter font`*, while output will be in `a plain typewriter font`. (The output for the above command was suppressed, as it would fill a couple of pages.)

Note that the up and down arrow keys may be used to scroll back through previously entered commands. Emacs users will be pleased to find that many

K.W. Broman, Ś. Sen, *A Guide to QTL Mapping with R/qtl*,
Statistics for Biology and Health, DOI 10.1007/978-0-387-92125-9_2,

of the Emacs key bindings may be used. (But be careful about Ctrl-p, which may lead you to print a page.)

2.1 Importing data

Importing QTL mapping data into R is accomplished with the `read.cross` function. Data may be read in a variety of formats. We strongly recommend the comma-delimited formats discussed in the next subsection, but MapMaker/QTL, Map Manager QTX, and QTL Cartographer formats may also be used. Sample data files in most of the formats are available at the R/qtl web site (`http://www.rqtl.org/sampledata`). The help file for `read.cross` contains the complete details on the file formats and the use of the function. The help file may be viewed by typing `?read.cross`; see Sec. A.5. Note that basic use of the `read.cross` function is described in Sec. 2.1.1 on the comma-delimited formats and is not repeated in the subsections on the other formats.

Before contemplating loading one's data into R, it must be assembled into one of the accepted formats. While the comma-delimited formats can be created in OpenOffice, Microsoft Excel, or other spreadsheet programs, a different format (or computer program) might be best for entering the data into the computer. (And ideally data should enter the computer directly from the measurement device, rather than be input by hand.) The reformatting of data files to conform to the requirements of specific software is a frequent task for geneticists, and hand manipulation of data files is time-consuming and error-prone. Thus we recommend that geneticists learn to program in a language like Perl, which will greatly simplify the task. While the up-front investment to learn Perl is large, the value such knowledge will provide over one's career is far larger.

2.1.1 Comma-delimited files

The recommended format for QTL mapping data to be imported into R/qtl is the comma-delimited format, `"csv"` (an abbreviation of "comma-separated values"). Several variations on this format will be described below. We begin by discussing the basic one.

In the basic `"csv"` format, all phenotype and genotype data, plus the genetic map of the typed markers, are combined into a single file with fields delimited by commas. The file may be constructed in a spreadsheet, such as OpenOffice or Microsoft Excel; an example is illustrated in Fig. 2.1. Be careful about the use of commas within the fields (though the use of quotation marks should prevent this from being a problem).

The initial columns are phenotypes (at least one phenotype must be included, such as a numeric index for each individual). Subsequent columns are markers. The first row contains the phenotype and marker names. The second

	A	B	C	D	E	F	G	H	I	J
1	pheno	sex	pgm	c1m1	c1m3	c1m4	c1m5	c2m1	c2m2	c2m3
2				1	1	1	1	2	2	2
3				8.3	49.0	59.5	89.0	1.0	15.0	45.0
4	0.093	f	0	B	B	–	H	B	B	B
5	0.177	f	0	H	H	H	H	H	H	H
6	–	f	0	H	B	A	A	H	–	B
7	0.230	f	0	B	H	H	–	A	A	A
8	0.228	f	0	B	–	H	H	H	B	B
9	0.279	f	0	B	B	A	H	A	H	H
10	0.419	f	0	H	H	H	H	A	B	B
11	0.427	f	0	–	A	B	B	B	H	H
12	0.282	f	0	–	A	B	B	A	A	A
13	0.400	f	0	H	B	A	A	–	H	H
14	0.521	f	0	B	H	B	B	H	H	H
15	0.385	f	0	H	B	A	A	B	H	–
16	0.518	f	0	–	H	H	H	H	H	B

Figure 2.1. Part of a data file in the `"csv"` format, as it might be viewed in a spreadsheet.

row must have empty fields in each of the phenotype columns. (This is quite rigid; even a space character will mess things up.) For the genotype columns, the second row should contain chromosome assignments. Numbers are best; character strings, such as “`Chr 1`” or “`six`” will make later data manipulation more cumbersome. Use “`X`” or “`x`” to identify the X chromosome.

An optional third row can contain the centiMorgan (cM) positions of the genetic markers. The fields in the phenotype columns should again be blank. Marker order is taken from the cM positions, if provided; otherwise it is taken from the column order.

Subsequent rows correspond to the individuals, with phenotypes followed by genotypes. Missing data should be indicated by “`NA`” or “`-`” or some other code. (It is always best to insert some code indicating missingness rather than leave some cells empty, as empty cells can be ambiguous: was the value missing or was an error in data entry made?) Multiple missing data codes may be used, but consistency between the phenotype and genotype data is required: a missing value code for the genotype data cannot be a legitimate phenotype and vice versa. No missing values are allowed in the chromosome identifiers or genetic map positions.

For a backcross, two genotype codes are to be used: one for homozygotes (e.g., `AA`) and one for heterozygotes (`AB`). For an intercross, five genotype codes may be used: the two homozygotes (`AA` and `BB`), the heterozygote (`AB`), and two further genotype codes to be used for dominant markers, such as `D` for “not `BB`” (i.e., `AA` or `AB`) and `C` for “not `AA`” (i.e., `AB` or `BB`), as used by the MapMaker software.

Consistency in genotype codes is required: one cannot use both `A` and `AA` to indicate a homozygous A genotype. Also note that spaces can mess things

Table 2.1. Possible intercrosses, and the appropriate code for the `pgm` "phenotype." In the crosses, females are always listed first, so A×B means a female A crossed to a male B.

Cross	Possible genotypes		pgm code
	Females	Males	
(A × B) × (A × B)	AA, AB	A-, B-	0
(B × A) × (A × B)	AA, AB	A-, B-	0
(A × B) × (B × A)	AB, BB	A-, B-	1
(B × A) × (B × A)	AB, BB	A-, B-	1

up: "`A `" is treated as different from "`A`". It is best to ensure that there are no spaces in the final data file.

X chromosome genotypes should be coded just like the autosomal genotype data; in particular, hemizygous males should be coded as if they were homozygous, rather than using separate codes for hemizygous and homozygous genotypes. If X chromosome genotype data are included, one of the phenotypes should indicate the sex of the individuals. This may be called "`sex`" or "`Sex`," and the sexes may be coded by `0`/`1` for females/males, or by the codes `f`/`m`, `F`/`M`, or `female`/`male`.

Further care is required for the X chromosome genotype data in an intercross, as the direction of the cross must be known. Four possible intercrosses may be performed, as shown in Table 2.1. In all cases, the males are hemizygous A or B at any one locus, but in the crosses (A×B)×(A×B) and (B×A)×(A×B), the females are either AA or AB, while in the crosses (A×B)×(B×A) and (B×A)×(B×A), the females are either AB or BB. Thus, the order of the cross producing the F_1 male is critical; for example, we wish to know whether the paternal grandmother was from strain A or B.

We thus require, for intercrosses, a "phenotype" column named `pgm` (for "paternal grandmother"), with codes `0` and `1` indicating which individuals came from which cross, as shown in Table 2.1.

If one includes a phenotype named "`id`," "`ID`," or "`Id`," it will be assumed to provide individual identifiers. These will be used in certain places to indicate the individuals (such as in `plot.geno`; see Sec. 3.5).

The specification of a file in the `"csv"` format is now complete. If the file was created in a spreadsheet program, such as OpenOffice or Microsoft Excel, you will need to use "Save as" and select the format "CSV (comma-delimited)" to create the actual file. The result will look something like that shown in Fig. 2.2. A complete example is provided at the R/qtl web site.

With our first file format understood, we now turn to the use of `read.cross` to load the data into R.

A list of the input arguments for `read.cross` may be viewed via the `args` function, as follows. (We often use `args` to get a quick reminder of the input to a function.) Remember that, if R/qtl is not yet loaded, one must use

```
pheno,sex,pgm,c1m1,c1m3,c1m4,c1m5,c2m1,c2m2,c2m3,c2m4,...
,,,1,1,1,1,2,2,2,2,2,2,2,3,3,3,3,3,3,4,4,4,4,5,5,5,5,5...
,,,8.3,49,59.5,89,1,15,45,68.9,80.9,87.4,99,0,11.2,39....
0.093,f,0,A,B,A,A,B,H,H,H,H,H,H,B,H,H,B,B,H,A,A,A,A,H,...
0.177,f,0,B,H,H,H,H,B,B,B,B,H,B,H,H,H,-,H,H,H,H,A,H,A,...
0.271,f,0,B,A,H,H,H,H,H,A,A,A,A,H,H,H,H,H,H,B,-,B,B,H,...
0.230,f,0,B,B,A,H,B,B,B,B,B,H,B,A,H,H,B,B,H,H,H,B,B,H,...
0.228,f,0,H,H,H,H,H,H,H,H,B,B,B,B,B,H,H,H,H,B,H,H,H,B,...
0.279,f,0,H,B,A,A,H,B,B,H,A,A,-,A,A,A,H,H,H,A,A,H,B,H,...
0.419,f,0,B,H,H,H,A,A,A,A,A,A,A,B,B,B,B,B,B,B,H,H,H,H,...
0.427,f,0,B,A,H,H,H,B,B,B,B,B,B,H,H,H,B,B,B,H,H,A,A,A,...
0.282,f,0,B,B,A,H,A,H,H,A,A,A,A,B,B,H,A,-,B,H,H,H,H,B,...
0.4,f,0,H,H,H,H,A,B,B,H,H,H,H,H,H,H,B,H,H,H,H,H,H,H,H,...
0.521,f,0,H,A,B,B,B,H,H,H,H,H,H,H,H,A,A,A,H,B,B,B,H,H,...
0.385,f,0,A,A,B,B,A,A,A,H,H,H,H,B,B,H,H,H,H,H,H,A,A,A,...
0.518,f,0,H,B,A,A,H,H,H,B,B,B,B,A,A,H,H,A,H,H,H,H,H,H,...
⋮
```

Figure 2.2. Part of a text file in the "csv" format. The terminal dots in each line are just to indicate that the file extends quite far to the right.

the `library` function to make it available. (Ignore the `NULL`; that's just a meaningless bit from the `args` function.)

```
> library(qtl)
> args(read.cross)

function (format = c("csv", "csvr", "csvs", "csvsr", "mm",
    "qtx", "qtlcart", "gary", "karl"), dir = "", file, genfile,
    mapfile, phefile, chridfile, mnamesfile, pnamesfile,
    na.strings = c("-", "NA"), genotypes = c("A", "H", "B", "D",
    "C"), alleles = c("A", "B"), estimate.map = TRUE,
    convertXdata = TRUE, ...)
NULL
```

This is, admittedly, rather forbidding, but not all of the arguments will be needed in all cases. Note that the `c` function is used to combine multiple items together into a vector.

The argument `format` will be used to indicate that we are reading data in the `"csv"` format. The possible formats are shown; the first listed is taken as the default. The argument `dir` is used to indicate the directory in which the file appears. By default, it is assumed that the file is in the current working directory. (For details on how to select or change the working directory, see

Sec. A.4.) The argument `file` will be used to give the name of the data file. The other file arguments are used for formats in which the data are split across multiple files.

The argument `na.strings` is used to indicate the set of missing data codes. By default, either "`-`" or "`NA`" will be treated as missing. Note that most things are case-sensitive, so "`na`" will be treated as different from "`NA`" and "`Na`". If all of these appear in the data file, all should be indicated via the `na.strings` argument.

The argument `genotypes` is used to indicate the genotype codes, and takes a vector of character strings. The order of the codes in the string is important. We often forget whether "`D`" stands for "not `AA`" or "not `BB`," and so we generally must refer to the help file for `read.cross`, where this is explained. Note, again, that the codes are case-sensitive, so "`a`" will be treated as different from "`A`."

The argument `alleles` is used to indicate custom names for the alleles (single-character names are best), so that if one does a mouse cross of BALB/c × DBA/2, one might want to use the codes `C` and `D` for the alleles. These will be used in certain plots (such as of phenotype against genotype) and summaries.

If the genetic map positions of the markers are not provided in the file and `estimate.map=TRUE`, the intermarker distances will be estimated, while if `estimate.map=FALSE`, a dummy map will be created. (If genetic map positions *are* provided, this argument will be ignored.) Estimation of the genetic map can sometimes be time-consuming, and so one may wish to use `estimate.map=FALSE`. One may later estimate the map with the function `est.map` and plug it into the data object with `replace.map`; see Sec. 3.4.3.

If marker positions are provided in the file, it is important that no two markers are placed at precisely the same position. If they are, this may be rectified with the function `jittermap`; see page 84.

The "`...`" at the end of the specification of `read.cross` is used to allow additional arguments to be specified; these are passed to the more basic R function `read.table`, which does the actual work of reading in the data. Its use will be explained further below.

There seems a lot to understand, but use of `read.cross` is generally not so tedious as it might appear, as most of the arguments to the function can be ignored. For example, suppose the data in Figures 2.1 and 2.2 are saved in one's working directory as the file `mydata.csv`. One could read this into R with the following.

```
> mydata <- read.cross("csv", "", "mydata.csv")
```

Note that "`<-`" is the *assignment operator*. The data are read from the `mydata.csv` file and combined into a single object (with a very special internal format, described in Sec. 2.6.1) and assigned to `mydata`. This will be a new object in our R workspace that we may manipulate and analyze. Type `ls()` or `objects()` to list the objects in your workspace.

Also note that arguments to functions in R may be specified by their position in the list, by their name, or they may be left unspecified (in which case the default values are assumed). Thus, in the code above, it is assumed that `format="csv"`, `dir=""`, and `file="mydata.csv"`, and we need not specify values for `na.strings`, `genotypes`, or `alleles`, as the default values suffice for our data. All of the following lines of code are equivalent.

```
> mydata <- read.cross("csv", , "mydata.csv")
> mydata <- read.cross("csv", file="mydata.csv")
> mydata <- read.cross(format="csv", file="mydata.csv")
> mydata <- read.cross(file="mydata.csv", format="csv")
> mydata <- read.cross(file="mydata.csv")
```

If the data file were in some location other than the R working directory, we would need to specify its location with the `dir` argument. The directory (or folder) hierarchy is indicated with forward slashes (/). In Windows, it is traditional to use backslashes (\), but these will not work in R, though double-backslashes (\\) may be used in place of forward slashes.

For example, if we were working on a Macintosh and our file was on the Desktop, we might use the following code. The tilde (~) denotes our home directory.

```
> mydata <- read.cross("csv", "~/Desktop", "mydata.csv")
```

If we were working in Windows and the file was located in `c:\My Data`, we could use the following code.

```
> mydata <- read.cross("csv", "c:/My Data", "mydata.csv")
```

If we had coded the genotype data differently, we would need to use the `genotypes` argument. Because of all of the intervening file name arguments, the `na.strings`, `genotypes`, and `alleles` arguments generally must be specified by name. For example, suppose missing data were coded "`na`" and that the genotypes were coded `BB`/`BC`/`CC`. Then the data would be read as follows.

```
> mydata <- read.cross("csv", "", "mydata.csv", na.strings="na",
+                      genotypes=c("BB","BC","CC"),
+                      alleles=c("B","C"))
```

We recommend downloading the example `"csv"` data file (`listeria.csv`) from the R/qtl web site and trying to load it into R. (The file is included with the R/qtl package, but it is in a spot that may be difficult to find.) If one has trouble importing one's own data, it is a good idea to try importing a file that is known to be correct, so one may determine whether the problem concerns some incompatibility in the file or an incomplete understanding of the use of `read.cross`.

Outside the United States, commas are sometimes used instead of periods in numbers, and so semicolons are sometimes used instead of commas in such CSV files. Files of this sort may also be read; one must make use of the flexibility in the `read.cross` function through the "`...`" in its specification, through

```
# The ch3c data
# File created by Karl W Broman, 7-19-06
# Intercross between C57BL/6J and A/J
# 100 females from the cross (AxB)x(AxB)
# 101 markers, including 10 on the X chromosome
pheno,sex,pgm,c1m1,c1m3,c1m4,c1m5,c2m1,c2m2,c2m3,c2m4,...
,,,1,1,1,1,2,2,2,2,2,2,2,3,3,3,3,3,3,4,4,4,4,5,5,5,5,5...
,,,8.3,49,59.5,89,1,15,45,68.9,80.9,87.4,99,0,11.2,39....
0.093,f,0,A,B,A,A,B,H,H,H,H,H,H,B,H,H,B,B,H,A,A,A,A,H,...
0.177,f,0,B,H,H,H,H,B,B,B,B,H,B,H,H,H,-,H,H,H,H,A,H,A,...
0.271,f,0,B,A,H,H,H,H,H,A,A,A,A,H,H,H,H,H,H,B,-,B,B,H,...
0.230,f,0,B,B,A,H,B,B,B,B,B,H,B,A,H,H,B,B,H,H,H,B,B,H,...
0.228,f,0,H,H,H,H,H,H,H,H,B,B,B,B,B,H,H,H,H,B,H,H,H,B,...
0.279,f,0,H,B,A,A,H,B,B,H,A,A,-,A,A,A,H,H,H,A,A,H,B,H,...
0.419,f,0,B,H,H,H,A,A,A,A,A,A,A,B,B,B,B,B,B,B,H,H,H,H,...
0.427,f,0,B,A,H,H,H,B,B,B,B,B,B,H,H,H,B,B,B,H,H,A,A,A,...
⋮
```

Figure 2.3. An example file in the "csv" format with comment lines included.

which further arguments are passed down to the more basic `read.table` function. That function allows arguments `sep`, for specifying the field separator, and `dec`, for specifying the character used for the decimal point.

Thus, if the `mydata.csv` file had used semicolons and commas rather than commas and periods, we would read it into R with the following code.

```
> mydata <- read.cross("csv", , "mydata.csv", sep=";", dec=",")
```

Note that these additional arguments *must* be specified by name.

One may include comments in an input file, to be ignored when it is imported, but useful to document its contents. A single symbol, such as `#`, may be used to indicate that the remainder of the line is to be ignored. The chosen symbol cannot appear anywhere in the data, and is indicated, in the call to `read.cross`, via the `comment.char` argument. (In R versions 2.3.1 and earlier, `comment.char="#"` was the default, but in R versions 2.4.0 and later, the default has become `comment.char=""`, and so no such commenting character is assumed.)

For example, the file in Fig. 2.3 contains initial comment lines, indicated by `#`. To read this file into R, we would use the following code.

```
> mydata <- read.cross("csv", , "mydata.csv", comment.char="#")
```

There are three related comma-delimited formats: `"csvr"`, `"csvs"`, and `"csvsr"`. These are primarily for the case of expression genetic data, in which

	A	B	C	D	E	F	G	H	I	J
1	pheno			0.093	0.177	–	0.230	0.228	0.279	0.419
2	sex			f	f	f	f	f	f	f
3	pgm			0	0	0	0	0	0	0
4	c1m1	1	8.3	B	H	H	B	B	B	H
5	c1m3	1	49.0	B	H	B	H	–	B	H
6	c1m4	1	59.5	–	H	A	H	H	A	H
7	c1m5	1	89.0	H	H	A	–	H	H	H
8	c2m1	2	1.0	B	H	H	A	H	A	A
9	c2m2	2	15.0	B	H	–	A	B	H	B
10	c2m3	2	45.0	B	H	B	A	B	H	B
11	c2m4	2	68.9	B	H	H	A	B	A	H
12	c2m5	2	80.9	B	B	A	A	B	A	H
13	c2m6	2	87.4	H	B	A	A	B	A	H
14	c2m7	2	99.0	B	B		–	B	A	H
15	c3m1	3	0.0	A	B	A	B	H	B	H
16	c3m2	3	11.2	H	B	–	B	H	B	H

Figure 2.4. Part of a data file in the "csvr" format, as it might be viewed in a spreadsheet.

QTL mapping is to be performed with the expression of all genes on a microarray, so that one has thousands or tens of thousands of phenotypes.

The "csvr" format is just like the "csv" format, but with rows and columns interchanged. (The "r" is for *rotate*, but the file is technically *transposed* rather than rotated.) In Fig. 2.4, the file from Fig. 2.1 is shown in the "csvr" format. All other aspects are the same as before, and the use of `read.cross` is unchanged, so such a file (call it "mydata_rot.csv") could be read in as follows.

```
> mydata <- read.cross("csvr", , "mydata_rot.csv")
```

Of course, other arguments, such as `genotypes`, may be used as before.

The "csvs" format is similar to the "csv" format, but with separate files for the phenotypes and the genotypes. The genotype data file must begin with a single column containing individual identifiers, followed by columns for each of the markers. As with the phenotype columns for the "csv" format, this initial column must have empty cells in the rows for the chromosome assignments and marker positions. The phenotype data file must contain a column with precisely the same name and contents, so that we can be sure that the phenotype and genotype data are appropriately aligned. An example of this format is display in Fig. 2.5.

To read data in the "csvs" format, one must specify the names of both files. This may be done via the `read.cross` arguments `genfile` and `phefile`, as follows. (We assume that both files are in the current working directory.)

```
> mydata <- read.cross("csvs", genfile="mydata_gen.csv",
+                      phefile="mydata_phe.csv")
```

Phenotype data file

	A	B	C	D
1	pheno	sex	pgm	id
2	0.093	f	0	1
3	0.177	f	0	2
4	–	f	0	3
5	0.230	f	0	4
6	0.228	f	0	5
7	0.279	f	0	6
8	0.419	f	0	7
9	0.427	f	0	8
10	0.282	f	0	9
11	0.400	f	0	10
12	0.521	f	0	11
13	0.385	f	0	12

Genotype data file

	A	B	C	D	E
1	id	c1m1	c1m3	c1m1	c1m3
2		1	1	1	1
3		8.3	49.0	8.3	49.0
4	1	B	B	B	B
5	2	H	H	H	H
6	3	H	B	H	B
7	4	B	H	B	H
8	5	B	–	B	–
9	6	B	B	B	B
10	7	H	H	H	H
11	8	–	A	–	A
12	9	–	A	–	A
13	10	H	B	H	B

Figure 2.5. Part of the genotype and phenotype data files for an example of the `"csvs"` format, as they might be viewed in a spreadsheet.

For the user's convenience, if the `phefile` argument was not specified, but the `file` and `genfile` arguments were, we assume that `file` and `genfile` are indicating the genotype and phenotype data files, respectively. This can simplify the code a bit. For example, suppose that we are working in a directory `MyProject/R`, and that the two data files are sitting in the directory `MyProject/Data`. The data could be imported as follows.

```
> mydata <- read.cross("csvs", "../Data", "mydata_gen.csv",
+                       "mydata_phe.csv")
```

The `"csvsr"` format is just like the `"csvs"` format, but with both files rotated as in the `"csvr"` format. We use `read.cross` in the same way as for the `"csvs"` format.

2.1.2 MapMaker/QTL

The format `"mm"` is for data in the format used by the MapMaker software. There are two files, a `.raw` file containing the genotype and phenotype data and a second file containing the genetic map information. Examples of these files are provided on the R/qtl web site.

The genetic map file may be in one of two formats. First, one may use a `.maps` file, produced by MapMaker/Exp. Second, one may create a space-delimited file, as illustrated in Fig. 2.6, with one row for each marker. The first column is the chromosome assignment, the second column is the marker name (which must match that used in the `.raw` file exactly), and an optional third column may contain the cM position of each marker.

Use of `read.cross` to read data in the `"mm"` format is similar to the case of the `"csvs"` format, discussed in the previous subsection. Specify the `.raw` file

```
1 D10M44 0.00
1 D1M3 1.00
1 D1M75 24.85
1 D1M215 40.41
1 D1M309 49.99
1 D1M218 52.80
1 D1M451 70.11
1 D1M504 70.81
1 D1M113 80.62
1 D1M355 81.40
1 D1M291 84.93
1 D1M209 92.68
1 D1M155 93.64
2 D2M365 0.00
2 D2M37 27.94
2 D2M396 47.11
⋮
```

Figure 2.6. The initial portion of a space-delimited file that may be used to indicate marker locations for the MapMaker (`"mm"`) format.

with the `file` argument and the genetic map file with the `mapfile` argument. (The format of the genetic map file is determined automatically.) Note that the `na.strings` and `genotypes` arguments are ignored with this format, as such codes are specified within the `.raw` file.

For the user's convenience, if the `mapfile` argument was not specified, but the `genfile` argument was, we assume that `genfile` indicates the genetic map file. This can simplify the code a bit. For example, suppose that we are working in a directory `MyProject/R`, and that the two data files are sitting in the directory `MyProject/Data`. The data could be imported as follows.

```
> mydata <- read.cross("mm", "../Data", "mydata.raw",
+                      "mydata.maps")
```

2.1.3 QTL Cartographer

The format `"qtlcart"` is for data in the format used by the QTL Cartographer software. There are two files, a `.cro` file containing the genotype and phenotype data and a `.map` file containing the genetic map. Examples of these files are provided on the R/qtl web site.

We use `read.cross` to read the QTL Cartographer files in a manner similar to that used for the MapMaker files. For example, suppose we are working in a directory MyProject/R and that the two data files are in the directory MyProject/Data; they could then be imported as follows.

```
> mydata <- read.cross("qtlcart", "../Data", "mydata.cro",
+                      "mydata.map")
```

2.1.4 Map Manager QTX

The format "qtx" is for data in the format used by the Map Manager QTX software. There is a single file, generally with extension .qtx, containing all of the genotype and phenotype data as well as the genetic markers' chromosome assignments and order. Genetic map positions for the markers are generally not included in the file, and so must be estimated. An example file is provided on the R/qtl web site.

Loading data from a .qtx file into R/qtl is simple. The na.strings and genotypes arguments need not be used, as such codes are included within the file. Suppose that we are working in the directory MyProject/R; to read the mydata.qtx from the directory MyProject/Data, type the following.

```
> mydata <- read.cross("qtx", "../Data", "mydata.qtx")
```

As the genetic map positions for the markers are generally not provided in the .qtx file, and so must be estimated from the data, the import of the data can be time consuming. One may wish to use estimate.map=FALSE in the call to read.cross, and then use est.map and replace.map to estimate the map and then plug it into the data. This process is described in more detail in Sec. 3.4.3, but let us briefly consider a simple example.

```
> mydata <- read.cross("qtx", "", "mydata.qtx",
+                      estimate.map=FALSE)
> themap <- est.map(mydata, error.prob=0.001)
> mydata <- replace.map(mydata, themap)
```

In the first line of code, we read in the data without estimating the intermarker distances, and so a dummy map is inserted into the mydata object. In the second line, we call est.map to estimate the genetic map, here assuming that genotypes may be in error with probability 0.1%. The result is placed in the object themap. In the final line of code, the replace.map function is used to replace the map within mydata, inserting themap in its place. The output is the same data but with a different map; we assign it back to mydata, writing over the original data. (We might have assigned it to an object with a different name, in which case both would appear in our R workspace.)

2.2 Exporting data

Data may be exported from R/qtl into several formats. This may be useful, for example, if one wishes to compare results from R/qtl to those from QTL Cartographer, or simulate data in R/qtl and analyze them in Cartographer.

The `write.cross` function is used for this purpose. The `cross` argument is the cross object to be exported. The `chr` argument may be used to indicate a subset of chromosomes that should exported. The `format` argument indicates the format to which the data should be written.

The `filestem` argument indicates the initial part of the file names. For example, with the `qtlcart` format, `.cro` and `.map` files will be created. If one uses `filestem="mydata"`, the files `"mydata.cro"` and `"mydata.map"` will be created.

The `filestem` can include a directory, so that the files may be written somewhere other than the current working directory. For example, if one wishes to save chromosomes 5 and 13 of the `listeria` data to a file in the `"csv"` format on the Desktop on a Macintosh computer, use the following code.

```
> data(listeria)
> write.cross(listeria, "csv", "~/Desktop/listeria", c(5, 13))
```

2.3 Example data

A variety of example data sets are included with R/qtl. A complete list may be obtained with the following.

```
> data(package="qtl")
```

Of particular interest are the `hyper` and `listeria` data, which will be used as the main examples in this book.

The `hyper` data set is from Sugiyama *et al.* (2001). (It was also discussed in Sec. 1.2.) This is a backcross using the C57BL/6J and A/J inbred mouse strains, with the F_1 mated back to the C57BL/6J strain. There are 250 male backcross individuals. Mice were given water containing 1% NaCl for two weeks; the phenotype is blood pressure (actually the average of 20 blood pressure measurements from each of 5 days).

The `listeria` data set is from Boyartchuk *et al.* (2001). This is an intercross using the C57BL/6ByJ and BALB/cByJ inbred mouse strains. There are 120 female intercross individuals (though only 116 were phenotyped). Mice were injected with *Listeria monocytogenes*; the phenotype is survival time (in hours). A large proportion of the mice (35/116) survived past the 240-hour time point and were considered to have recovered from the infection; their phenotype was recorded as 264.

A number of further example data sets will be used in this book. (For a summary of all data sets considered in the book, see Appendix C.) These have been compiled into an R package, R/qtlbook (known in R as the `qtlbook` package). It may be obtained from the website for the book (`http://www.rqtl.org/book`) and from the Comprehensive R Archive Network (CRAN, `http://cran.r-project.org`).

Additional example data may be obtained at the QTL Archive at The Jackson Laboratory (`http://cgd.jax.org/nav/qtlarchive1.htm` or use Google to search for "QTL Archive"). Most of the data sets are available in the `"csv"` format. One must register to access the data. As stated at the QTL Archive, "The authors of the datasets retain individual ownership of the data. We request, as a courtesy to the authors, that you alert them in advance of any publications that result from reanalysis of these data or obtain permission prior to redistribution of data or results."

2.4 Data summaries

All of the data read by `read.cross` (including genotypes, phenotypes, and the genetic map) will be stored in a single object. (This object is stored in a quite complex form; see Sec. 2.6.1.) A number of functions are provided to get summary information about the object.

The most important function is `summary.cross`. In addition to providing a brief summary of the cross, it performs an extensive series of checks of the integrity of the data (for example, that there are the same number of individuals in the phenotype data as in the genotype data).

The data object for a QTL mapping experiment is assigned a "class" `"cross"`. R includes some simple object-oriented features, so that one may use the generic functions `summary` and `plot` on an object, and the relevant summary or plot is made.

For example, the following code loads the `listeria` data and displays a brief summary.

```
> data(listeria)
> summary(listeria)

    F2 intercross

    No. individuals:    120

    No. phenotypes:     2
    Percent phenotyped: 96.7 100

    No. chromosomes:    20
        Autosomes:      1 2 3 4 5 6 7 8 9 10 11 12 13 14 15 16
                        17 18 19
        X chr:          X

    Total markers:      133
    No. markers:        13 6 6 4 13 13 6 6 7 5 6 6 12 4 8 4 4 4
                        4 2
    Percent genotyped:  88.5
```

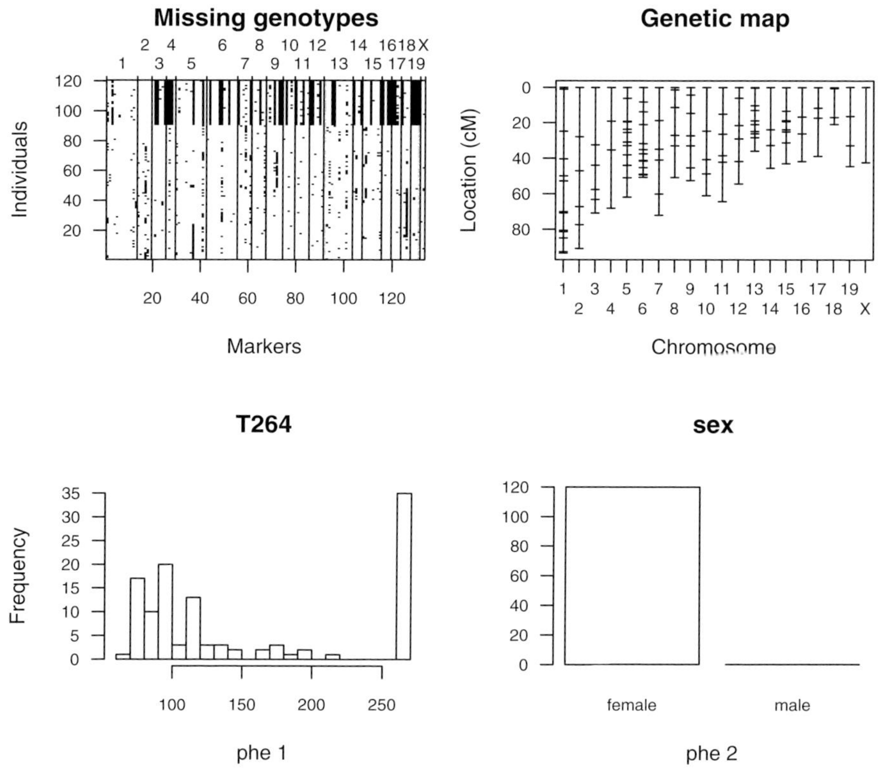

Figure 2.7. The summary plot of the `listeria` data provided by the `plot.cross` function, including the pattern of missing genotype data (upper left; black pixels indicate missing data), the genetic map of the typed markers (upper right), a histogram of the phenotype (lower left), and a bar plot of the sexes (lower right).

```
    Genotypes (%):        CC:26.2  CB:48.9  BB:24  not BB:0
                          not CC:0.9
```

We see that this is an intercross with 120 individuals, that there are two phenotypes, and 20 chromosomes containing 133 markers, and with genotype completion of 88.5%.

In the above code, the generic `summary` function sees that `listeria` has class `"cross"` and passes it to the `summary.cross` function, which provides the actual summary.

Similarly, the following code provides a summary plot of the `listeria` data, and in this case the generic `plot` function passes `listeria` to the `plot.cross` function, which makes the plot (shown in Fig. 2.7).

```
> plot(listeria)
```

The individual panels in Fig. 2.7 may be obtained with the following code.

```
> plot.missing(listeria)
> plot.map(listeria)
> plot.pheno(listeria, 1)
> plot.pheno(listeria, 2)
```

The `plot.missing` function creates the plot with the pattern of missing genotype data. It takes an argument `reorder` which can be used to order the individuals according to their phenotype. The genetic map is obtained with `plot.map`. The function `plot.pheno` plots a phenotype, either as a histogram (using the R function `hist`) or as a bar plot (using the R function `barplot`), depending on the nature of the phenotype.

Finally, there are a variety of other functions for getting additional small pieces of information about a cross object. They are largely self-explanatory.

```
> nind(listeria)

[1] 120

> nphe(listeria)

[1] 2

> totmar(listeria)

[1] 133

> nchr(listeria)

[1] 20

> nmar(listeria)

 1  2  3  4  5  6  7  8  9 10 11 12 13 14 15 16 17 18 19  X
13  6  6  4 13 13  6  6  7  5  6  6 12  4  8  4  4  4  4  2
```

The function `nmar` gives the numbers of markers on individual chromosomes.

2.5 Simulating data

One can simulate QTL mapping data in R/qtl with the `sim.cross` function; it can simulate only additive QTL models. These basic facilities are described in the next subsection. More complex QTL models may also be simulated by making use of the QTL genotype data, which are stored in the object output by `sim.cross`. This will be described in the following subsection. Computer simulations are particularly useful for exploring the power to detect QTL and the precision of localization of QTL. For further details, see Sec. 6.6.

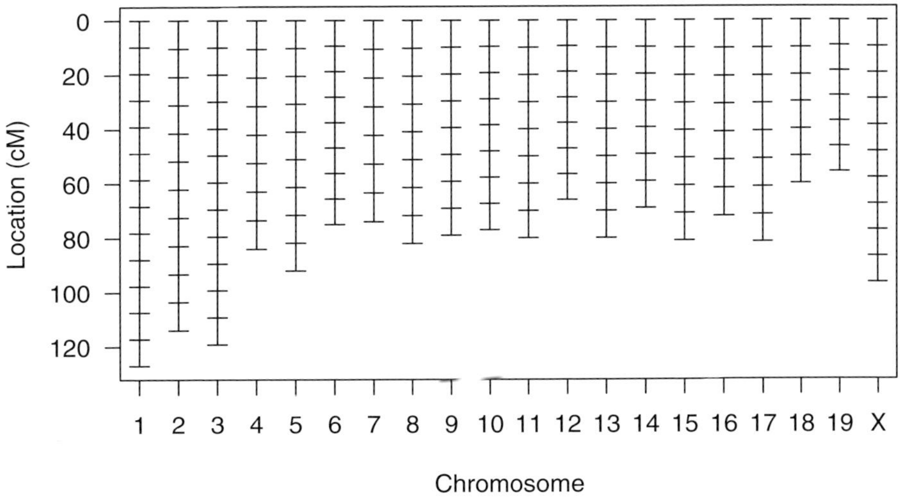

Figure 2.8. A genetic map, with approximately 10 cM marker spacing, modeled after the mouse genome and contained in the `map10` data set in R/qtl.

2.5.1 Additive models

The `sim.cross` function may be used to simulate a backcross or intercross with an additive QTL model. It requires, as input, a genetic map of markers. Such a map must be stored in a specific and rather complicated form (see Sec. 2.6.2), and so we first describe how to create such a map.

First, an example map, modeled after the mouse genome and having approximately evenly spaced markers (at ~10 cM) is provided with R/qtl in the data set `map10`. To access the object and plot the map, type the following.

```
> data(map10)
> plot(map10)
```

The plot is shown in Fig. 2.8. The marker spacing varies slightly across chromosomes so that the lengths of the chromosomes match those of the mouse genome.

Second, one may extract the genetic map from a QTL mapping data set with the `pull.map` function. For example, the following code extracts the map from the `listeria` data.

```
> data(listeria)
> listmap <- pull.map(listeria)
```

Finally, one may use `sim.map` to generate a map, with equally spaced markers or with markers placed randomly. Important arguments to `sim.map` include `len` (the cM lengths of the chromosomes), `n.markers` (the numbers of markers on the chromosomes), `anchor.tel` (indicates whether the ends of the chromosomes should be forced to have markers), `include.x` (whether the final chromosome should be designated to be the X chromosome, versus all chromosomes being autosomes), and `eq.spacing` (whether markers should be spaced evenly).

For example, to create a map with a single autosome of length 200 cM and having markers equally spaced at 20 cM, type the following.

```
> mapA <- sim.map(200, 11, include.x=FALSE, eq.spacing=TRUE)
```

To create a map with 19 autosomes and an X chromosome, chromosomes all of length 100 cM, and each containing 10 randomly positioned markers, though ensuring one marker at each end of each chromosome, we would type the following.

```
> mapB <- sim.map(rep(100, 20), 10)
```

A similar map, but without anchoring the telomeres, would be obtained as follows.

```
> mapC <- sim.map(rep(100, 20), 10, anchor.tel=FALSE)
```

Finally, to get a map with four autosomes of lengths 50, 75, 100, and 125 cM, respectively, and with equally spaced markers at a 5 cM spacing, type the following.

```
> L <- c(50, 75, 100, 125)
> mapD <- sim.map(L, L/5+1, eq.spacing=TRUE, include.x=FALSE)
```

Note that one can use the `summary.map` function to get a short summary of a genetic map; it works much like the `summary.cross` function described in Sec. 2.4. We can get a summary of the `mapD` object, created above, as follows.

```
> summary(mapD)

        n.mar length ave.spacing max.spacing
1          11     50           5           5
2          16     75           5           5
3          21    100           5           5
4          26    125           5           5
overall    74    350           5           5
```

With a genetic map in hand, we can now turn to the simulation of the actual data. The following code simulates data for a backcross of 100 individuals, with complete and error-free genotype data, and markers placed according to the genetic map in `map10`.

```
> simA <- sim.cross(map10, n.ind=100, type="bc")
```

We would simulate an intercross in the same way, using `type="f2"`.

If QTL are to be simulated, we must specify the model via the `model` argument, which should be a matrix with three columns for a backcross and four columns for an intercross. The first column in the matrix gives the chromosomes on which the QTL sit and the second column gives their cM positions.

The third column contains the additive effect of each QTL: in a backcross, the difference between the phenotype averages in heterozygotes and homozygotes; in an intercross, half the difference between phenotype averages for the homozygotes. In an intercross, there must be a fourth column giving the dominance effect for each QTL (the difference between the average phenotype for the heterozygotes and the midpoint between the average phenotypes for the homozygotes).

Phenotypes are simulated from a normal distribution with residual variance $\sigma^2 = 1$. Thus, in a backcross, if there is one QTL with additive effect a, the proportion of the phenotypic variance explained by the QTL (i.e., the heritability due to the QTL) will be $a^2/4/(a^2/4 + 1)$. In an intercross with one QTL exhibiting no dominance, the proportion of the phenotypic variance explained is $a^2/2/(a^2/2 + 1)$.

Let us first simulate a backcross with two additive QTL, each responsible for 8% of the phenotypic variance. Place the first at 50 cM on chromosome 1 and the second at 65 cM on chromosome 14. We must first find the additive effects that correspond to 8% phenotypic variance. Since the QTL are unlinked and have the same size effect, we need $(a^2/4)/[2(a^2/4) + 1] = 0.08$. Solving for a, we obtain $a = \sqrt{4 \times 0.08/(1 - 2 \times 0.08)}$.

```
> a <- 2 * sqrt(0.08 / (1 - 2 * 0.08))
> mymodel <- rbind(c(1, 50, a), c(14, 65, a))
> simB <- sim.cross(map10, type="bc", n.ind=200, model=mymodel)
```

We use the `c` function to combine the chromosome, position and effect of each QTL into a vector, and then `rbind` to combine the two into a matrix (`rbind` makes them rows in the matrix).

As a further example, we simulate an intercross of 250 individuals with three QTL, two having no dominance but with effects in the opposite directions and a third being strictly dominant. Let's have the first two QTL be linked on chromosome 3 at positions 40 cM and 65 cM, and place the third on chromosome 4 at 5 cM. For simplicity, let's set the effects at 0.5.

```
> mymodel2 <- rbind(c(3, 40, 0.5, 0), c(3, 65, -0.5, 0),
+                   c(4, 5, 0.5, 0.5))
> simC <- sim.cross(map10, type="f2", n.ind=250, model=mymodel2)
```

By default, there are no errors in the genotype data. Errors can be included at random via the `error.prob` argument. Genotype data are also, by default, complete. The genotype data can be missing at random with some probability via the `missing.prob` argument. And so we can repeat our backcross simulation with 1% genotyping errors and 5% missing data as follows.

```
> simD <- sim.cross(map10, type="bc", n.ind=200, model=mymodel,
+                   error.prob=0.01, missing.prob=0.05)
```

Random missing genotype data is rather artificial. For more realistic missing data, we can simulate an intercross of the same size as the `listeria` data and apply the missing data observed in that data set. This is not so simple, due to the complexity of the cross data objects and the need for a loop over chromosomes, and so the following code has little chance of being understood by the novice.

```
> data(listeria)
> listmap <- pull.map(listeria)
> simE <- sim.cross(listmap, type="f2", n.ind=nind(listeria),
+                   model=mymodel2)
> for(i in 1:nchr(simE))
+   simE$geno[[i]]$data[ is.na(listeria$geno[[i]]$data) ] <- NA
```

By default, simulations are performed assuming no crossover interference at meiosis. One may also simulate the crosses under the χ^2 model or the Stahl model. (See Sec. 2.7 for references.) The χ^2 model has a single parameter, m, which is a non-negative integer; $m = 0$ corresponds to no interference. With $m > 0$, it is assumed that, on the four-strand bundle at meiosis, chiasmata and intermediate points are thrown down at random (according to a Poisson process), and that every $(m + 1)$st point is a chiasma. No chromatid interference is assumed, so that the particular strands involved in each chiasma are at random, independent between chiasmata. As a result, the crossovers on a random meiotic product may be obtained by "thinning" the chiasmata independently with probability 1/2. (That is, each chiasma has 1/2 chance of being a crossover on the random product, with independence between chiasmata.) In the Stahl model, chiasmata arise according to two independent mechanisms, one following a χ^2 model and the other exhibiting no interference; the observed chiasma locations are the superposition of the two processes. There is one additional parameter, p, giving the proportion of chiasmata to come from the mechanism exhibiting no interference.

We can simulate under the χ^2 model and the Stahl model via the arguments `m` and `p` to `sim.cross`. By default, `m=0` (in which case `p` is irrelevant), indicating no crossover interference. The mouse exhibits strong crossover interference with $m \approx 10$. We can repeat our previous simulation, but with recombination according to a $\chi^2(m = 10)$ model as follows.

```
> simF <- sim.cross(map10, type="f2", n.ind=250, model=mymodel2,
+                   m=10)
```

We can simulate from the Stahl model, with $m = 10$ and $p = 0.1$, as follows.

```
> simG <- sim.cross(map10, type="f2", n.ind=250, model=mymodel2,
+                   m=10, p=0.1)
```

2.5.2 More complex models

The simulations in the previous section were restricted to strictly additive QTL models and with residual variation following a normal distribution with variance $\sigma^2 = 1$. However, the QTL genotype data are stored as a matrix within the output of `sim.cross`; with these data one may simulate data from essentially any QTL model.

First, let us simulate two QTL exhibiting epistasis. Consider a backcross of 200 individuals, with a QTL located at 25 cM on chromosome 4 and another at 45 cM on chromosome 5. Assume that an effect is seen only if an individual is homozygous at both QTL, in which case the phenotype is reduced by one unit.

We begin by simulating QTL having no effect, just so that their genotypes may be obtained, but so that the simulated phenotype will follow a normal(0,1) distribution, independent of genotype. We then modify the phenotype for individuals who are homozygous at both QTL. This requires a bit of mucking about in the cross data object.

```
> data(map10)
> nullmodel <- rbind(c(4, 25, 0), c(5, 45, 0))
> episim <- sim.cross(map10, type="bc", n.ind=200,
+                     model=nullmodel)
> qtlg <- episim$qtlgeno
> wh <- qtlg[,1]==1 & qtlg[,2]==1
> episim$pheno[wh, 1] <- episim$pheno[wh, 1] - 1
```

In the fifth line, we pull out the QTL genotype data. (The columns are the QTL; the rows are the individuals.) In the sixth line, we identify the individuals that are homozygous at both QTL. (Internally, in a backcross, `1` and `2` correspond to the homozygous and heterozygous genotypes, respectively. In an intercross, `1` and `3` are the two homozygous genotypes and `2` is the heterozygous genotype.)

We might create a binary version of this phenotype by thresholding at 1. (Individuals with quantitative phenotype > 1 become affected; the others are unaffected.) We can paste this into the simulated data as a second phenotype.

```
> binphe <- as.numeric(episim$pheno[,1] > 1)
> episim$pheno$affected <- binphe
```

There will now be a second phenotype named "`affected`" with 1 and 0 indicating affected and unaffected, respectively.

Finally, we might assign sexes to the individuals at random, and include a sex difference in the phenotype and even a difference in the effect of the QTL in the two sexes (a QTL $\times$ sex interaction). We'll create a third phenotype with these features, and place "sex" in the data as a fourth phenotype. Here, `0` and `1` correspond to females and males, respectively.

```
> sex <- sample(0:1, nind(episim), replace=TRUE)
> phe3 <- rnorm(nind(episim), 0, 1)
> phe3[wh & sex==0] <- phe3[wh & sex==0] - 1.5
> phe3[wh & sex==1] <- phe3[wh & sex==1] - 0.5
```

```
> episim$pheno$pheno3 <- phe3
> episim$pheno$sex <- sex
```

We use the R function `sample` to sample with replacement from the vector (0, 1), and `rnorm` to simulate standard normal data. The epistasis pattern for the two QTL is as before, but the effects are different in the two sexes. We reuse the `wh` object, created above, that indicated the individuals who were homozygous at both QTL.

2.6 Internal data structure

In this section, we describe the internal data structures used by R/qtl for cross and genetic map objects and the R syntax required to get access to the data. Other data structures (such as those produced by the `scanone` and `scantwo` functions) will be described in later chapters. This section is quite technical and will require a reasonably detailed understanding of R, and so it should probably be skipped initially. The choice of data structures required some balance between ease of programming and simplicity for the user interface. The syntax for references to certain pieces of the internal data can be quite complicated.

2.6.1 Experimental cross

We describe the internal data structure used by R/qtl for QTL mapping data; we will look at the data set `hyper` as an example. First, the object has a "class," which indicates that it corresponds to data for an experimental cross, and gives the cross type. By having class `"cross"`, the functions `plot` and `summary` know to send the data to `plot.cross` and `summary.cross`.

```
> data(hyper)
> class(hyper)
```

```
[1] "bc"    "cross"
```

As you can see, the class is a two-element vector containing first a character string indicating the cross type (`"bc"` or `"f2"`) and second `"cross"` to indicate that it is an experimental cross.

Every cross object is a list with two components, one containing the genotype data and genetic maps and the other containing the phenotype data.

```
> names(hyper)
```

```
[1] "geno"  "pheno"
```

The phenotype data is simply a matrix (more strictly a data frame) with rows corresponding to individuals and columns corresponding to phenotypes. We look at the phenotypes for the first five individuals as follows.

```
> hyper$pheno[1:5,]

     bp  sex
1 109.6 male
2 109.8 male
3 110.1 male
4 110.6 male
5 115.0 male
```

The first phenotype is the blood pressure of each mouse; the second phenotype indicates their sex. (In this case, all mice are male.) The phenotypes can be either numeric or factors. The sex phenotype can be coded `0/1`, `f/m`, `F/M`, or `female/male` for female/male; in all but the first case, it must be a factor.

The genotype data is a list with components corresponding to chromosomes. Each chromosome has a name and a class. The class for a chromosome is `"A"` or `"X"`, for autosomes or the X chromosome, respectively.

```
> names(hyper$geno)

 [1] "1"  "2"  "3"  "4"  "5"  "6"  "7"  "8"  "9"  "10" "11"
[12] "12" "13" "14" "15" "16" "17" "18" "19" "X"

> sapply(hyper$geno, class)

  1   2   3   4   5   6   7   8   9  10  11  12  13  14  15
"A" "A" "A" "A" "A" "A" "A" "A" "A" "A" "A" "A" "A" "A" "A"
 16  17  18  19   X
"A" "A" "A" "A" "X"
```

Each component of `geno` is itself a list with two components, `data` (containing the marker genotype data) and `map` (containing the positions of the markers, in cM). The genotype data are coded `1/2` for homozygotes and heterozygotes in a backcross, and `1/2/3/4/5` for the genotypes AA/AB/BB/not BB/not AA in an intercross.

```
> names(hyper$geno[[3]])

[1] "data" "map"

> hyper$geno[[3]]$data[91:94,]

   D3Mit164 D3Mit6 D3Mit11 D3Mit14 D3Mit44 D3Mit19
91        2      1       1       1       1       1
92        1      1       1       1       1       1
93       NA      2      NA      NA      NA      NA
94       NA      2      NA      NA      NA      NA

> hyper$geno[[3]]$map

D3Mit164   D3Mit6  D3Mit11  D3Mit14  D3Mit44  D3Mit19
     2.2     17.5     37.2     44.8     57.9     66.7
```

On the X chromosome, all individuals are coded with genotypes `1/2`. We use the phenotypes `sex` and `pgm`, if they are available, to recode these as AA/AB/BB/AY/BY before later analysis. The `1/2` codes simplify the use of the HMM algorithms (as in `calc.genoprob`, to calculate genotype probabilities), as all individuals may be treated as a backcross.

That completes the description of the raw data. However, other information may exist in a cross object, as when one runs `calc.genoprob`, `sim.geno`, or `calc.errorlod`, the output is the input cross object with the derived data attached to each component (the chromosomes) of the `geno` component.

```
> names(hyper$geno[[3]])

[1] "data" "map"

> hyper <- calc.genoprob(hyper, step=10, error.prob=0.01)
> names(hyper$geno[[3]])

[1] "data" "map"  "prob"

> hyper <- sim.geno(hyper, step=10, n.draws=2, error.prob=0.01)
> names(hyper$geno[[3]])

[1] "data"  "map"   "prob"  "draws"

> hyper <- calc.errorlod(hyper, error.prob=0.01)
> names(hyper$geno[[3]])

[1] "data"     "map"      "prob"     "draws"    "errorlod"
```

The structure of the individual components that were added is relatively self-explanatory.

Finally, when one runs `est.rf`, a matrix containing the pairwise recombination fractions and LOD scores is added to the cross object.

```
> names(hyper)

[1] "geno"  "pheno"

> hyper <- est.rf(hyper)
> names(hyper)

[1] "geno"  "pheno" "rf"
```

The `hyper$rf` object is a matrix. Values on the diagonal are the number of individuals that were genotyped for the corresponding marker. Values above the diagonal are LOD scores for a test of linkage; values below the diagonal are estimated recombination fractions.

```
> hyper$rf[1:4,1:4]
```

```
         D1Mit296 D1Mit123 D1Mit156 D1Mit178
D1Mit296  92.0000  11.4201   3.1422   0.6321
D1Mit123   0.1413  92.0000   9.9274   0.6321
D1Mit156   0.3043   0.1630 250.0000   2.9045
D1Mit178   0.1667   0.1667   0.2449  49.0000
```

The function `clean.cross` may be used to remove the intermediate results from a cross object (such as those created with `calc.genoprob` and `est.rf`), as follows.

```
> hyper <- clean(hyper)
> names(hyper)
```

```
[1] "geno"  "pheno"
```

```
> names(hyper$geno[[3]])
```

```
[1] "data" "map"
```

2.6.2 Genetic map

A genetic map object, as produced by `sim.map` or as extracted from a cross object with `pull.map`, also has a somewhat complex form. We will look at the data set `map10`, a genetic map modeled after the mouse genome. Such a map object has class `"map"` so that `plot` and `summary` will call `plot.map` and `summary.map`, respectively.

```
> data(map10)
> class(map10)
```

```
[1] "map"
```

The map is a list whose components are the individual chromosomes. Each chromosome has class either `"A"` or `"X"` according to whether it is an autosome or the X chromosome.

```
> names(map10)
```

```
 [1] "1"  "2"  "3"  "4"  "5"  "6"  "7"  "8"  "9"  "10" "11"
[12] "12" "13" "14" "15" "16" "17" "18" "19" "X"
```

```
> sapply(map10, class)
```

```
  1   2   3   4   5   6   7   8   9  10  11  12  13  14  15
"A" "A" "A" "A" "A" "A" "A" "A" "A" "A" "A" "A" "A" "A" "A"
 16  17  18  19   X
"A" "A" "A" "A" "X"
```

The individual chromosomes are vectors specifying the marker locations in cM, with names being the marker names.

```
> map10[[15]]

D15M1 D15M2 D15M3 D15M4 D15M5 D15M6 D15M7 D15M8 D15M9
 0.00 10.12 20.25 30.38 40.50 50.62 60.75 70.88 81.00
attr(,"class")
[1] "A"
```

2.7 Further reading

Broman and Heath (2007) discuss the management and manipulation of genetic data. They emphasize the need for biologists to learn to program, and the value of the Perl programming language for geneticists. While they focus on human linkage data, the general principles apply to all genetic data

Useful Perl books include *Learning Perl* (Schwartz *et al.*, 2008) for beginners, *Programming Perl* (Wall *et al.*, 2000) as a reference, and *Perl Cookbook* (Christiansen and Torkington, 2003) for its recipes encompassing many common tasks. These books, plus a couple of others, may be purchased together on a CD for a very good price: the *Perl CD Bookshelf*, available from O'Reilly Media.

Regarding the χ^2 model for crossover interference, see Zhao *et al.* (1995). The Stahl model was described in Copenhaver *et al.* (2002).

3

Data checking

Our ability to map the loci contributing to variation in a trait depends critically on the quality and integrity of the data. Odd mapping results can often be traced to errors in the genotype data, the genetic maps, or the phenotype data. Thus, the first order of business, following data import, should be to identify and correct errors in the data.

A variety of data diagnostics are provided in R and R/qtl. We illustrate these below using real but anonymized data. The process of checking data can be quite interesting detective work. The features that should be studied are generally well characterized, but in many cases it can be tricky to identify the primary cause of a particular problem.

3.1 Phenotypes

We first take a look at the phenotype data. We look for individuals with unusual phenotypes. These may be truly unusual individuals, but they may also indicate errors in data entry and so deserve careful follow-up. We also look for systematic problems in the phenotype data (such as drifts in the measurements over time or between batches).

We begin by considering the example data, `ch3a`. We use the `library` function to load the `qtl` and `qtlbook` packages (the latter contains the example data used in this book), and use the `data` function to make the `ch3a` data set available to us.

```
> library(qtl)
> library(qtlbook)
> data(ch3a)
```

These data have five related phenotypes; Fig. 3.1 contains histograms of the phenotypes. Note that the histograms are generally skewed; this may influence our choice of QTL mapping method, or we may seek to transform

K.W. Broman, Ś. Sen, *A Guide to QTL Mapping with R/qtl*,
Statistics for Biology and Health, DOI 10.1007/978-0-387-92125-9_3,

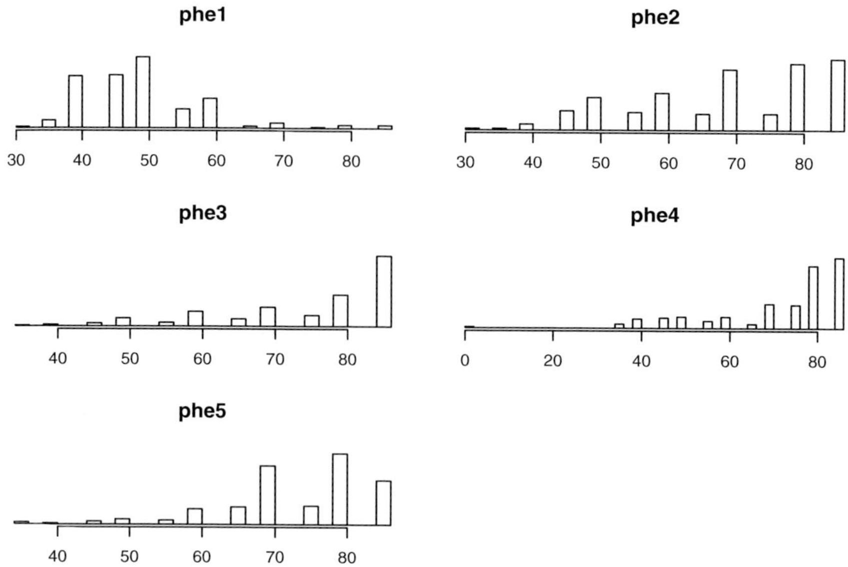

Figure 3.1. Histograms of the phenotypes from the `ch3a` data.

the phenotypes. More importantly, though, note that there is one individual whose fourth phenotype is 0, considerably lower than the other individuals.

Figure 3.1 may be produced with the following code.

```
> par(mfrow=c(3,2))
> for(i in 1:5)
+   plot.pheno(ch3a, pheno.col=i)
```

The function `par` is used to modify graphics parameters; we create three rows and two columns of plots with `mfrow=c(3,2)`. The function `c` is used to combine multiple items together into a vector.

We step through the five phenotypes using a `for` loop. (The `plot.pheno` function is called five times, with `i` taking the values 1, 2, . . . , 5, sequentially.) The distribution of a phenotype may displayed with the `plot.pheno` function, which will create either a histogram or a bar plot, according to whether the phenotype is numeric or categorical.

The unusual individual is rather difficult to see in Fig. 3.1; it is more clear in scatterplots of the phenotypes against one another, displayed in Fig. 3.2. Each panel contains the data for one phenotype plotted against the data for another phenotype. The individual with 0 at the fourth phenotype now stands out.

Figure 3.2 was created with the following code.

```
> pairs(jitter( as.matrix(ch3a$pheno) ), cex=0.6, las=1)
```

Since the phenotypes are discrete, we use `jitter` to add a bit of noise so that individual points may be distinguished. The code `ch3a$pheno` is used to

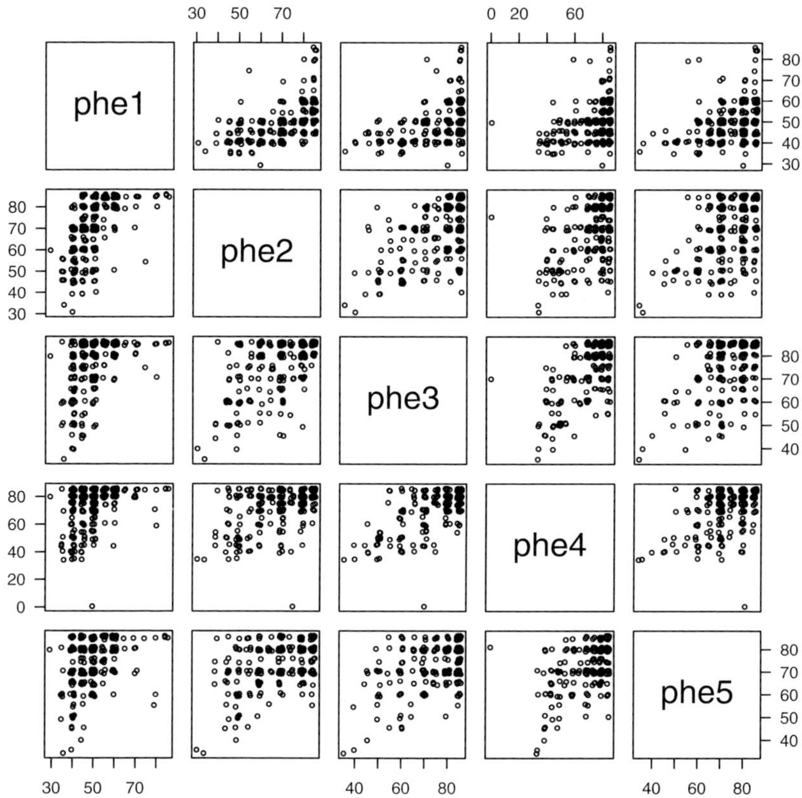

Figure 3.2. Scatterplots of the phenotypes against one another for the `ch3a` data.

pull out the phenotype data, which must be converted to a numeric matrix with `as.matrix` in order to use the `jitter` function. The function `pairs` creates the set of scatterplots; `cex=0.6` is used to change the size of the points and `las=1` is used to change the orientation of the y-axis labels.

It is best to go back to the primary data to determine whether the 0 is a true phenotype or whether it is a data entry error. In the latter case, we may set the phenotype to be missing as follows.

```
> ch3a$pheno[ch3a$pheno == 0] <- NA
```

Any 0 phenotypes will then be replaced with `NA` and therefore are treated as missing.

To complete our exploration of the phenotype data, we plot the individuals' phenotypes against their index, which may correspond to the order in which they were measured. The left panel in Fig. 3.3 contains a plot of the average of the five phenotypes for each individual, in the order they appeared in the

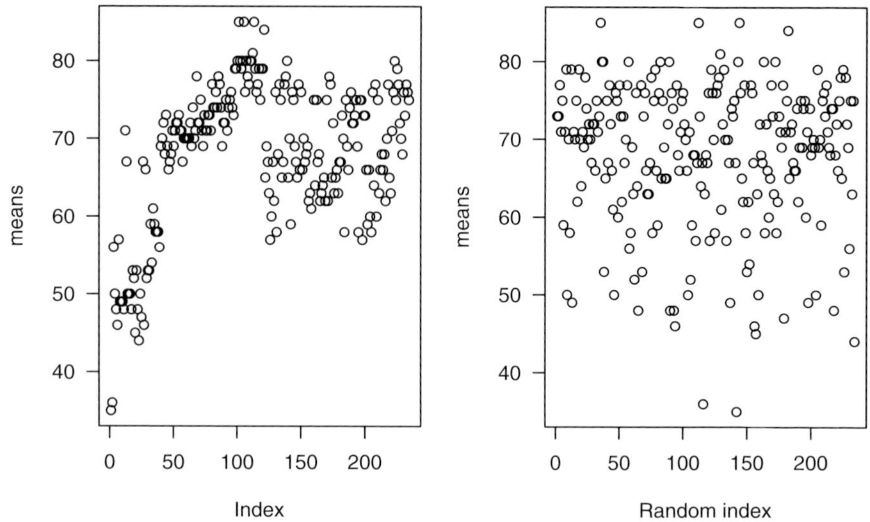

Figure 3.3. Plots of the average phenotype against index (left panel) and against a randomized index (right panel) for the `ch3a` data.

data. The right panel contains the same individual phenotype averages, but in a random order.

There is a clear pattern in the average phenotype that is not seen in the case that the data have been randomized (which we include just to emphasize the point). If this truly represents systematic changes in the phenotype over the course of the measurements, it is cause for considerable concern, as it indicates an important source of uncontrolled (and nongenetic) variation.

Figure 3.3 was produced with the following code. The function `apply` is used to obtain the row averages in the phenotype data, and `sample` is used to randomize the order of these averages.

```
> par(mfrow=c(1,2), las=1, cex=0.8)
> means <- apply(ch3a$pheno, 1, mean)
> plot(means)
> plot(sample(means), xlab="Random index", ylab="means")
```

3.2 Segregation distortion

It is important to check for segregation distortion at all markers (i.e., that the genotypes appear in the expected proportions), as apparent distortion may indicate genotyping problems. For example, consider the example data, `ch3b`. We use the function `geno.table` to inspect the genotype frequencies at each marker. The last column in the output is a p-value for a χ^2 test of Mendelian proportions (1:2:1 in an intercross).

```
> data(ch3b)
> gt <- geno.table(ch3b)
> gt[ gt$P.value < 1e-7, ]
```

```
       chr missing AA  AB  BB not.BB not.AA AY BY   P.value
c4m7     4       0 31   0 113      0      0  0  0 2.829e-52
c6m9     6       0 28   0 116      0      0  0  0 2.374e-55
c7m3     7      40 62   2  40      0      0  0  0 1.257e-23
c10m8   10      64 36   4  40      0      0  0  0 6.950e-15
c11m3   11      26 10   0 108      0      0  0  0 1.070e-61
c13m1   13      18 99  27   0      0      0  0  0 1.923e-43
c13m4   13      51 43  13  37      0      0  0  0 2.241e-11
c14m1   14      57 10  77   0      0      0  0  0 1.979e-12
c14m2   14       0 61  79   4      0      0  0  0 8.048e-11
c14m5   14      18 45   1  80      0      0  0  0 1.900e-31
c16m2   16      12  0 104  28      0      0  0  0 8.293e-13
c16m6   16      15 32   1  96      0      0  0  0 1.149e-41
c18m5   18      38  1   1 104      0      0  0  0 2.379e-66
c19m3   19       9  0 134   1      0      0  0  0 3.500e-29
```

We see 14 markers, on 11 chromosomes, showing quite extreme distortion. For example, markers c4m7 and c6m9 had no heterozygotes, while marker c19m3 had almost all heterozygotes. It is likely that these problems are due to genotyping errors.

As a further example, let us look at the `listeria` data from Boyartchuk *et al.* (2001), which is included with R/qtl. Some markers show segregation distortion, but the distortion is much less severe than was observed in the `ch3b` data.

```
> data(listeria)
> gt <- geno.table(listeria)
> p <- gt$P.value
> gt[ !is.na(p) & p < 0.01, ]
```

```
        chr missing CC CB BB not.BB not.CC   P.value
D6M284    6       0 27 76 17      0      0 0.0060967
D6M254    6       0 18 77 25      0      0 0.0053804
D12M46   12      33 34 33 20      0      0 0.0083345
D13M99   13       0 48 49 23      0      0 0.0007282
D13M233  13      14 41 43 22      0      0 0.0050294
D13M106  13       0 45 55 20      0      0 0.0036066
D13M147  13       0 45 55 20      0      0 0.0036066
```

Note that `is.na` returns TRUE or FALSE according to whether the values are missing or not, respectively, and `!` is the logical "not."

Many markers on chromosome 13 show a reduced frequency of B alleles, which may indicate real segregation distortion (e.g., that there is a locus

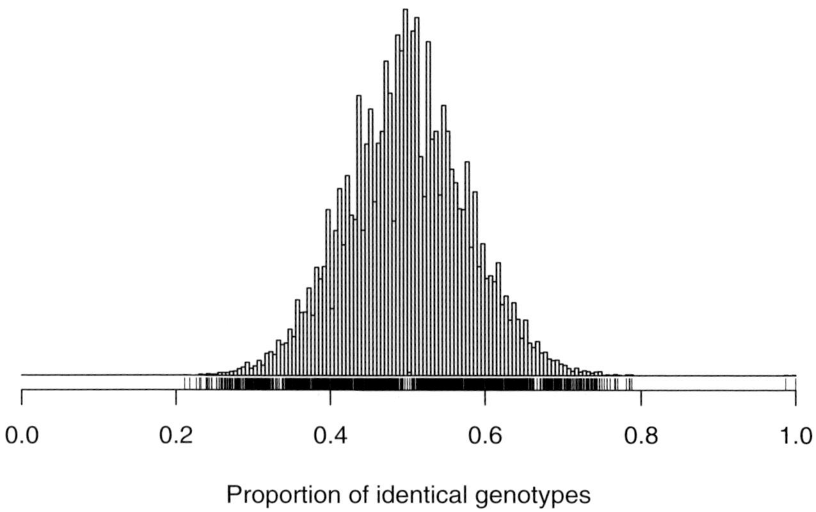

Figure 3.4. Histogram of the proportion of markers with identical genotypes for each pair of individuals in the `ch3a` data.

on that chromosome that is associated with early mortality), rather than genotyping errors.

3.3 Compare individuals' genotypes

We have occasionally found it useful to compare the genotype data for each pair of individuals from a cross, to identify pairs that have unusually similar genotypes. These may indicate sample mix-ups of some kind.

For example, Fig. 3.4 contains a histogram of the proportion of markers with identical genotypes for each pair of individuals in the `ch3a` data. There are two pairs of individuals that have very similar genotype data.

Figure 3.4 was created with the following code. The `comparegeno` function returns a matrix whose (i, j)th element is the proportion of markers at which individuals i and j have the same genotype. The function `hist` creates the actual histogram; the argument `breaks` defines the number of bins (and can be used to define the precise breakpoints for the bins). The function `rug` is used to create, underneath the histogram, line segments at the individual data points, so that the two outliers may be clearly seen.

```
> data(ch3a)
> cg <- comparegeno(ch3a)
> hist(cg, breaks=200,
+      xlab="Proportion of identical genotypes")
> rug(cg)
```

With the following code, we can identify the pairs of individuals with very similar genotype data.

```
> which(cg > 0.9, arr.ind=TRUE)
```

```
     row col
[1,] 138   5
[2,]  55  12
[3,]  12  55
[4,]   5 138
```

Individuals 5 and 138 have identical genotypes at all 86 markers at which they were both typed; individuals 12 and 55 have the same genotype at 75/76 markers. Real backcross individuals shouldn't show such similarity in their genotypes, and so these individuals' data should be viewed with suspicion.

3.4 Check marker order

It is critical that one check that markers are placed on the correct chromosomes and in the correct order, as the incorrect placement of markers on the map can destroy the results of the QTL analysis. Even if marker positions are based on a high-quality physical map, mislabeling of markers can occur, and so marker labels may not match the true markers that were genotyped.

3.4.1 Pairwise recombination fractions

The first thing to do is to estimate, for each pair of markers, the recombination fraction between them, r, and calculate a LOD score for the test of $r = 1/2$. Markers on different chromosomes should not appear linked, and for markers on the same chromosome, the estimated recombination fraction should be smaller for more closely linked markers.

We again consider a set of real but anonymized data, included in the `qtlbook` package. We use `est.rf` to estimate the recombination fractions between all pairs of markers. It inserts the results back into the cross object, and so we assign the results back to the object, `ch3c`.

```
> data(ch3c)
> ch3c <- est.rf(ch3c)
```

```
Warning message:
In est.rf(ch3c) : Alleles potentially switched at markers
  c1m3 c1m4 c7m1 c7m2
```

Before proceeding further, note the warning message produced by `est.rf`. There may be markers whose alleles got switched (A for B and B for A). These potential problems are identified by looking for markers whose LOD scores are larger for cases with $\hat{r} > 0.5$ rather than $\hat{r} < 0.5$. The function `checkAlleles` gives slightly more detail; the last column in the output is the difference between the largest LOD score corresponding to an estimated recombination fraction > 0.5 and the largest LOD score corresponding to an estimated recombination fraction < 0.5. Note that markers that are tightly linked to a problem marker will also show up in this table.

```
> checkAlleles(ch3c)

   marker chr index diff.in.max.LOD
2    c1m3   1     2          21.378
3    c1m4   1     3          15.010
32   c7m1   7     1           6.691
33   c7m2   7     2          10.769
```

There appear to be problems on chromosomes 1 and 7. Let us look in more detail at the genotype data for the markers on chr 1. We use `pull.map` to display the map for that chromosome, so that we can see the other marker names, and then use `geno.crosstab` to create tables of genotypes at one marker against genotypes at another marker.

```
> pull.map(ch3c, 1)

c1m1 c1m3 c1m4 c1m5
 8.3 49.0 59.5 89.0

> geno.crosstab(ch3c, "c1m3", "c1m4")

    c1m4
c1m3 - AA AB BB
  -  7  0  0  0
  AA 0  0  3 19
  AB 0  0 38  7
  BB 0 22  3  1

> geno.crosstab(ch3c, "c1m3", "c1m5")

    c1m5
c1m3 - AA AB BB
  -  7  0  0  0
  AA 0  2 11  9
  AB 0  9 24 12
  BB 0 12 12  2

> geno.crosstab(ch3c, "c1m4", "c1m5")

    c1m5
c1m4 - AA AB BB
  -  7  0  0  0
```

```
AA 0 11 10  1
AB 0 11 28  5
BB 0  1  9 17
```

It looks like marker 2 ("c1m3") is the problem: for that marker, relative to markers c1m4 and c1m5, the double-recombinant classes are more common than the nonrecombinant ones, while the table of two-locus genotypes for markers c1m5 and c1m5 looks okay.

To fix the problem, we pull out the genotypes for chromosome 1 using the function pull.geno, swap the alleles (replacing 1's with 3's and vice versa), and then put the new data back.

```
> g <- pull.geno(ch3c, 1)
> g[,"c1m3"] <- 4 - g[,"c1m3"]
> ch3c$geno[[1]]$data <- g
```

By a similar approach, we find that it is marker 2 ("c7m2") on that is the problem one on chr 7. We fix it as follows.

```
> g <- pull.geno(ch3c, chr=7)
> g[,"c7m2"] <- 4 - g[,"c7m2"]
> ch3c$geno[[7]]$data <- g
```

If we now rerun est.rf and checkAlleles, we'll find there are no further problems of this form.

```
> ch3c <- est.rf(ch3c)
> checkAlleles(ch3c)

No apparent problems.
```

We now return to the recombination fractions themselves, and our assessment of marker placement. The function plot.rf is used to plot the pairwise recombination fractions and LOD scores. We use the option alternate.chrid=TRUE so that the individual chromosome IDs may be more easily distinguished.

```
> plot.rf(ch3c, alternate.chrid=TRUE)
```

The results are displayed in Fig. 3.5; the estimated recombination fractions between markers are in the upper left, and the LOD scores are in the lower right. Red indicates pairs of markers that appear to be linked (low $\hat{r}$ or high LOD), and blue indicates pairs that are not linked (high $\hat{r}$ or low LOD).

There are a number of red points in the lower right, indicating markers on different chromosomes that appear linked. In particular, there appear to be problems on chromosomes 1, 7, 12, 13, and 15. With the following code, we plot the results for just those chromosomes (see Fig. 3.6).

```
> plot.rf(ch3c, chr=c(1, 7, 12, 13, 15))
```

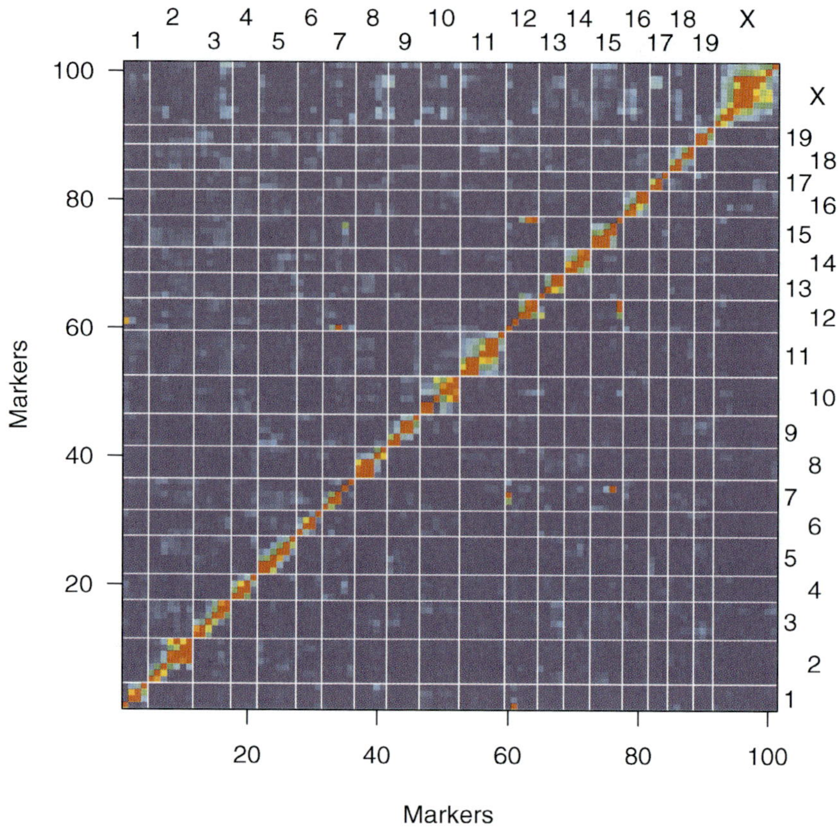

Figure 3.5. Estimated recombination fractions (upper left) and LOD scores (lower right) for all pairs of markers in the `ch3c` data.

As a further indication of this problem, it is valuable to use the available genotype data to reestimate the intermarker distances of the genetic map. This is done, and the map plotted, with the following code. Note that the `nm` object will have class `"map"`, and so `plot(nm)` is equivalent to `plot.map(nm)`. Also note that with `error.prob=0.001`, we assume a 0.1% genotyping error rate.

```
> nm <- est.map(ch3c, error.prob=0.001)
> plot(nm)
```

The estimated map, shown in Fig. 3.7, indicates clear problems on chromosomes 7, 12 and 15: enormous map expansion occurs as a result of markers that do not belong on those chromosomes. There may also be problems on chromosomes 10 and 13. The results in Fig. 3.6 indicate that the fourth marker on chr 7 belongs on chr 15, the first marker on chr 12 belongs on chr 7, the

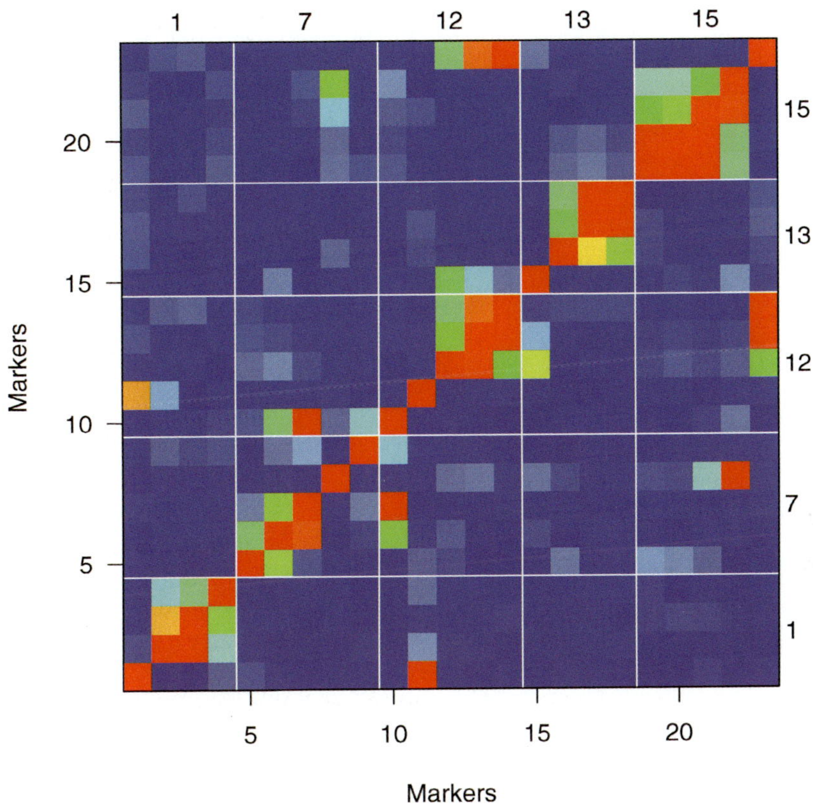

Figure 3.6. Estimated recombination fractions (upper left) and LOD scores (lower right) for pairs of markers on selected chromosomes in the `ch3c` data.

second marker on chr 12 belongs on chr 1, the first marker on chr 13 belongs on chr 12, and the fifth marker on chr 15 belongs on chr 12.

It is questionable whether we should move these markers to the positions that they appear to be linked, or just omit them. The ideal solution would be to retype these markers starting with new material. If we want to use the available data for an initial analysis, and so seek to fix these problems in marker positions, R/qtl does have some limited facilities for moving markers between chromosomes.

We first need to identify the names of the markers, which can be accomplished with the function `find.marker`, using the argument `index` to specify the markers by their numeric indices within the chromosomes. We then use the `movemarker` function to move the markers to the chromosomes that they appear to be belong. (The markers are moved to the end of the chromosome, and so we will later need to fix the marker order on those chromosomes.)

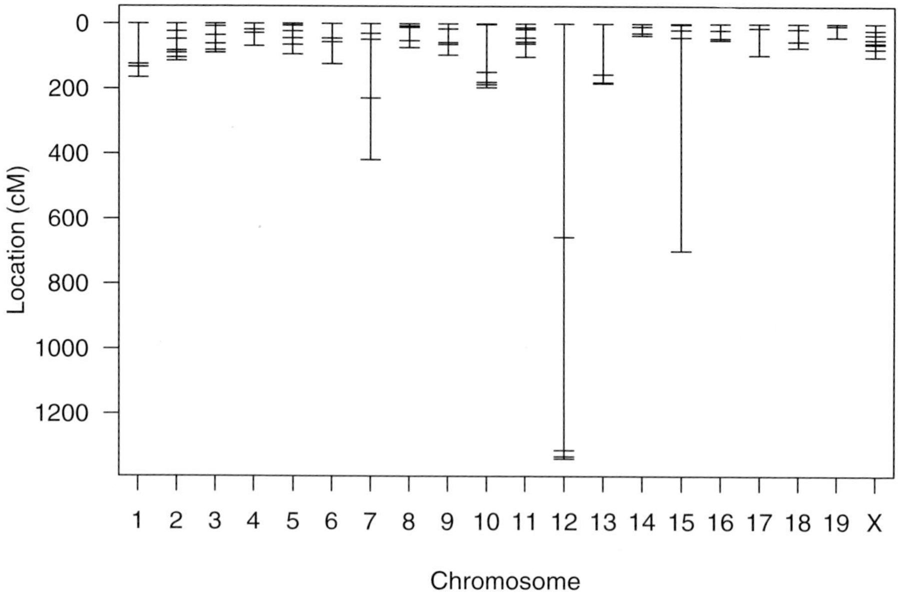

Figure 3.7. Genetic map, as estimated from the ch3c data.

```
> ch3c <- movemarker(ch3c, find.marker(ch3c, 7,index=4), 15)
> ch3c <- movemarker(ch3c, find.marker(ch3c,12,index=2),  1)
> ch3c <- movemarker(ch3c, find.marker(ch3c,12,index=1),  7)
> ch3c <- movemarker(ch3c, find.marker(ch3c,13,index=1), 12)
> ch3c <- movemarker(ch3c, find.marker(ch3c,15,index=5), 12)
```

We needed to be careful to move the second marker on chr 12 before moving the first marker; if we had first moved the initial marker, the second marker would no longer be in the second position.

We can then use `est.rf` to plot the recombination fractions and LOD scores for the relevant chromosomes again. The results are in Fig. 3.8. The markers now appear to be on the correct chromosomes, though there remain some problems with the order of markers within the chromosomes. (We will address that issue in Sec. 3.4.2.)

One last point on pairwise linkages: some strategies for selective genotyping can lead to odd results in the pairwise recombination fractions. For example, consider the `hyper` data. At most markers, only individuals with extreme phenotypes were genotyped, and at some markers, only individuals showing recombination events at the surrounding markers were genotyped. (The pattern of missing genotype data is diplayed in Fig. 1.7 on page 10). The latter strategy leads to somewhat odd results in the pairwise recombination fractions and LOD scores, as in estimating the recombination fraction between a pair of markers, we consider only the data on individuals who were typed at both markers.

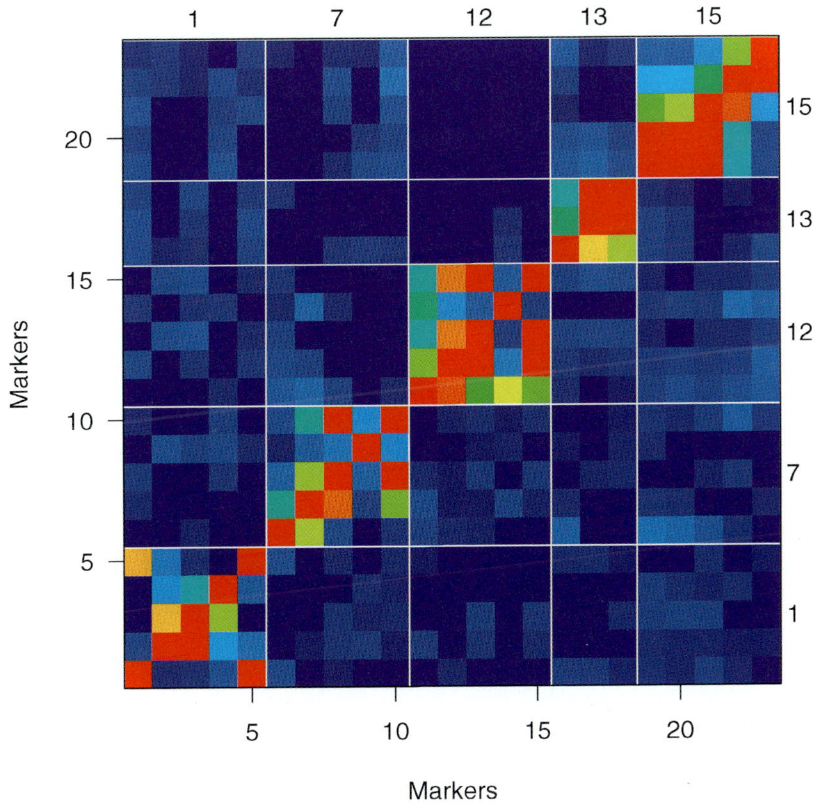

Figure 3.8. Estimated recombination fractions (upper left) and LOD scores (lower right) for pairs of markers on selected chromosomes in the `ch3c` data, after some problems with marker positions have been fixed.

To calculate the recombination fractions for the `hyper` data set, we do the following. The results are displayed in Fig. 3.9.

```
> data(hyper)
> hyper <- est.rf(hyper)
> plot.rf(hyper, alternate.chrid=TRUE)
```

Note that, while the LOD scores in the lower right triangle indicate no linkages between nonsyntenic markers, there are some pairs with low estimated recombination fractions (red pixels in the upper left triangle), as some markers were typed on very few individuals. The checkerboard patterns on chr 1, 4, 11, and 15, might indicate problems with marker order, but are really due to the strategy of typing only recombinant individuals at some markers.

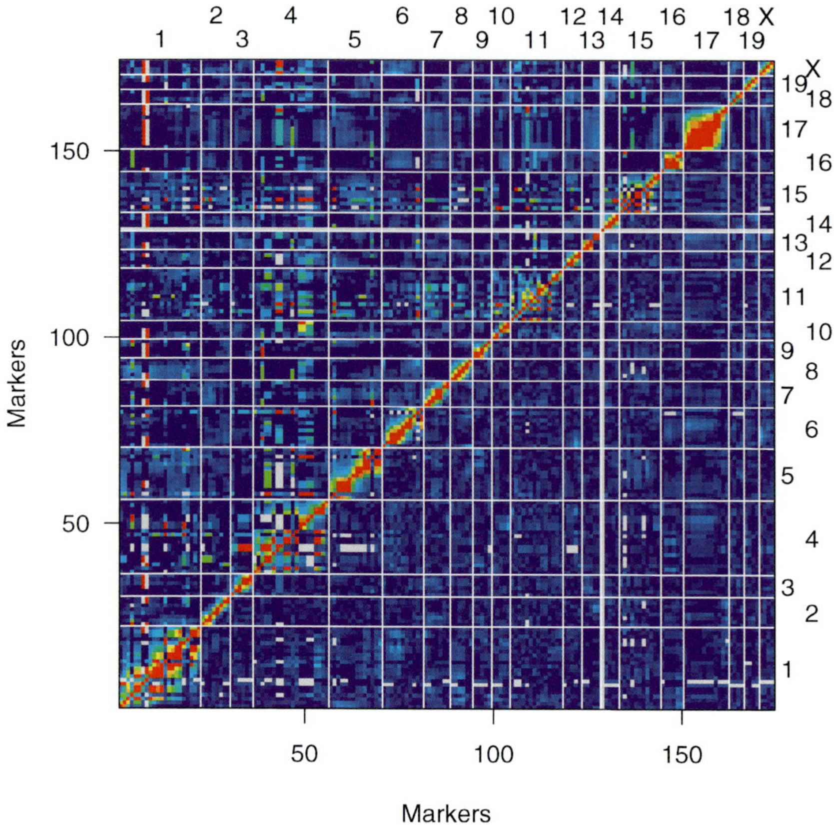

Figure 3.9. Estimated recombination fractions (upper left) and LOD scores (lower right) for all pairs of markers in the `hyper` data.

3.4.2 Rippling marker order

One can check the order of markers on a chromosome using the `ripple` function (whose name was taken from a similar function in MapMaker/EXP).

We would like to compare all possible orderings of markers on a chromosome, but with even a moderate number of markers, the number of possible marker orders is too large for such an exhaustive evaluation. If there are n markers on a chromosome, there are $n!/2$ possible marker orders, where $n! = n \times (n-1) \times (n-2) \times \cdots \times 2 \times 1$. In the case of 10 markers, there are 1,814,400 possible marker orders. Thus, in `ripple` we consider a sliding window of markers and consider all possible orders of the markers within the window, keeping the order of markers outside the window fixed.

The `ripple` function can use two methods for comparing orders: maximum likelihood and minimal obligate crossovers. Maximum likelihood (in which one considers the probability of the observed genotype data given a particular order, and then chooses the marker order for which this probability is maximized) is generally preferred, but is considerably more computationally intensive. It is simpler to count the number of obligate crossovers in the data, for a given order, and then choose the order for which the number of obligate crossovers is minimized. (In a backcross, one is simply counting crossovers, though with the assumption that no double crossovers occur between typed markers.) Counting crossovers is hundreds of times faster than maximum likelihood, and the results are remarkably similar. Thus we generally recommend using the crossover count method with a large window, followed by maximum likelihood with a much smaller window.

The key arguments for the `ripple` function include the `cross`, the specified chromosome (`chr`; just one chromosome is considered at a time), the `window` size, and the `method` (either `"countxo"` or `"likelihood"`). For the likelihood method, one may also specify an assumed genotyping error rate with `error.prob`.

Let us return to the `ch3c` data; we had ensured that all markers were on the correct chromosomes, but saw that some chromosomes showed clear problems in marker order. We first look at chromosome 1, for which there are just five markers. We'll first count crossovers, and look at all possible marker orders. We do so as follows.

```
> rip <- ripple(ch3c, 1, 5)
```

```
    60 total orders
```

The result (assigned to the object `rip`) contains the number of obligate crossovers for each possible marker order. We can get a summary of the results as follows.

```
> summary(rip)
```

```
                      obligXO
Initial  1 2 3 4 5        197
1        1 5 2 3 4        124
2        4 3 2 1 5        134
3        2 3 4 1 5        146
4        1 5 4 3 2        146
5        1 5 3 2 4        152
... [ 16  additional rows] ...
```

The first row is the original marker order (i.e., that which is in the data). Other marker orders are sorted by the number of obligate crossovers (which appears in the final column), but only the first few are displayed. Note that by moving marker 5 from one side of the chromosome to the position between markers 1 and 2, the number of obligate crossovers is reduced from 197 to 124. (Marker 5 was originally on chromosome 12, and when we used `movemarker` to move it to chromosome 1, we just placed it at the end.)

We can adopt the second order (with the minimal number of obligate crossovers) using the `switch.order` function, whose main arguments are `cross`, `chr`, and `order`. The `order` object should be a vector of integers indicating the new marker order; we may insert the second row from the output of `ripple` directly, to save a bit of effort, even though it is one value longer than the number of markers. The `switch.order` function also takes an argument `error.prob`: the assumed genotyping error rate in the estimation of the genetic map for the new order. Thus, we can switch the marker order on chromosome 1 with the following.

```
> ch3c <- switch.order(ch3c, 1, rip[2,])
```

With the markers in their new order, we will now run `ripple` again using the likelihood method but with a smaller window size, to see if the likelihood approach is inconsistent with the approximate method. We assume a genotyping error rate of 0.001.

```
> rip <- ripple(ch3c, 1, 3, method="likelihood",
+                error.prob=0.001, verbose=FALSE)
> summary(rip)

                      LOD  chrlen
Initial  1 2 3 4 5    0.0   117.3
1        2 1 3 4 5   -1.2   173.7
```

When one uses `method="likelihood"`, the `ripple` function gives some indication of its progress; we suppress this by using `verbose=FALSE`.

The second-to-last column in the summary is a LOD score ($\log_{10}$ likelihood ratio) comparing the original order to the alternative order (or orders); a negative value (as here) indicates that the original order has higher likelihood. The last column gives the estimated genetic length of the chromosome with the different marker orders; the best marker order is generally that giving the shortest chromosome length. We see that no further change is needed.

We would now use the same approach with all other chromosomes. While one could type the above commands repeatedly, a more detailed knowledge of R can save quite a bit of effort. For example, we can use a `for` loop to run `ripple` for each chromosome, one at a time, as follows. (We include chromosome 1 again, to make the code simpler.)

```
> rip <- vector("list", nchr(ch3c))
> names(rip) <- names(ch3c$geno)
> for(i in names(ch3c$geno))
+   rip[[i]] <- ripple(ch3c, i, 7, verbose=FALSE)
```

The first line creates a “list” that will contain the output. The second line assigns it the names of the chromosomes in the data.

Now things get a bit hairy. We use `sapply` to pull out, for each chromosome, the difference in the number of obligate crossovers between the initial order and the best of the other orders.

```
> dif.nxo <- sapply(rip, function(a) a[1,ncol(a)]-a[2,ncol(a)])
> dif.nxo
```

```
  1   2   3   4   5   6   7   8   9  10  11  12  13  14  15
-10  16  -7 -23  -8 -22  44 -10 -12  45 -12  78  -5 -10  -4
 16  17  18  19   X
-10 -11  -5  -9   2
```

It is probably best to not seek a complete understanding of this code at the moment. We assigned the results to the object `dif.nxo`, and then typed the name of the object to print the results. The positive numbers indicate that an order other than that in the data showed a decrease in the number of obligate crossovers.

We can now loop through all of the chromosomes and switch the order of markers whenever the alternate order showed an improvement.

```
> for(i in names(ch3c$geno)) {
+   if(dif.nxo[i] > 0)
+     ch3c <- switch.order(ch3c, i, rip[[i]][2,])
+ }
```

Again, the code is complex, but consider the effort saved; we hope this encourages the reader to learn a bit more R.

Since for some chromosomes, we did not look at all possible marker orders, it is a good idea to repeat the process to see if any further improvement may be found.

```
> for(i in names(ch3c$geno))
+   rip[[i]] <- ripple(ch3c, i, 7, verbose=FALSE)
> dif.nxo <- sapply(rip, function(a) a[1,ncol(a)]-a[2,ncol(a)])
```

We can now look to see whether any of the values in `dif.nxo` are positive (indicating that an alternate order is better).

```
> any(dif.nxo > 0)
```

```
[1] FALSE
```

Finally, we go back through all of the chromosomes with `ripple`, this time using `method="likelihood"` and a window size of three markers.

```
> for(i in names(ch3c$geno))
+   rip[[i]] <- ripple(ch3c, i, 3, method="likelihood",
+                      error.prob=0.001, verbose=FALSE)
> lod <- sapply(rip, function(a) a[2, ncol(a)-1])
```

The object `lod` contains the LOD scores for the best alternate order relative to that in the data. The following prints those that are positive (indicating that the alternate order is an improvement).

```
> lod[lod > 0]
```

```
    X
1.655
```

The X chromosome shows some improvement, and so we look at those results more closely.

```
> summary(rip[["X"]])

                                    LOD chrlen
Initial  1 2 3 4 5 6 7 8 9 10       0.0   96.9
1        1 2 3 5 4 6 7 8 9 10       1.7  102.2
2        1 2 3 5 6 4 7 8 9 10       1.7  102.2
3        1 2 3 4 5 7 6 8 9 10       0.0   96.9
```

Switching markers 4 and 5 increases the likelihood by a factor of $10^{1.7} \approx 50$, but leads to a longer chromosome. It is questionable whether the marker order should be switched or not, as a LOD score of 1.7 is not exceptionally strong evidence for the alternate order. Both orders might be considered in later QTL analyses, and if there is evidence for a QTL in the region, we will want to look especially carefully at the marker order.

Finally, we may take another look at the pairwise recombination fractions, at least for chromosomes that were shown in Fig. 3.8 to have some problems. Calls to `est.rf` and `plot.rf` produce the results in Fig. 3.10, as follows.

```
> ch3c <- est.rf(ch3c)
> plot.rf(ch3c, chr=c(1, 7, 12, 13, 15))
```

The results are just what we want: red along the diagonal, fading to blue off the diagonal.

3.4.3 Estimate genetic map

Now that we have the markers in what we believe are the correct orders, we finish this section by estimating the intermarker distances from the observed data; we further compare the results to the map that was included with the data. The function `est.map` does the map estimation. We can use `plot.map` to plot a single genetic map or to plot two maps against each other. Moreover, if a cross object is input to this function, it pulls out the genetic map from the data and plots the map. So with the following, we can estimate the genetic map for the `ch3c` data in its final form and plot it against the map in the data (see Fig. 3.11).

```
> nm <- est.map(ch3c, error.prob=0.001, verbose=FALSE)
> plot.map(ch3c, nm)
```

For many chromosomes, the estimated map is identical to the one within the `ch3c` data set. That is because we had moved some markers around, and if marker order is modified, the intermarker distances must be estimated with

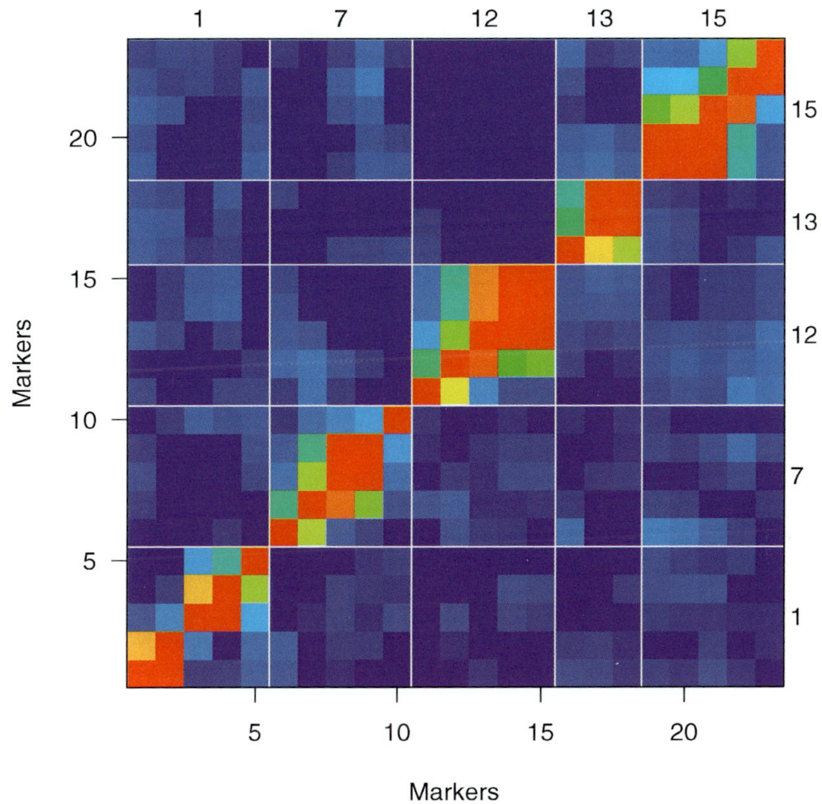

Figure 3.10. Estimated recombination fractions (upper left) and LOD scores (lower right) for pairs of markers on selected chromosomes in the `ch3c` data, after some problems with marker positions have been fixed.

the available data. Several chromosomes exhibit considerable map expansion (e.g., chromosome 6): the estimated map is quite a bit longer than the map in the data. This may indicate the presence of genotyping errors.

One may wish, at this point, to replace the map within the `ch3c` with that estimated from the data. Reference genetic maps are often based on a rather small number of individuals. (For example, the original MIT mouse genetic map was based on an intercross with just 46 individuals.) One's own data often contains many more individuals, and so may produce a more accurate map. The only caveat is that reference genetic maps generally contain a much more dense set of markers, which (as described in the next section) provides greater ability to detect genotyping errors. Thus reference genetic maps may be based on cleaner genotype data.

To replace the genetic map in the `ch3c` data with that estimated from the data, we use the function `replace.map`, as follows.

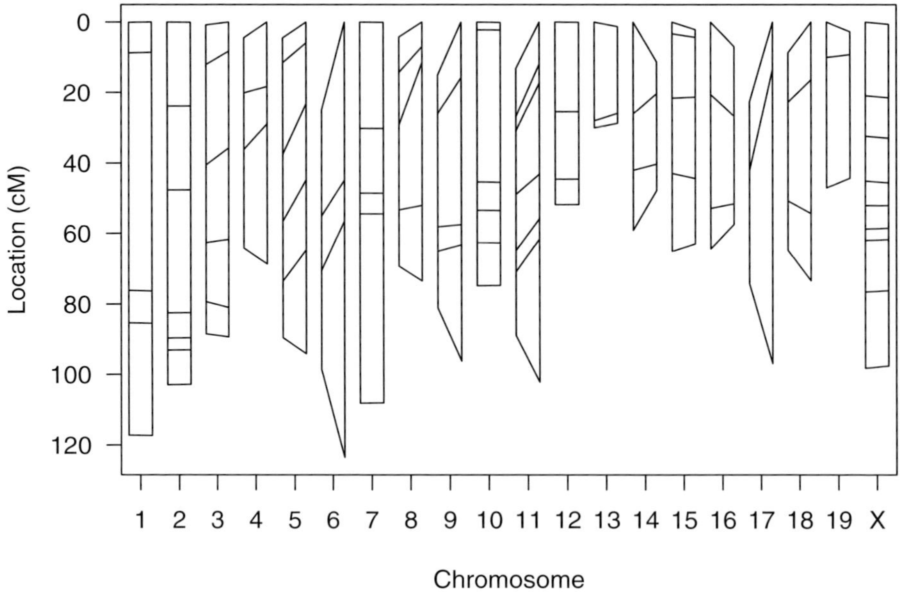

Figure 3.11. The genetic map in the `ch3c` data (after considerable revisions in marker order) plotted against the map estimated from the data. For each chromosome, the line on the left is the map in the data, and the line on the right is the map estimated from the data; line segments connect the positions for each marker.

```
> ch3c <- replace.map(ch3c, nm)
```

3.5 Identifying genotyping errors

Our ability to map QTL relies on high-quality genotype data; errors in the genotype data will lessen our ability to detect QTL. Genotyping errors may appear as apparent tight double crossovers. Meiosis generally exhibits strong crossover interference, and so crossovers will not occur too close together. Thus, if the genotype at a single marker is out of phase with the surrounding markers, it is likely in error. This requires dense marker genotype data; with sparse markers, one cannot be sure whether the apparent tight double crossover is an error or is a true double crossover.

Detection of genotyping errors is facilitated by the calculation of genotyping error LOD scores. For each individual at each marker, we calculate a $\log_{10}$ likelihood ratio comparing the hypothesis that the particular genotype is in error to the hypothesis that it is correct; the likelihood uses the genotype data at all other markers on the chromosome. (For further details on this calculation, see Sec. D.7.) These LOD scores serve largely to ease the identification of unusually tight double crossovers.

To calculate the genotyping error LOD scores, we use the `calc.errorlod` function. One must specify an assumed genotyping error rate, and the results can be sensitive to this rate. The results are especially sensitive to the genetic map. Thus, before calculating the error LOD scores, we may first wish to replace the map in the data with that estimated from the data. In the following we do that plus calculate the error LOD scores for the `hyper` data.

```
> data(hyper)
> newmap <- est.map(hyper, error.prob=0.01)
> hyper <- replace.map(hyper, newmap)
> hyper <- calc.errorlod(hyper)
```

The `top.errorlod` function prints information about the genotypes with error LOD score above a specified cut off (indicated by the argument `cutoff`). An argument `chr` may be used to give results for a selected subset of the chromosomes. In the following, we look at the genotypes with error LOD scores > 5. We save the results to the object `top`, and then type the name of the object to print the results.

```
> top <- top.errorlod(hyper, cutoff=5)
> top

  chr id    marker errorlod
1  16 50 D16Mit171   16.000
2  16 54 D16Mit171   16.000
3  16 81   D16Mit5    8.915
4  16 24   D16Mit5    8.915
5  16 71   D16Mit5    8.915
6  16 34   D16Mit5    8.915
7  13 42  D13Mit78    8.000
8  13 42 D13Mit148    7.881
```

There are a number of genotypes indicated to be likely in error. The `id` column is a numeric index here, but if a phenotype named "`id`" or "`ID`" had been included in the data, such labels would be used.

We can look more closely at the problem genotypes with the function `plot.geno`, which plots the genotype data for a single chromosome, for selected individuals. We pull out the individual IDs from the result of `top.errorlod`, and plot their genotype data for chromosome 16. We use the `cutoff` argument to indicate that genotypes with error LOD score > 5 should be flagged.

```
> plot.geno(hyper, 16, top$id[top$chr==16], cutoff=5)
```

The results are shown in Fig. 3.12.

A small number of genotyping errors will not have much influence on the results, and so one should be concerned only if an inordinate number of possible errors are seen (though this may also indicate a problem with marker

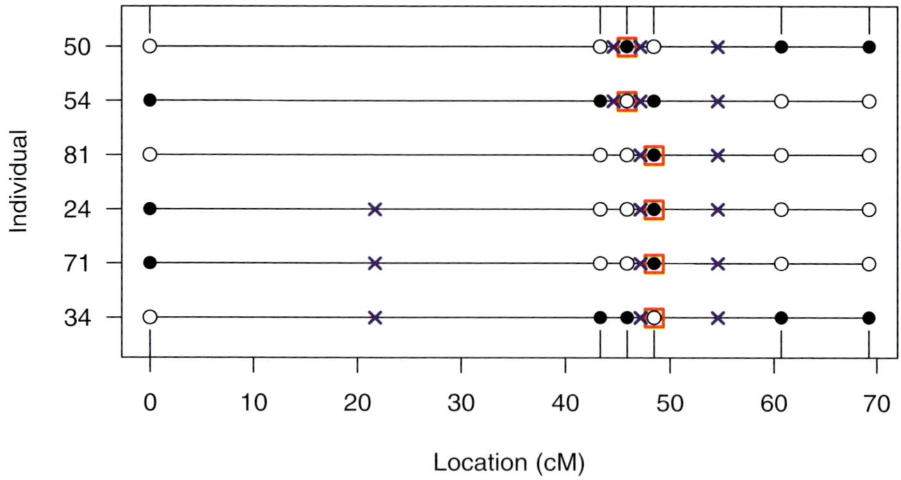

Figure 3.12. Chromosome 16 genotypes for selected individuals in the `hyper` data. Open and closed circles are homozygous and heterozygous genotypes, respectively. Possible genotyping errors are flagged with red squares; inferred crossovers are indicated with blue ×'s.

order). However, if one sees evidence for a QTL in the region, it may be valuable to take a second look at the raw genotype information or to rerun the genotypes on some markers.

3.6 Counting crossovers

Another useful diagnostic is to count the number of crossovers implied by the genotype data in each individual. Individuals with an unusually small or large number of crossovers should be viewed with suspicion. This may be an indication of either poor quality DNA or sample mix-ups.

The function `countXO` may be used to count the number of observed crossovers for each individual. One may use the `chr` argument to focus on a selected set of chromosomes. By default, we count the total number of crosssovers across the genome. If `countXO` is called with the argument `bychr=TRUE`, the function returns a matrix containing the numbers of crossovers on the individual chromosomes.

Let us again consider the `hyper` data. We may count crossovers and plot them as follows (see Fig. 3.13).

```
> nxo <- countXO(hyper)
> plot(nxo, ylab="No. crossovers")
```

Note that these counts are the minimal number of crossovers required to explain the observed genotype data. We see a large shift in the distribution

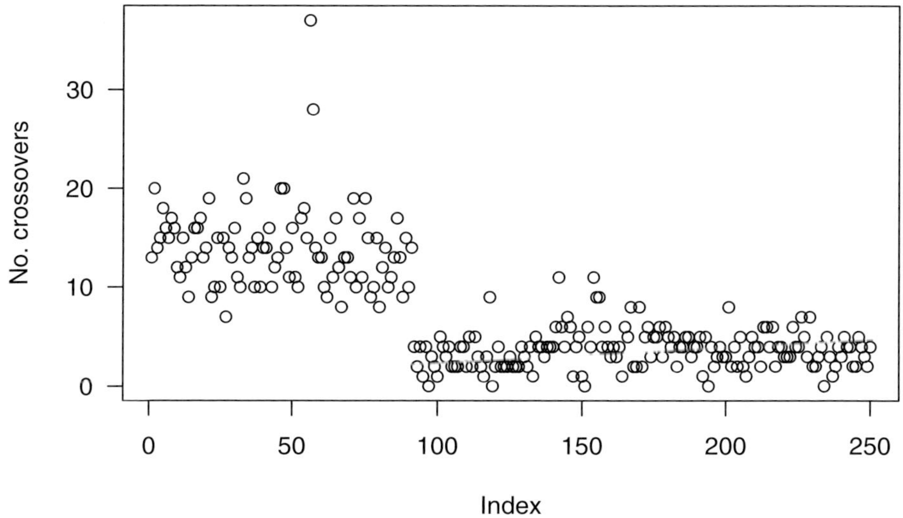

Figure 3.13. The observed number of crossovers for each individual in the `hyper` data.

between the first 92 individuals and the remaining 158 individuals, due to the selective genotyping of the data. (The initial 92 individuals were genotyped at markers across the genome; the remaining individuals were typed only on selected chromosomes that exhibited evidence for a QTL.)

Particularly interesting are the two individuals with > 25 crossovers.

```
> nxo[nxo>25]

56 57
37 28
```

The 56th individual exhibited 37 crossovers. (The first row in the above output contains labels with the indexes of the individuals; the second row contains the crossover counts.) This is considerably larger than was seen in others (see Fig. 3.13). The initial 92 individuals showed an average of 14 crossovers. The remaining 158 individuals (genotyped only on selected chromosomes) had an average of 4 crossovers.

```
> mean(nxo[1:92])

[1] 13.84

> mean(nxo[-(1:92)])

[1] 3.741
```

If we pull out the crossover counts for each chromosome for individual 56, we can identify the chromosomes that are particularly problematic.

```
> countXO(hyper, bychr=TRUE)[56,]
```

```
 1  2  3  4  5  6  7  8  9 10 11 12 13 14 15 16 17 18 19  X
 2  0  1  1  5  7  2  2  1  2  4  1  0  1  4  1  1  1  1  0
```

The genotype data for chromosome 6 (for which this individual shows seven apparent crossovers) are particularly suspicious, and deserve further investigation.

3.7 Missing genotype information

As a final diagnostic, we compute the proportion of missing genotype information at positions along the genome, given the available marker data. This can help us to identify regions where further markers might be added. In addition, as we will see in the next chapter, standard interval mapping to identify regions harboring QTL can occasionally give spurious evidence for linkage in regions of low genotype information, and so an evaluation of the proportion of missing information in regions of inferred QTL can help us to identify such problems.

Consider a fixed position in the genome and let g_i denote the genotype of individual i at that site. We first calculate $p_{ij} = \Pr(g_i = j \mid \boldsymbol{M}_i)$, where $\boldsymbol{M}_i$ denotes the multipoint marker genotype data for individual i. We consider two methods for defining the proportion of missing genotype information. First, we correspond the possible genotypes with integers (1 and 2 for a backcross, and 1, 2, and 3 for an intercross), and calculate the conditional variance of the genotypes, given the available marker genotype data, $\sum_i \text{var}(g_i|\boldsymbol{M}_i)$, and look at the ratio of this variance to the variance in the case of no genotype information ($n/4$ for a backcross and $n/2$ for an intercross). If there is complete genotype information (e.g., at a fully typed marker), we obtain a ratio of 0; if there is no genotype information, we obtain a ratio of 1.

In the second method, we use the information theoretic concept of entropy, $-\sum_i \sum_j p_{ij} \log_2 p_{ij}$, where we take $0 \log 0 = 0$. We again take the ratio of this quantity to the value for the case of no genotype information (n for a backcross and $3n/2$ for an intercross). Again, if there is complete genotype information, we obtain a ratio of 0; if there is no genotype information, we obtain a ratio of 1.

These quantities may be calculated and plotted with `plot.info`. For example, a plot of the missing genotype information in the `hyper` data appears in Fig. 3.14, which was created with the following.

```
> plot.info(hyper, col=c("blue", "red"))
```

We use `col` to indicate that the entropy and variance versions of the results should be plotted in blue and red, respectively.

The proportion of missing genotype information is effectively 0 at the fully typed markers. For several chromosomes, the minimal missing information is

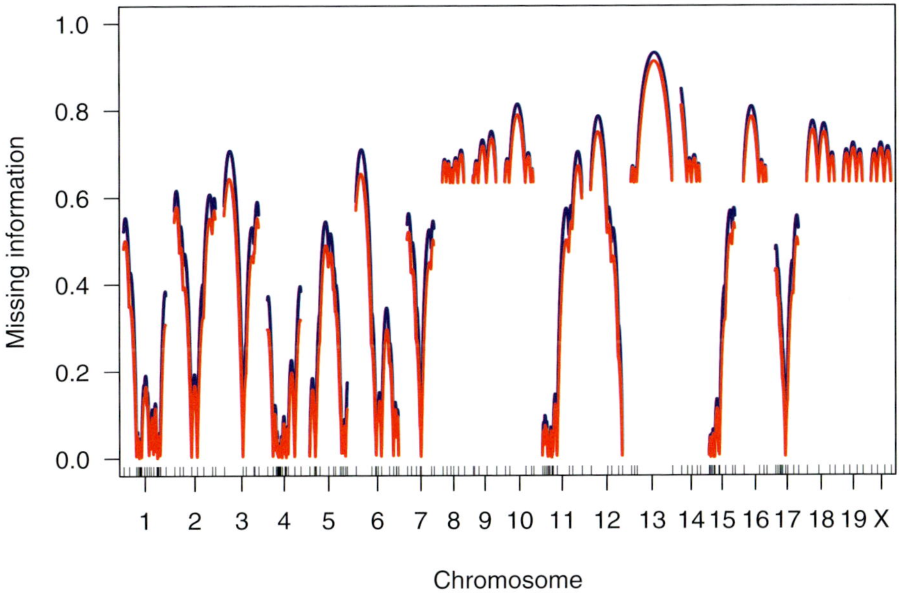

Figure 3.14. The proportion of missing genotype information in the `hyper` data. Results by the entropy and variance versions are shown in blue and red, respectively.

about 63%, as only the 92 individuals (out of 250) with extreme phenotypes were genotyped.

The detailed results of `plot.info` may be saved in an object. We can get the results just at the markers by rerunning `plot.info` with `step=0` and then show the results just for chromosome 14 as follows. (The argument `step` indicates the density of the grid, in cM, at which the missing inforamtion is to be calculated. The default value is `step=1`; use of `step=0` indicates that the calculation should be performed only at the markers.)

```
> z <- plot.info(hyper, step=0)
> z[ z[,1]==14, ]
```

```
          chr   pos misinfo.entropy misinfo.variance
D14Mit48   14  0.00          0.8475           0.8098
D14Mit14   14 16.40          0.6355           0.6335
D14Mit37   14 29.05          0.6331           0.6324
D14Mit7    14 43.68          0.6344           0.6333
D14Mit266  14 52.97          0.6355           0.6336
```

The result is a matrix with four columns: the chromosome and cM position followed by the missing information results by the entropy and variance methods, respectively.

A related function of interest is `nmissing`, which returns the number of missing genotypes for each individual or each marker (according to whether

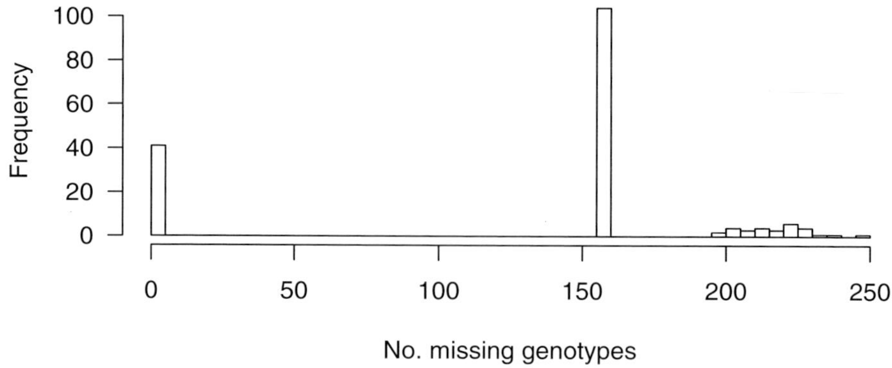

Figure 3.15. Histogram of the number of individuals with missing genotypes at the markers in the `hyper` data.

the argument `what` is `"ind"` or `"mar"`, respectively). For example, we can get a histogram of the number of missing genotypes at the markers in the `hyper` data as follows. The results are in Fig. 3.15.

```
> hist(nmissing(hyper, what="mar"), breaks=50)
```

About 40 markers were typed on essentially everyone; over 100 were typed on only the 92 individuals with extreme phenotypes. The remaining 29 markers were typed only on a few individuals.

There is also a function `ntyped` which provides the opposite of `nmissing`: the number of typed markers for each individual, or the number of typed individuals for each marker.

3.8 Summary

Success in QTL mapping requires high-quality data. Given the time and expense in gathering data, reasonable effort should be devoted to identifying and correcting errors in the data prior to QTL analysis.

Histograms, scatterplots and time-course plots can assist in the identification of gross errors in the phenotype data, or of real but odd individuals deserving careful consideration in later analysis.

The assessment of genotype data begins with the inspection of the segregation patterns. Problem markers are often revealed by departures from the 1:1 and 1:2:1 patterns expected in a backcross and intercross, respectively.

While the availability of sequence-based marker maps in many organisms has eliminated much of the effort that was once required to establish marker order, the genetic maps of typed markers should still be carefully inspected. Errors in marker labels are not uncommon, and these may be revealed by an inspection of pairwise linkages and the reestimation of intermarker genetic distances.

Finally, it can be useful to study the pattern of crossovers to identify genotyping errors that are revealed by apparent tight double crossovers. The calculation of genotyping error LOD scores simplies this effort. However, the presence of a small number of genotyping errors will not have much influence on later results, and so this aspect of the detective work is generally not critical.

3.9 Further reading

Surprisingly little has been written on the detective work involved in identifying and resolving errors in QTL mapping data. Of some relevance is Broman (1999), concerning cleaning human genotype and pedigree data. The genotyping error LOD scores were developed by Lincoln and Lander (1992).

4

Single-QTL analysis

The most commonly used method for QTL analysis is interval mapping, in which one posits the presence of a single QTL and considers each point on a dense grid across the genome, one at a time, as the location of the putative QTL. A central issue concerns the treatment of missing genotype information: at a position between genetic markers, genotype data are not available and must be inferred on the basis of the available marker genotype data. Several methods are available; we describe the most popular. These methods all have analogs for the fit of multiple-QTL models, which will be discussed in Chap. 8 and 9. We further discuss the establishment of statistical significance in such single-QTL genome scans, and the special treatment that is required for the X chromosome. But first, in order to introduce the basic ideas in QTL mapping, we describe an even simpler method, sometimes called marker regression.

Each section will begin with a bit of theory, followed by R code for performing the analyses with R/qtl. The R code is cumulative through the chapter; the results in one section may rely on code executed in a previous section.

4.1 Marker regression

The simplest method for the analysis of QTL mapping data is to consider each marker individually, split the individuals into groups, according to their genotypes at the marker, and compare the groups' phenotype averages. While this method can seldom be recommended for use in practice, it provides a valuable framework for thinking about QTL mapping and for describing some of the essential issues in QTL mapping.

Consider, for example, the `hyper` data, described in Sec. 2.3. The blood pressure phenotype is plotted against the genotype at markers D4Mit214 and D12Mit20 in Fig. 4.1. At D4Mit214, the homozygous individuals exhibit a larger average phenotype than the heterozygotes, indicating that this marker is linked to a QTL. At D12Mit20, on the other hand, the two genotype groups

K.W. Broman, Ś. Sen, *A Guide to QTL Mapping with R/qtl*,
Statistics for Biology and Health, DOI 10.1007/978-0-387-92125-9_4,

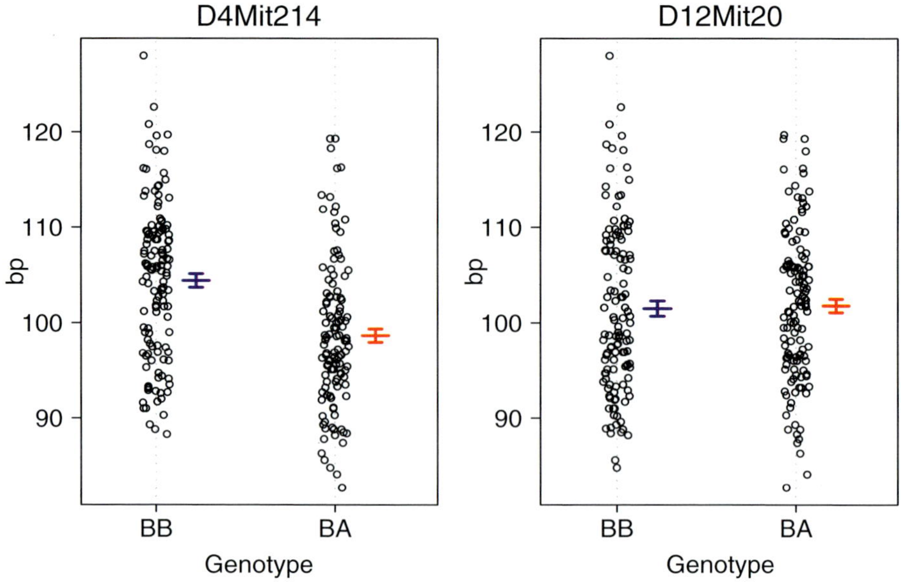

Figure 4.1. Dot plots of the blood pressure phenotype against the genotype at two selected markers, for the `hyper` data. Confidence intervals for the average phenotype in each genotype group are shown.

show similar phenotypes, and so D12Mit20 is not indicated to be linked to a QTL.

In a backcross, we test for linkage of a marker to a QTL by a t test; in an intercross, we would use analysis of variance (ANOVA), which gives an F statistic. Traditionally, evidence for linkage to a QTL is measured by a LOD score: the $\log_{10}$ likelihood ratio comparing the hypothesis that there is a QTL at the marker to the hypothesis that there is no QTL anywhere in the genome.

The LOD score at a marker is calculated as follows. First, consider the null hypothesis of no QTL, in which case, with y_i denoting the phenotype for individual i, $y_i \sim N(\mu, \sigma^2)$ (i.e., the phenotypes follow a single normal distribution, independent of the genotypes). We consider the likelihood function $L_0(\mu, \sigma^2) = \Pr(\text{data} \mid \text{no QTL}, \mu, \sigma^2) = \prod_i \phi(y_i; \mu, \sigma^2)$, where ϕ is the density of the normal distribution. We take as estimates of μ and σ^2 the values for which the likelihood is maximized (such estimates are called the maximum likelihood estimates, MLEs). For this model, the MLE of μ is simply the phenotype average, $\bar{y}$, and the MLE of σ^2 is RSS_0/n, where $\text{RSS}_0 = \sum_i (y_i - \bar{y})^2$ is the null residual sum of squares and n is the sample size. The $\log_{10}$ likelihood for the null hypothesis is obtained by plugging in the MLEs; with a bit of algebra, it reduces to $-\frac{n}{2} \log_{10} \text{RSS}_0$.

Under the alternative hypothesis, that there is a QTL at the marker under test, we assume that $y_i | g_i \sim N(\mu_{g_i}, \sigma^2)$, where g_i is the genotype of individual i at the marker, μ_{AA} and μ_{AB} are the phenotype averages for the two genotype groups, and σ^2 is the residual variance (assumed to be the same in the two groups). The likelihood function is $L_1(\mu_{AA}, \mu_{AB}, \sigma^2) = \Pr(\text{data} \mid \text{QTL at marker}, \mu_{AA}, \mu_{AB}, \sigma^2) = \prod_i \phi(y_i; \mu_{g_i}, \sigma^2)$. We again estimate the parameters by maximum likelihood: the values for which L_1 achieves its maximum. The MLEs for the μ_i are simply the phenotype averages within the two genotype groups. The MLE of σ^2 is the pooled estimate, RSS_1/n, where $\text{RSS}_1 = \sum_i (y_i - \hat{\mu}_{g_i})^2$ is the residual sum of squares under the alternative. The $\log_{10}$ likelihood for the alternative hypothesis is obtained by plugging in the MLEs; it reduces to $-\frac{n}{2}\log_{10} \text{RSS}_1$.

Finally, the LOD score is the difference between the $\log_{10}$ likelihood under the alternative hypothesis and the $\log_{10}$ likelihood under the null hypothesis, and so we obtain the following.

$$\text{LOD} = \frac{n}{2} \log_{10}\left(\frac{\text{RSS}_0}{\text{RSS}_1}\right)$$

Note that individuals with missing genotype data at the marker must be omitted from the estimation under the alternative, and so, in the calculation of the LOD score, such individuals are omitted from both the alternative and null likelihoods.

The LOD score is equivalent to the F statistic from ANOVA. (And note that the t statistic in a backcross is the signed square root of the F statistic that would be obtained from ANOVA.) Let df denote the degrees of freedom (df=1 for a backcross and df=2 for an intercross); the connection between the F statistic and the LOD score is as follows.

$$\begin{aligned} F &= \left(\frac{\text{RSS}_0 - \text{RSS}_1}{\text{RSS}_1}\right)\left(\frac{n - \text{df} - 1}{\text{df}}\right) \\ &= \left(\frac{\text{RSS}_0}{\text{RSS}_1} - 1\right)\left(\frac{n - \text{df} - 1}{\text{df}}\right) \\ &= \left(10^{\frac{2}{n}\text{LOD}} - 1\right)\left(\frac{n - \text{df} - 1}{\text{df}}\right) \end{aligned}$$

The inverse of this formula is also of interest.

$$\text{LOD} = \frac{n}{2} \log_{10}\left[F\left(\frac{\text{df}}{n - \text{df} - 1}\right) + 1\right]$$

Note further that the estimated proportion of the phenotypic variance explained by the QTL (i.e., the estimated heritability due to the QTL) is $(\text{RSS}_0 - \text{RSS}_1)/\text{RSS}_0$. Thus, the estimated percent variance explained by the QTL is $1 - 10^{-\frac{2}{n}\text{LOD}}$.

Large LOD scores indicate evidence for the presence of a QTL, but in considering the statistical significance of LOD scores, we must take account of the multiple tests performed. Discussion of this issue is deferred to Sec. 4.3.

The key advantage of marker regression is its simplicity: we just perform a t test or ANOVA at each marker. Thus, no special software is required, and one may easily incorporate covariates (such as sex) or extend the analysis to more complex models (such as for the treatment of censored survival times).

A key disadvantage is that one must omit individuals with missing marker genotypes. Further, one cannot inspect positions between markers, and one obtains rather poor information about QTL location. Also, the apparent effect of a QTL is attenuated by its incomplete linkage to a marker. For example, consider a backcross with a single QTL, and let μ_{AA} and μ_{AB} denote the phenotype averages for the two QTL genotypes, so that the effect of the QTL is $\Delta = \mu_{AB} - \mu_{AA}$. If the recombination fraction between a marker and the QTL is r, then the individuals with marker genotype AA will consist of a fraction $(1-r)$ with QTL genotype AA and a fraction r with QTL genotype AB. Thus the average phenotype for individuals that are AA at the marker will be $\mu_{AA}(1-r) + \mu_{AB}r$. Similarly, the average phenotype for individuals that are AB at the marker will be $\mu_{AB}(1-r)+\mu_{AA}r$. As a result, the difference between the phenotype averages for the two marker genotype groups will be $[\mu_{AB}(1-r)+\mu_{AA}r] - [\mu_{AA}(1-r)+\mu_{AB}r] = \Delta(1-2r)$, and so the apparent effect of the QTL is reduced by a factor $(1-2r)$ due to its incomplete linkage to the marker.

The most important disadvantage of marker regression is that we consider the presence of a single QTL. Thus we have limited ability to separate linked QTL and no ability to assess possible interactions among QTL. This disadvantage is shared with all of the methods described in this chapter.

Example

Let us now turn to the actual analysis with R/qtl, using the `hyper` data as an example. We first load R/qtl and the data.

```
> library(qtl)
> data(hyper)
```

First note that the dot plots of phenotype against genotype, in Fig. 4.1, were created with the `plot.pxg` function, as follows.

```
> par(mfrow=c(1,2))
> plot.pxg(hyper, "D4Mit214")
> plot.pxg(hyper, "D12Mit20")
```

The fit of single-QTL models is accomplished with the function `scanone`. The use of the argument `method="mr"` indicates to use marker regression (i.e., ANOVA or a t test at each marker). For the `hyper` data, we type the following.

```
> out.mr <- scanone(hyper, method="mr")
```

The result, saved in `out.mr`, is a matrix with three columns: chromosome, cM position, and LOD score. We can look at the results for the chromosome 12 as follows.

```
> out.mr[ out.mr$chr == 12, ]

          chr  pos       lod
D12Mit37   12  1.1 0.3610905
D12Mit110  12 16.4 0.0009559
D12Mit34   12 23.0 0.0005335
D12Mit118  12 40.4 0.0003868
D12Mit20   12 56.8 0.0136116
```

The output of `scanone` has class `"scanone"`, and so use of the functions `plot` and `summary` with the `out.mr` object will create a plot or summary via `plot.scanone` and `summary.scanone`, respectively. For example, we may pull out the single largest LOD score from each chromosome with `summary.scanone`; we show just those having LOD > 3 with the `threshold` argument.

```
> summary(out.mr, threshold=3)

         chr  pos  lod
D1Mit14    1 82.0 3.52
D4Mit214   4 21.9 6.86
```

The function `max.scanone` may be used to pick out the single biggest LOD score, as follows.

```
> max(out.mr)

         chr  pos  lod
D4Mit214   4 21.9 6.86
```

A plot of the LOD scores for chromosomes 4 and 12 is obtained with the following. The result appears in Fig. 4.2.

```
> plot(out.mr, chr=c(4, 12), ylab="LOD score")
```

The argument `chr` is used to select the chromosomes to plot; by default all chromosomes will be plotted. The argument `ylab` is used to change the y-axis label.

The jagged appearance of the LOD curve for chromosome 4 is due to the pattern of missing marker genotype data. Recall that in the marker regression method, we split the individuals into groups according to their genotype at a marker. If an individual's genotype is missing, that individual must be omitted, as we do not know the genotype group in which it should be placed. At some markers in the `hyper` data, only recombinant individuals were genotyped (see Fig. 1.7 on page 10), and so these markers, in isolation, will provide little evidence for linkage to a QTL.

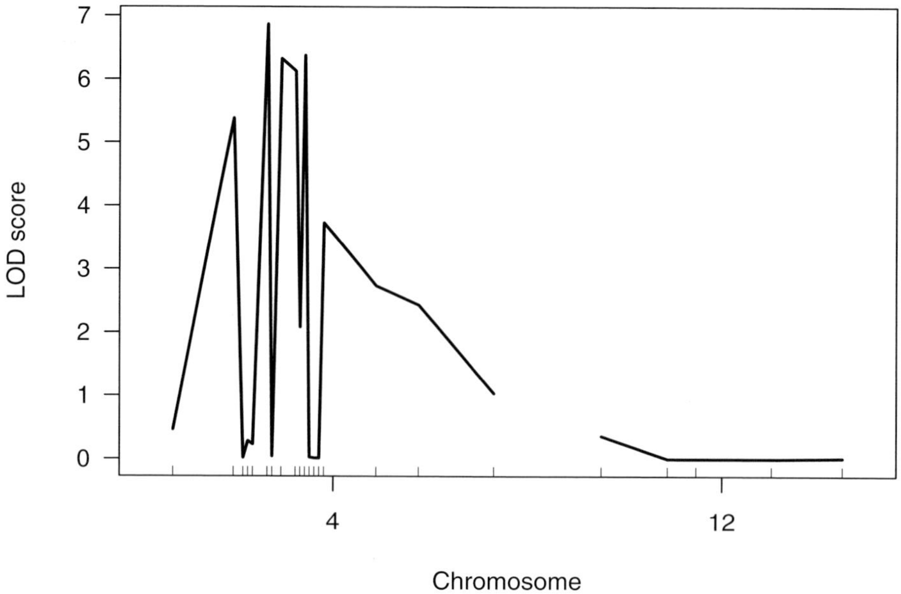

Figure 4.2. LOD scores for each marker on chromosomes 4 and 12 for the `hyper` data, calculated by marker regression.

4.2 Interval mapping

In this section, we consider several variants on interval mapping. Interval mapping improves on the marker regression method by taking account of missing genotype data at a putative QTL. The various interval mapping methods differ in their treatment of the missing genotype data. Standard interval mapping uses maximum likelihood estimation under a mixture model, while the Haley–Knott regression methods use approximations to the mixture model. The multiple imputation method uses the same mixture model but with multiple imputation in place of maximum likelihood.

4.2.1 Standard interval mapping

In standard interval mapping, we again assume the presence of a single QTL, but now we consider a grid of positions along the genome as the possible locations for the QTL. Consider a particular fixed position for the QTL, and let g_i denote the QTL genotype for individual i. As with the marker regression method, we assume that $y_i|g_i \sim N(\mu_{g_i}, \sigma^2)$, but now the QTL genotype is generally not known.

For each individual, we may calculate $p_{ij} = \Pr(g_i = j|\boldsymbol{M}_i)$, where $\boldsymbol{M}_i$ denotes the multipoint marker genotype data for individual i. For example, consider a backcross and two markers separated by a recombination fraction

Table 4.1. Conditional probabilities for the QTL genotypes in a backcross, given the genotypes at two flanking markers.

Marker genotype		QTL genotype	
Left	Right	AA	AB
AA	AA	$(1 - r_{1Q})(1 - r_{2Q})/(1 - r_{12})$	$r_{1Q}r_{2Q}/(1 - r_{12})$
AA	AB	$(1 - r_{1Q})r_{2Q}/r_{12}$	$r_{1Q}(1 - r_{2Q})/r_{12}$
AB	AA	$r_{1Q}(1 - r_{2Q})/r_{12}$	$(1 - r_{1Q})r_{2Q}/r_{12}$
AB	AB	$r_{1Q}r_{2Q}/(1 - r_{12})$	$(1 - r_{1Q})(1 - r_{2Q})/(1 - r_{12})$

of r_{12}. Suppose that our putative QTL sits in the intervening interval, and let r_{iQ} denote the recombination fraction between marker i and the QTL. If we assume no crossover interference and no genotyping errors, and if an individual is typed at both markers, the QTL genotype probabilities given the marker genotypes are as in Table 4.1. In general, one may use algorithms for hidden Markov models (HMMs) for these sorts of calculations, as one can then allow for the presence of genotyping errors and more simply deal with partially informative genotypes (such as the case of dominant markers in an intercross). For further details, see Sec. 1.3.1 and Appendix D.

Given the marker data, an individual's phenotype follows a *mixture* of normal distributions, with known mixing proportions (the p_{ij}). That is, the density function for the phenotype of individual i is $\sum_j p_{ij}\phi(y_i; \mu_j, \sigma^2)$, where ϕ is the density of the normal distribution and the sum is over the possible QTL genotypes.

For example, consider a backcross containing a single QTL, and consider two markers separated by 20 cM, with the QTL placed in the intervening interval, 7 cM from the left marker. The phenotype distributions, conditional on the genotypes at the two markers, are shown in Fig. 4.3.

Except in the case of rare double recombination events, all individuals who are AA at both markers (top panel) will also be AA at the QTL, and so their phenotypes will follow a normal distribution centered at μ_{AA}. Similarly, individuals who are AB at both markers (lower panel) will generally also be AB at the QTL, and so their phenotypes will follow a normal distribution centered at μ_{AB}.

Individuals who are AA at the left marker but AB at the right marker will consist of some individuals who are AA at the QTL (i.e., their recombination event occurred to the right of the QTL) and some who are AB at the QTL (having recombined to the left of the QTL). Similarly, individuals who are AB at the left marker but AA at the right marker will consist of some individuals who are AB at the QTL (having recombined to the right of the QTL) and some who are AA at the QTL (having recombined to the left of the QTL). The dashed curves in the central panels in Fig. 4.3 are the distributions of the groups with common QTL genotype; the solid curves are the mixture distributions.

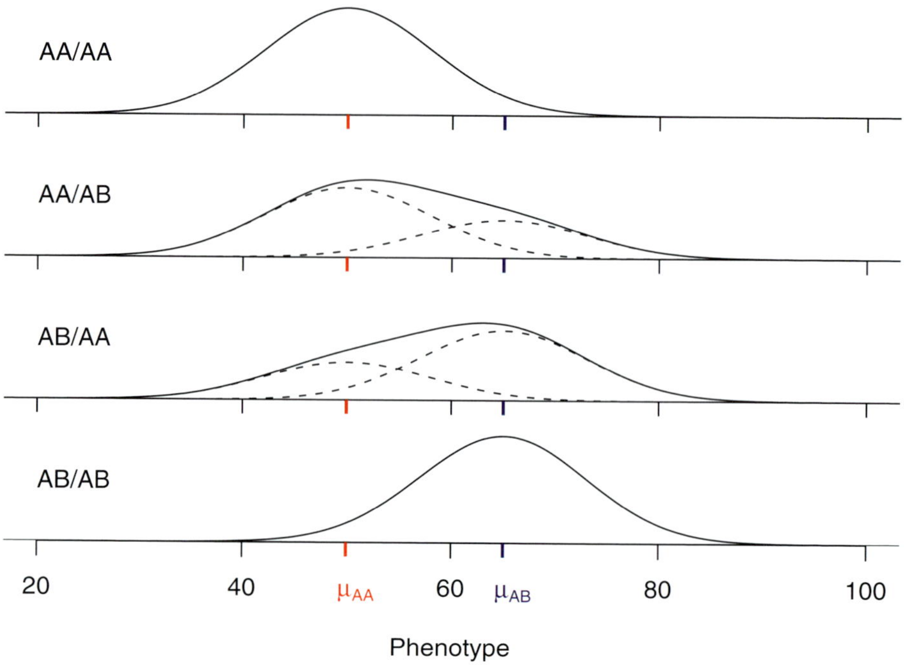

Figure 4.3. Phenotype distributions conditional on the genotype at two markers, in the case of a backcross containing a single QTL. The markers are separated by 20 cM and the QTL sits within the marker interval, 7 cM from the left marker. The dashed curves are the distributions of the groups with common QTL genotype; the solid curves are the mixture distributions.

We estimate the μ_j and σ by maximum likelihood; that is, we take as our estimates those values for which the observed data is most probable. The likelihood function is $L(\boldsymbol{\mu}, \sigma) = \prod_i \sum_j p_{ij}\phi(y_i; \mu_j, \sigma^2)$, where the sum is over the possible QTL genotypes. The MLEs cannot be obtained in closed form; an iterative algorithm is necessary. We use a form of the EM algorithm. We begin with initial estimates $\hat{\mu}_j^0$ and $\hat{\sigma}^{(0)}$.

In the E-step at iteration s, we calculate the conditional probability that an individual is in QTL genotype group j given its marker data, phenotype, and our current estimates of the μ_j and σ.

$$
\begin{aligned}
w_{ij}^{(s)} &= \Pr(g_i = j | \boldsymbol{M}_i, y_i, \hat{\mu}_j^{(s-1)}, \hat{\sigma}^{(s-1)}) \\
&= \frac{p_{ij}\phi(y_i; \hat{\mu}_j^{(s-1)}, \hat{\sigma}^{(s-1)})}{\sum_k p_{ik}\phi(y_i; \hat{\mu}_k^{(s-1)}, \hat{\sigma}^{(s-1)})}
\end{aligned}
$$

In the M-step, we update our estimates of the μ_j and σ, treating the $w_{ij}^{(s)}$ as weights.

$$\hat{\mu}_j^{(s)} = \sum_i w_{ij}^{(s)} y_i / \sum_i w_{ij}^{(s)}$$

$$\hat{\sigma}^{(s)} = \sqrt{\frac{\sum_{ij} w_{ij}^{(s)} (y_i - \hat{\mu}_j^{(s)})^2}{n}}$$

Iterations are repeated until the estimates converge (i.e., until the estimates stop changing). The EM algorithm has the advantage that the likelihood is nondecreasing across iterations. It may be that the algorithm converges to a local maximum, but with relatively dense markers and relatively complete marker genotype data, the likelihood is well behaved and the EM algorithm will converge to the global maximum. If there were multiple modes in the likelihood surface, one would want to use multiple initial estimates for the EM algorithm, but because the likelihood is well behaved in this context, we generally start the algorithm by taking $w_{ij}^{(0)} = p_{ij}$ and then doing the M-step. This is equivalent to using Haley–Knott regression (Sec. 4.2.2) to get the initial estimates.

Once the maximum likelihood estimates of the μ_j and σ have been obtained, a LOD score is calculated as follows.

$$\text{LOD} = \log_{10}\left(\frac{\prod_i \sum_j p_{ij} \phi(y_i; \hat{\mu}_j, \hat{\sigma}^2)}{\prod_i \phi(y_i; \hat{\mu}_0, \hat{\sigma}_0^2)} \right)$$

where $\hat{\mu}_0$ and $\hat{\sigma}_0$ are the average and SD of the y_i, so that the denominator of the LOD score is the likelihood under the null hypothesis that there is no QTL anywhere in the genome.

In standard interval mapping, the EM algorithm is performed at each position on a grid of putative QTL locations along the genome, while the estimates and likelihood under the null hypothesis are calculated just once.

The key advantage of interval mapping, relative to marker regression, is that one takes appropriate account of missing genotype information. Thus, we need not omit individuals with missing genotype at a marker, and we may inspect positions between markers. We thus obtain a more clear understanding of QTL location. (The smooth LOD curves are also nicer to look at. Before the development of interval mapping, papers reporting the results of QTL mapping contained long tables of p-values; the plot of LOD curves was an important advance.) Further, we may obtain an improved estimate of the effect of a QTL, as such an estimate may be obtained at its estimated position, rather than using the genotypes at the nearest genetic marker.

Disadvantages of interval mapping include increased computation time, the need for specialized software, and difficulty in generalizing the method for more complex models or for the inclusion of environmental and other covariates. These disadvantages are no longer of much importance, as interval mapping requires only a couple of seconds of computer time, a variety of relevant computer programs are available, and a variety of extensions of interval mapping have been implemented.

The most important disadvantage of interval mapping is that we are still considering only a single-QTL model, and so we have limited ability to separate linked QTL and no ability to assess possible interactions among QTL. With complete marker genotype data, interval mapping and marker regression give precisely the same results at the markers. Interval mapping seems fancy, but it is little different, conceptually, from simply performing ANOVA at each marker.

Example

Having completed our discussion of the theory underlying standard interval mapping, we now turn to the calculations in R/qtl. We first must calculate the conditional genotype probabilities, $p_{ij} = \Pr(g_i = j|\boldsymbol{M}_i)$. This is done with the function `calc.genoprob`. The argument `step` is used to define the density (in cM) of the grid on which these probabilities will be calculated; this will determine the density at which interval mapping is performed. The argument `error.prob` allows the probabilities to be calculated assuming a given rate of genotyping errors.

```
> hyper <- calc.genoprob(hyper, step=1, error.prob=0.001)
```

In `calc.genoprob`, one may also use the argument `off.end` to calculate the probabilities to some distance past the terminal markers on the chromosome, so that interval mapping will be performed past the terminal markers, but this can lead to the artifacts in the results (for example, a QTL near the end of the chromosome may show a mirror image past the terminal marker), and so we generally use `off.end=0` (the default).

We should emphasize that it is important, for analyses with R/qtl, that no two markers are placed at precisely the same position. If there are markers that coincide, a warning will be produced by the function `summary.cross`. Marker positions may be moved apart slightly with the function `jittermap`, as follows. Note that this should be done *prior* to the call to `calc.genoprob`.

```
> hyper <- jittermap(hyper)
```

Interval mapping is performed with the `scanone` function, using the argument `method="em"` (for "EM algorithm").

```
> out.em <- scanone(hyper, method="em")
```

Standard interval mapping is the default method, and so `method="em"` can be omitted, as follows.

```
> out.em <- scanone(hyper)
```

The form of the results was described in the previous section. We can plot the results for all chromosomes as follows; the results appear in Fig. 4.4.

```
> plot(out.em, ylab="LOD score")
```

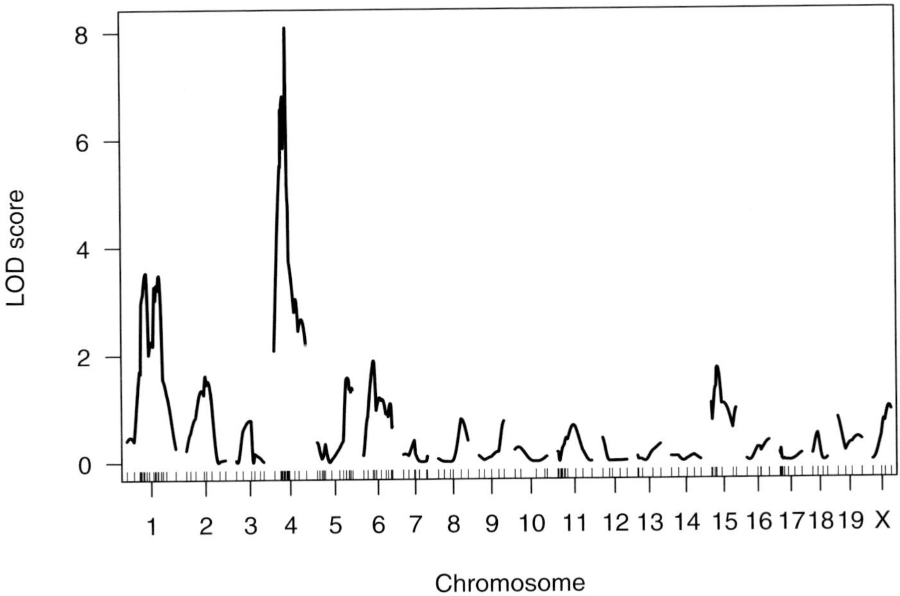

Figure 4.4. LOD scores by standard interval mapping for the `hyper` data.

We can also use `plot.scanone` to plot both the interval mapping results and those by marker regression (obtained in the previous section) together in one figure. This may be done in a couple of different ways. First, we can send both results to the `plot.scanone` function. We plot the results for chromosomes 4 and 12 as follows; see Fig. 4.5.

```
> plot(out.em, out.mr, chr=c(4, 12), col=c("blue", "red"),
+       ylab="LOD score")
```

We can produce the same figure by first plotting the interval mapping results and then adding the marker regression results using `add=TRUE`.

```
> plot(out.em, chr=c(4, 12), col="blue", ylab="LOD score")
> plot(out.mr, chr=c(4, 12), col="red", add=TRUE)
```

Finally, we can create a black-and-white plot with different line types, using the `lty` argument. Use `lty=1` for a solid line, `lty=2` for a dashed line, and `lty=3` for a dotted line. We give it a vector with two plot types; the first line type is used for the first result sent to the plot; the second line type is for the second result.

```
> plot(out.em, out.mr, chr=c(4, 12), col="black", lty=1:2,
+       ylab="LOD score")
```

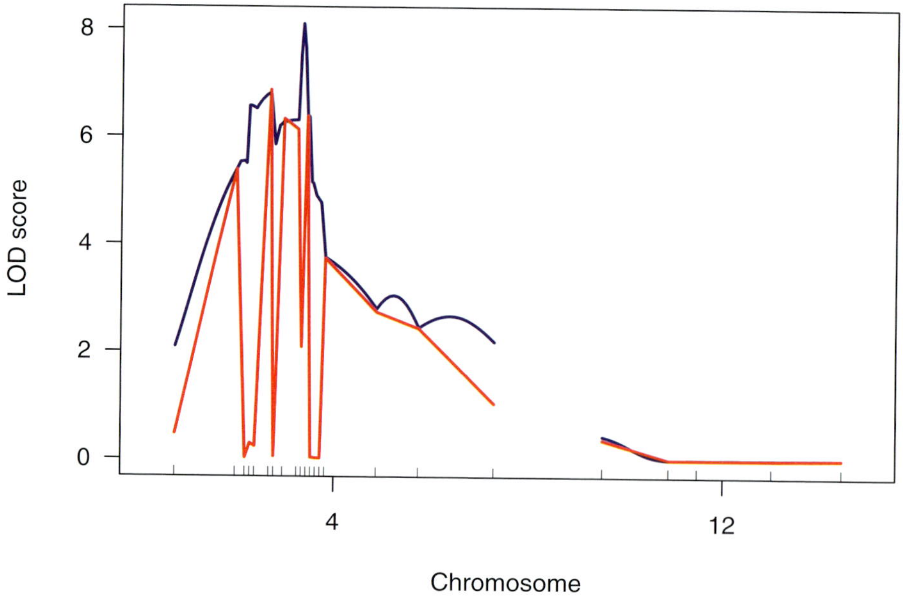

Figure 4.5. LOD scores for selected chromosomes for the `hyper` data by standard interval mapping (blue) and marker regression (red).

4.2.2 Haley–Knott regression

Haley–Knott regression provides a fast approximation of the results of standard interval mapping. In standard interval mapping, we assume that $y_i|g_i \sim N(\mu_{g_i}, \sigma^2)$, where y_i is the phenotype of individual i and g_i is its (unobserved) QTL genotype. We further calculate $p_{ij} = \Pr(g_i = j|\boldsymbol{M}_i)$, where $\boldsymbol{M}_i$ is the marker genotype data for individual i. Recall that $y_i|\boldsymbol{M}_i$ follows a mixture of normal distributions.

Note that $\mathrm{E}(y_i|\boldsymbol{M}_i) = \sum_j p_{ij}\mu_j$ and so the conditional phenotype average, given the available marker data, is linear in the μ_j. This suggests that the μ_j might be estimated by linear regression of the y_i on the p_{ij}.

This is Haley–Knott regression. At each position on our grid across the genome, we calculate the p_{ij} and then regress the phenotype on this matrix. In doing so, we pretend that $y_i|\boldsymbol{M}_i \sim N(\sum_j p_{ij}\mu_j, \sigma^2)$. That is, we replace the normal mixture with a single normal distribution, though with the correct mean function. We may thus calculate a LOD score as

$$\mathrm{LOD} = \frac{n}{2}\log_{10}\left(\frac{\mathrm{RSS}_0}{\mathrm{RSS}_1}\right)$$

where RSS_0 is the null residual sum of squares and RSS_1 is the residual sum of squares from the regression of the y_i on the p_{ij}.

Haley–Knott regression can be much faster than standard interval mapping, as an iterative algorithm is not needed; one performs a single regression at each position. However, the treatment of missing genotype information is less than ideal, and so its approximation of standard interval mapping can be poor in regions of low genotype information (such as widely spaced or incompletely genotyped markers). The approximation is especially poor in the case of selective genotyping (in which only individuals with extreme phenotypes are genotyped). We will discuss this issue further in Sec. 4.2.5, below.

Example

To perform Haley–Knott regression, we again need the genotype probabilities calculated by `calc.genoprob`, though we do not need to run the function again here, as it was executed in order to get the results by standard interval mapping. We again use the `scanone` function, and use `method="hk"` for Haley–Knott regression, as follows.

```
> out.hk <- scanone(hyper, method="hk")
```

We plot the results with those of standard interval mapping with the following. We look only at chromosomes 1, 4, and 15, so that the differences may be more clearly seen; the result appears in Fig. 4.6.

```
> plot(out.em, out.hk, chr=c(1,4,15), col=c("blue","red"),
+        ylab="LOD score")
```

The results from standard interval mapping and Haley–Knott regression are seen to be quite similar, though they deviate from each other at the ends of the chromosomes, particularly at the distal end of chromosome 15. The discrepancies occur in regions of missing genotype information, and are particularly affected by the selective genotyping strategy used for these data. The terminal markers on chromosomes 1 and 4, and the distal markers on chromosome 15, were genotyped at only the 92 individuals with most extreme phenotypes.

R/qtl contains a function `-.scanone` for subtracting two sets of LOD scores from each other, provided that they conform exactly (that they come from the same cross and that calculations were performed at the same density). We thus may look at the differences in the LOD scores from the two methods as follows. The result appears in Fig. 4.7.

```
> plot(out.hk - out.em, chr=c(1,4,15), ylim=c(-0.5, 1.0),
+        ylab=expression(LOD[HK] - LOD[EM]))
> abline(h=0, lty=3)
```

We used `abline` to add a dotted horizontal line at 0, and we used the function `expression` to get a fancy y-axis label. Type `?plotmath` to read about the possibilities with `expression`, and consider the following code. (We omit the resulting figure and any explanation; hopefully the interested reader can figure this out.)

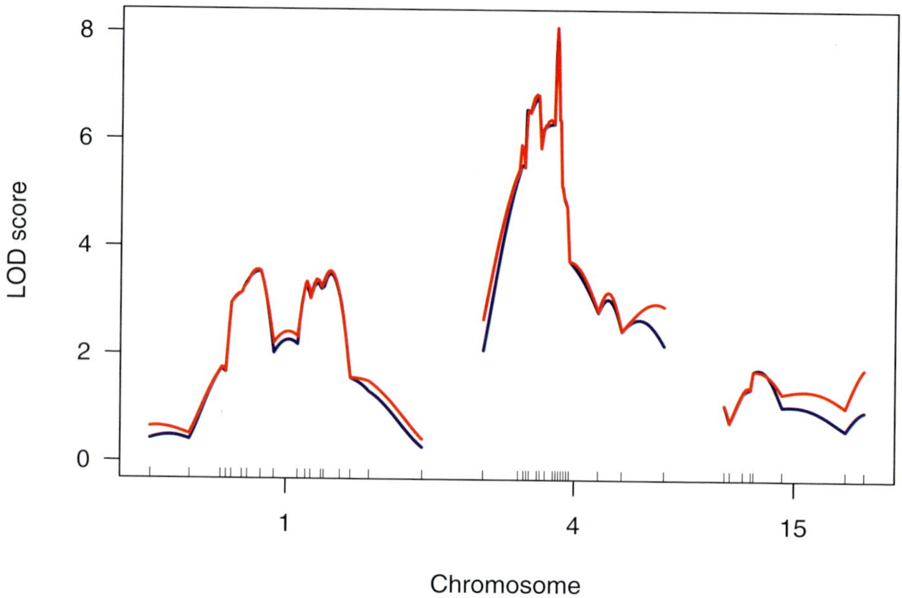

Figure 4.6. LOD scores for selected chromosomes for the `hyper` data by standard interval mapping (blue) and Haley–Knott regression (red).

```
> plot(rnorm(100), rnorm(100), xlab=expression(hat(mu)[0]),
+        ylab=expression(alpha^beta),
+        main=expression(paste("Plot of ", alpha^beta,
+             " versus ", hat(mu)[0])))
```

4.2.3 Extended Haley–Knott regression

An improved version of Haley–Knott regression may be obtained by also considering the variances. In Haley–Knott regression, we used the fact that $\mathrm{E}(y_i|\boldsymbol{M}_i) = \sum_j p_{ij}\mu_j$ and made the approximation $y_i|\boldsymbol{M}_i \sim N(\sum_j p_{ij}\mu_j, \sigma^2)$. That is, we approximate the mixture distribution with a single normal distribution with the correct mean, but with constant variance independent of genotype.

In the extended Haley–Knott regression method, we note that

$$\begin{aligned}
\mathrm{var}(y_i|\boldsymbol{M}_i) &= \mathrm{var}[\mathrm{E}(y_i|g_i)|\boldsymbol{M}_i] + \mathrm{E}[\mathrm{var}(y_i|g_i)|\boldsymbol{M}_i] \\
&= \mathrm{var}(\mu_{g_i}|\boldsymbol{M}_i) + \mathrm{E}(\sigma^2|\boldsymbol{M}_i) \\
&= \textstyle\sum_j p_{ij}[\mu_j - \sum_k p_{ik}\mu_k]^2 + \sigma^2
\end{aligned}$$

Write $m_i(\boldsymbol{\mu}) = \sum_j p_{ij}\mu_j$ and $v_i(\boldsymbol{\mu}, \sigma^2) = \sum_j p_{ij}[\mu_j - m_i(\boldsymbol{\mu})]^2 + \sigma^2 = \sum_j p_{ij}\mu_j^2 - (\sum_j p_{ij}\mu_j)^2 + \sigma^2$. In the extended Haley–Knott method, we assume

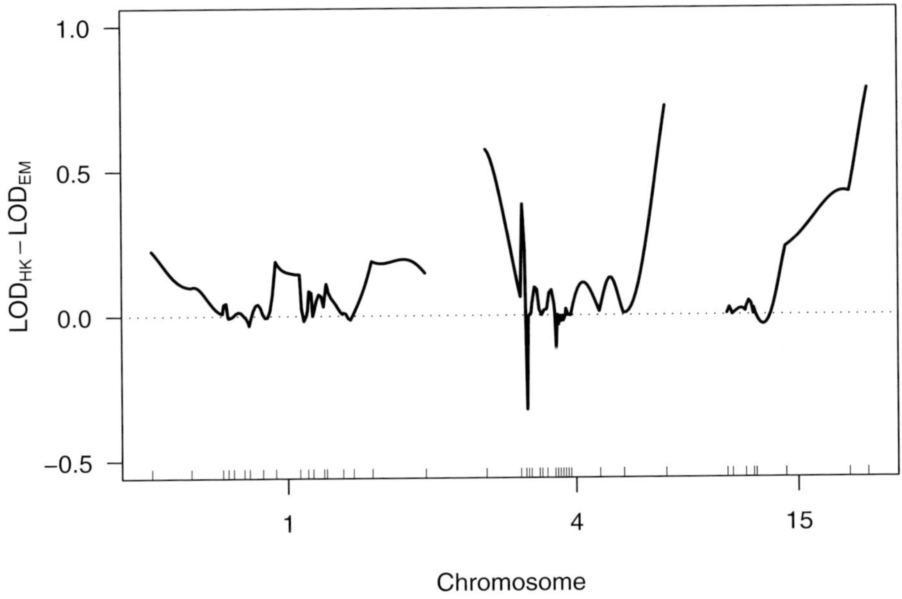

Figure 4.7. Differences in the LOD scores from Haley–Knott regression and standard interval mapping for selected chromosomes from the `hyper` data.

that $y_i|\boldsymbol{M}_i \sim N[m_i(\boldsymbol{\mu}), v_i(\boldsymbol{\mu}, \sigma^2)]$. That is, we replace the mixture distribution with a single normal distribution but with the correct mean and variance functions.

We estimate the μ_j and σ^2 by maximum likelihood (though with this approximate normal model). This requires an iterative method, though it generally converges more quickly than the EM algorithm under the mixture model.

The extended Haley–Knott method is not as fast as Haley–Knott regression, but it provides an improved approximation and is still somewhat faster than standard interval mapping. Most importantly, the extended Haley–Knott method is more robust than standard interval mapping; see Sec. 4.2.5.

Example

As with standard interval mapping and Haley–Knott regression, we need the genotype probabilities calculated by `calc.genoprob`, though we do not need to run the function again here. We use the `scanone` function with `method="ehk"` for the extended Haley–Knott method, as follows.

```
> out.ehk <- scanone(hyper, method="ehk")
```

We can plot the three interval mapping methods together with the following. The results appear in Fig. 4.8.

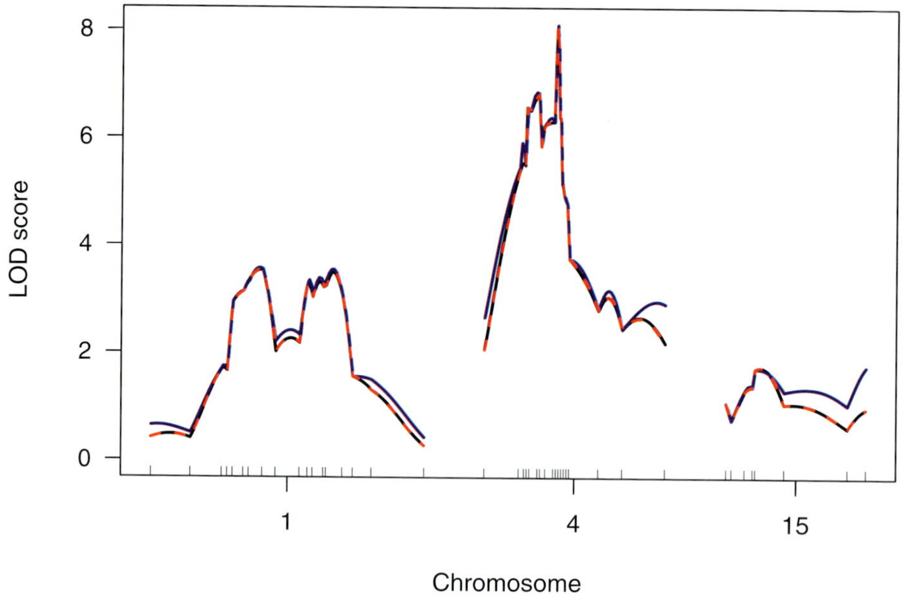

Figure 4.8. LOD scores for selected chromosomes for the `hyper` data by standard interval mapping (black), Haley–Knott regression (blue), and the extended Haley–Knott method (red, dashed).

```
> plot(out.em, out.hk, out.ehk, chr=c(1,4,15), ylab="LOD score",
+      lty=c(1,1,2))
```

The colors black, blue, and red are used by default. The argument `lty` is used to define line types (1 is solid; 2 is dashed). As we will see, the results by standard interval mapping and the extended Haley–Knott method are almost indistinguishable; we plot the extended Haley–Knott results with dashed curves so that the interval mapping results may still be seen.

Note how much more closely the results from the extended Haley–Knott method follow the results of standard interval mapping. The black curves are completely covered by the red curves, as standard interval mapping and the extended Haley–Knott method give results that are almost indistinguishable.) Up to three results may be plotted with a single call to `plot.scanone`. Alternatively, we may use `add=TRUE`, to obtain this plot.

```
> plot(out.em, chr=c(1,4,15), ylab="LOD score")
> plot(out.hk, chr=c(1,4,15), col="blue", add=TRUE)
> plot(out.ehk, chr=c(1,4,15), col="red", lty=2, add=TRUE)
```

To more clearly see the differences among the results, we can again plot the differences between the LOD scores. The results appear in Fig. 4.9.

```
> plot(out.hk - out.em, out.ehk - out.em, chr=c(1, 4, 15),
+      col=c("blue", "red"), ylim=c(-0.5, 1),
```

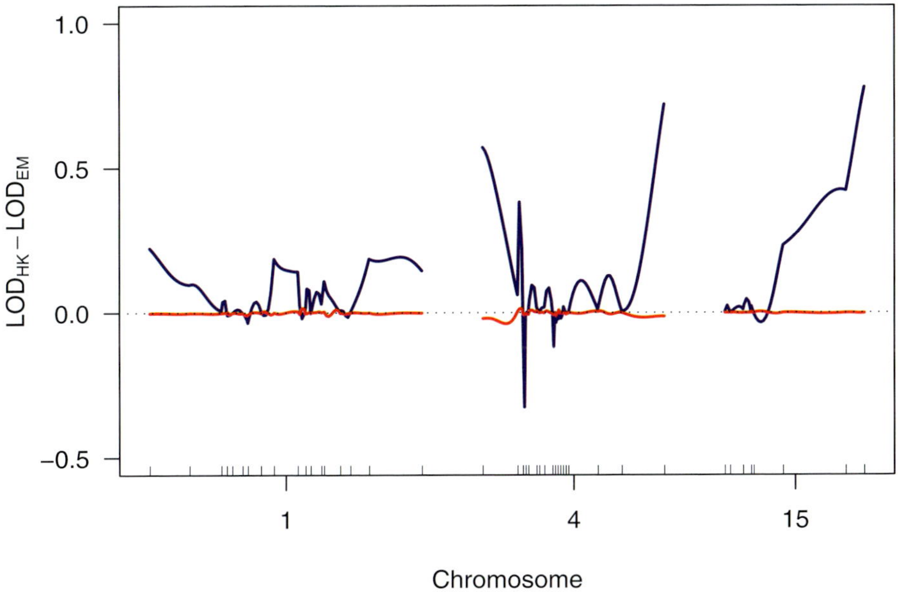

Figure 4.9. Differences in the LOD scores from Haley–Knott regression (in blue) and the extended Haley–Knott method (in red) from the LOD scores of standard interval mapping, for selected chromosomes from the `hyper` data.

```
+       ylab=expression(LOD[HK] - LOD[EM]))
> abline(h=0, lty=3)
```

4.2.4 Multiple imputation

As we discussed in Sect. 1.3, the QTL mapping problem can be split into two parts: the missing data problem and the model selection problem. The multiple imputation approach dispenses with the missing data problem by filling in all missing genotype data, even at sites between markers (on a grid along the chromosomes). With complete genotype data, the fit of a QTL model reduces to ANOVA (for single-QTL models) or multiple regression (for multiple-QTL models).

The only wrinkle is that such imputations must be done multiple times, with the final result being a combination of the results from the multiple imputations. Moreover, the combination of the single-imputation results can be complicated. The imputations are simple and model fit with each imputation is simple, but the combination of the imputations can be difficult.

Genotypes are imputed randomly, but conditional on the observed marker genotype data: we simulate from the joint genotype distribution given the observed data. For example, consider Fig. 4.10, which illustrates the imputation of a single backcross individual's genotype data. The observed genotype data

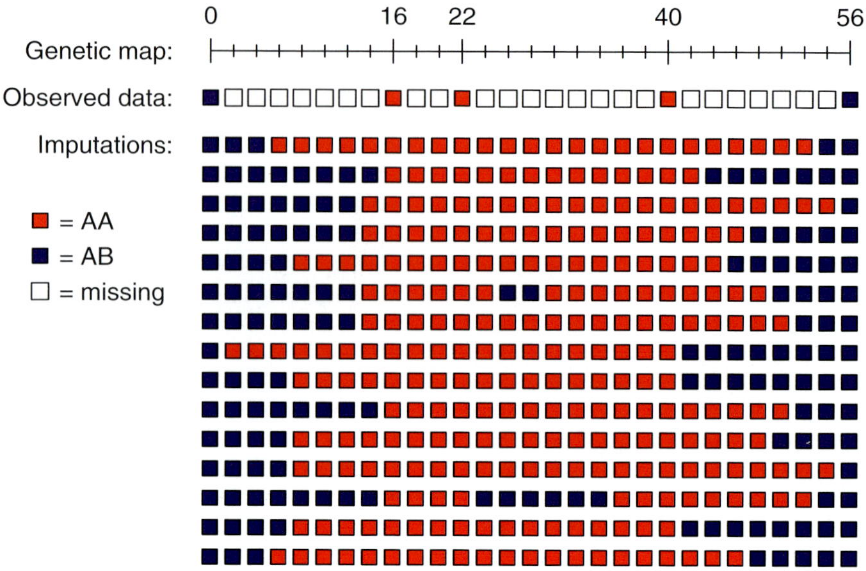

Figure 4.10. Illustration of multiple imputations for a single backcross individual. Red and blue squares correspond to homozygous and heterozygous genotypes, respectively, while open squares indicate missing data. The marker genotype data for the individual is shown at the top, below a genetic map (in cM) for the chromosome; multiple imputations of the genotype data, at the markers and at intervening positions at 2 cM steps along the chromosome, are shown below.

at five genetic markers is shown at the top, followed by the imputed genotypes for 15 different imputations. Note that the positions of the recombination events vary among the imputations, and a couple of imputations exhibit double-crossovers between markers. The imputed genotypes at the markers match those observed, as these data were simulated assuming no genotyping errors (an assumption that may be relaxed).

Such multiple imputations would be obtained on all individuals. With a given set of imputed data, we can simply perform a t test or ANOVA at each position, as with such complete genotype data on all individuals, we know how to split the individuals into genotype groups. Following tradition, we express the results as LOD scores.

As a further illustration, consider Fig. 4.11, which contains imputation results for chromosome 4 of the `hyper` data. The LOD curve from each of 16 imputations are shown in gray, and a combined LOD curve from a total of 64 imputations is shown in black. At markers with complete genotype data, all LOD curves coincide, as all imputations give the same set of data. In regions with less genotype information, the LOD curves from the individual imputations deviate from one another.

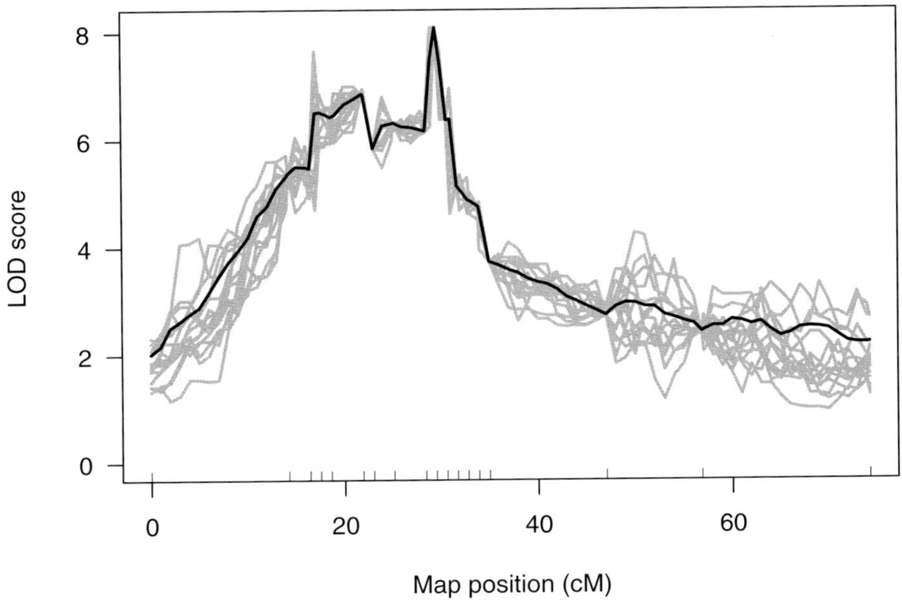

Figure 4.11. Illustration of the imputation results for chromosome 4 of the `hyper` data. The individual LOD curves from 16 imputations are shown in gray; the combined LOD score from a total of 64 imputations is shown in black.

For the normal model, the combined LOD score is the average of the LOD scores from the individual imputations, though on the 10^{LOD} scale. To obtain a more stable estimate of this average, we use a trimmed mean. In the case of m imputations, we trim the lowest and highest $\log_2(m)/2$ LOD scores. (More precisely, we trim $\lfloor \log_2(m)/2 \rfloor$ from each end, where $\lfloor x \rfloor$ is the greatest integer $\leq x$. We generally take the number of imputations to be a power of 2, like 64 or 128: $m = 2^k$ for some positive integer k.) Moreover, we assume that the LOD scores at a particular position, across imputations, approximately follow a lognormal distribution, and we use the fact that, if $L = \ln(W) \sim N(\mu, \sigma^2)$, so that W is lognormally distributed, $\text{E}(W) = \exp(\mu + \sigma^2/2)$.

To be precise, the combined LOD scores that we calculate by the multiple imputation method are not truly LOD scores. Strictly speaking, this is a Bayesian method, and the combined scores are the log posterior distribution (LPD) of QTL location, but the results are generally similar to the LOD scores from standard interval mapping.

The multiple imputation approach is intensive in both computation time and memory use. While it is more robust than standard interval mapping, it has little advantage over the extended Haley–Knott method for single-QTL models, particularly because of the large up-front cost to obtain the imputations. The multiple imputation approach has greatest value for the fit and exploration of multiple-QTL models (see Chap. 9).

Example

To perform multiple imputation in R/qtl, we first perform the imputations using the `sim.geno` function. This function is similar to `calc.genoprob`, though it has an additional argument, `n.draws`, through which the number of imputations is specified.

```
> hyper <- sim.geno(hyper, step=1, n.draws=64, error.prob=0.001)
```

The imputed genotypes can take up an *enormous* amount of memory, especially if `step` is small, `n.draws` is large, and there are many individuals in the cross. In such cases, you will need a computer with a lot of RAM. You may wish to perform initial analyses with a rather coarse `step` size and with fewer imputations, reserving the refined `step` and larger `n.draws` for the fit of later multiple-QTL models (see Chap. 9), in which case you can focus on just those chromosomes that appear to harbor a QTL. If time and memory are not as issue, more imputations are always better (just as a smaller `step` size is always better). With relatively complete genotype data and relatively dense markers, very few imputations will be required; the more sparse the genotype information, the more imputations should be used. Repeating the analysis with independent imputations can give a good indication of the need for a larger number of imputations. If the results are hardly distinguishable, the chosen number of imputations is sufficient.

The analysis is again accomplished with `scanone`, with `method="imp"` for the multiple imputation method. For example, we type the following to perform interval mapping by multiple imputation with the `hyper` data.

```
> out.imp <- scanone(hyper, method="imp")
```

We plot the results for selected chromosomes with those from standard interval mapping (Sec. 4.2.1) via the following; the results appear in Fig. 4.12.

```
> plot(out.em, out.imp, chr=c(1,4,15), col=c("blue", "red"),
+        ylab="LOD score")
```

It is again worthwhile to look at the differences between the LOD curves. See Fig. 4.13.

```
> plot(out.imp - out.em, chr=c(1,4,15), ylim=c(-0.5, 0.5),
+        ylab=expression(LOD[IMP] - LOD[EM]))
> abline(h=0, lty=3)
```

4.2.5 Comparison of methods

In this section, we summarize the relative advantages and disadvantages of the various single-QTL mapping methods that we have discussed in this chapter. For a quick summary, see Table 4.2 on page 102.

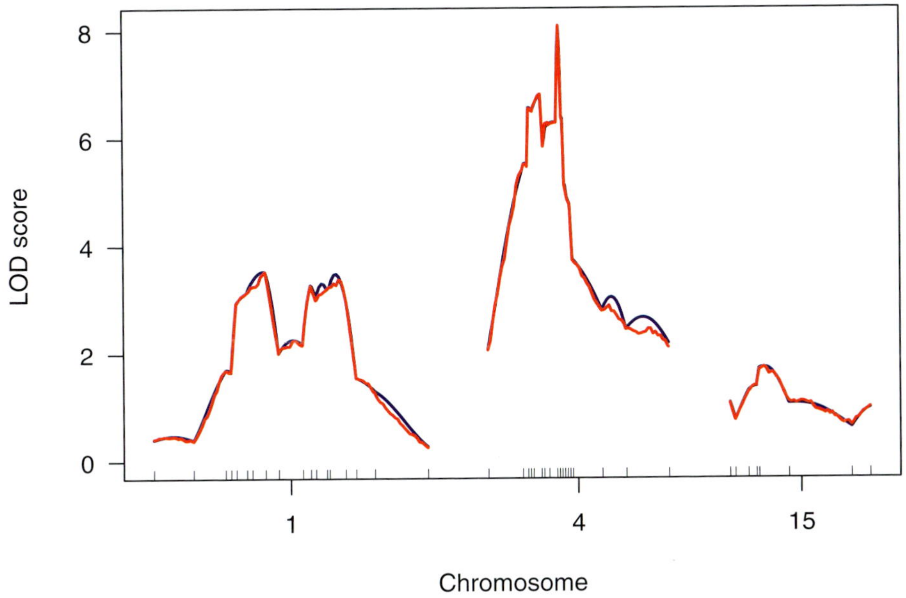

Figure 4.12. LOD scores for selected chromosomes for the `hyper` data by standard interval mapping (blue) and multiple imputation (red).

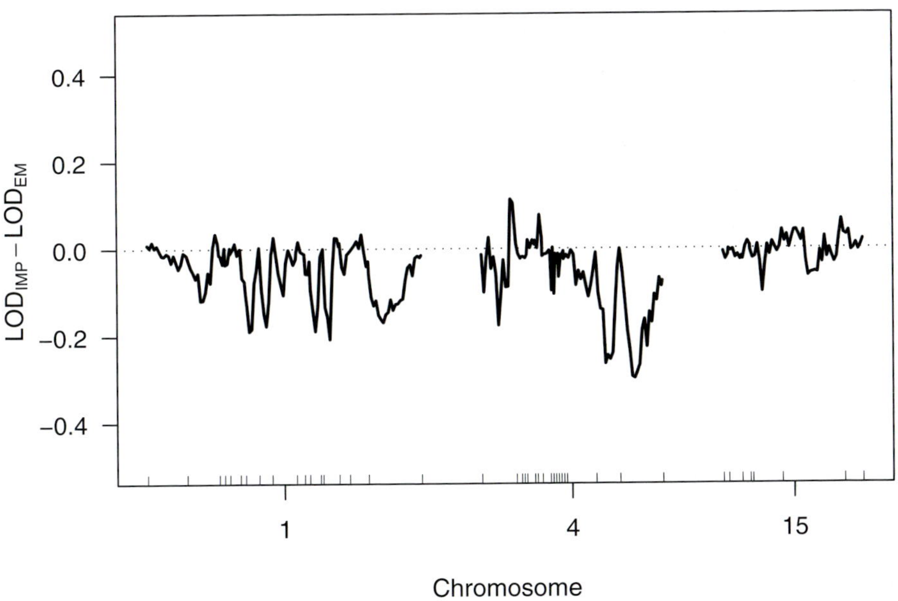

Figure 4.13. Differences in the LOD scores from multiple imputation and standard interval mapping for selected chromosomes from the `hyper` data.

Marker regression is not recommended for use in practice, except in the case of dense markers with complete genotype data (as is sometimes the case with data on recombinant inbred lines), because individuals with missing genotype at a marker must be omitted from the analysis, and one cannot inspect positions between markers. The interval mapping methods take account of missing genotype data.

Standard interval mapping, which uses maximum likelihood estimation in a normal mixture model, is generally the preferred method, but it can give artifacts. Recall that the LOD score is the $\log_{10}$ likelihood ratio, comparing the hypothesis of a single QTL at the position under test to the null hypothesis of no QTL anywhere in the genome. If the phenotype distribution exhibits multiple modes, so that it would be better approximated by a mixture of normal distributions rather than a single normal distribution, one can get spuriously large LOD scores in regions of low genotype information (such as a large gap between typed markers). At a typed marker, one is constrained by the observed genotypes, but in a region with low genotype information, the likelihood under the alternative essentially concerns the fit of a mixture model.

For example, consider the `listeria` data. The phenotype is time-to-death following infection with *Listeria monocytogenes*, but a large number of individuals recovered from infection, and so their phenotype was censored (and recorded as 264 hours); see Fig. 2.7 on page 35. The markers are relatively dense, and the genotype data relatively complete, and so application of standard interval mapping with these data do not cause problems. If we omit most of the markers on a chromosome, we can illustrate our point. Chromosome 1 was typed at 13 markers; let us omit all but the terminal markers. We use the function `markernames` to pull out the marker names for chromosome 1 and `drop.markers`, to drop all but the first and last markers.

```
> data(listeria)
> mar2drop <- markernames(listeria, chr=1)[2:12]
> listeria <- drop.markers(listeria, mar2drop)
```

If we now perform standard interval mapping, we will see a large LOD peak on chromosome 1, in the middle of the large interval with no genotype data (see Fig. 4.14). The other interval mapping methods do not exhibit this problem; we consider just the Haley–Knott methods here. We use the argument `chr=1` in `scanone` so that only chromosome 1 is analyzed.

```
> listeria <- calc.genoprob(listeria, step=1, error.prob=0.001)
> out1.em <- scanone(listeria, chr=1)
> out1.hk <- scanone(listeria, chr=1, method="hk")
> out1.ehk <- scanone(listeria, chr=1, method="ehk")
> plot(out1.em, out1.hk, out1.ehk, ylab="LOD score",
+      lty=c(1,1,2))
```

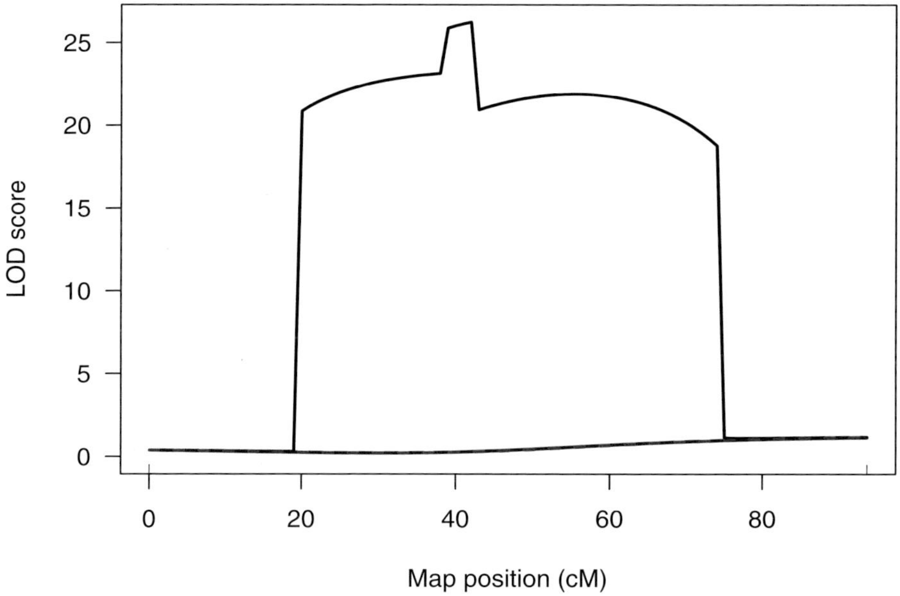

Figure 4.14. LOD scores by standard interval mapping (black), Haley–Knott regression (blue) and the extended Haley–Knott method (red, dashed) for chromosome 1 of the `listeria` data, when the genotype data for all but the terminal markers have been omitted.

The discontinuities in the LOD curve from standard interval mapping concern multiple modes in the likelihood surface. At a typed marker, one is constrained by the observed genotype data. At the center of the large interval, there is essentially no genotype data, and so one is fitting a pure mixture model.

The Haley–Knott method is more robust than standard interval mapping; however, the approximation used in Haley–Knott regression, in which we regress the phenotype on conditional genotype probabilities, can be poorly behaved with appreciable missing genotype data, particularly in the case of selective genotyping. In the selective genotyping strategy (discussed further in Chap. 6), only individuals with extreme phenotypes (for example, the individuals in the upper and lower 10% of the phenotype distribution) are genotyped. In the case of an inexpensive phenotype, this greatly reduces the cost of a study yet provides nearly equivalent power for QTL detection.

Consider a backcross with genotypes coded as 0 and 1. In Haley–Knott regression, an individual with no genotype data will be treated as if its genotype were 1/2 (and were known to be 1/2). This results in a slightly inflated estimate of the effect of a QTL but also a greatly reduced estimate of the residual variation, and so can give inflated LOD scores.

In the extended Haley–Knott method, individuals with no genotype data are also treated as having genotype 1/2, but their residual variance is adjusted, so they are given very little weight in the estimation. Standard interval mapping and the multiple imputation method explicitly take account of the fact that the individuals without genotypes have an equal chance of being homozygous and heterozygous. As a result, these other methods give remarkably similar LOD curves, irrespective of whether the individuals without genotype data are included in the analysis.

If one omitted individuals with absolutely no genotype data, Haley–Knott regression will provide a good approximation to standard interval mapping. However, often (as with the `hyper` data) selective genotyping is followed by complete genotyping at certain markers, and so no individual is completely lacking in genotype data.

Consider the `hyper` data as an example. The relationship between blood pressure and the genotype at marker D4Mit214 is shown in the left panel Fig. 4.15. The regression line of phenotype on genotype (equivalent to a t test) is shown. In the right panel, the genotype data for all but the 92 individuals with extreme phenotypes have been omitted. The blue line is the regression line when only the genotyped individuals are considered. The red line is the regression line obtained when those not genotyped are placed intermediate between the two genotyped groups, as is done in Haley–Knott regression. The regression lines in the right panel are more steep than that in the left panel, but the biggest difference is that, in the right panel, with the ungenotyped individuals placed in the center, the variation around the regression line is artificially reduced.

Let us look more closely at the LOD curves that will be obtained by the different methods. In the `hyper` data, further genotyping was performed in regions showing initial evidence for a QTL; we will omit these genotypes, to convert these data to the "pure" selective genotyping form.

First, we identify the markers that have mostly missing data (those that were typed only on recombinant individuals) and use `drop.markers` to remove them from the data set. We assign the revised data to a new object, `hyper.rev`.

```
> data(hyper)
> nt.mar <- ntyped(hyper, "mar")
> mar2drop <- names(nt.mar[nt.mar < 92])
> hyper.rev <- drop.markers(hyper, mar2drop)
```

Next, we eliminate the genotype data for the individuals with intermediate phenotype, who may be identified by their missing genotype data. This requires a loop over the chromosomes; the genotype data for the relevant individuals is replaced with `NA` (for "missing").

```
> nm.ind <- nmissing(hyper.rev)
> ind2drop <- nm.ind > 0
```

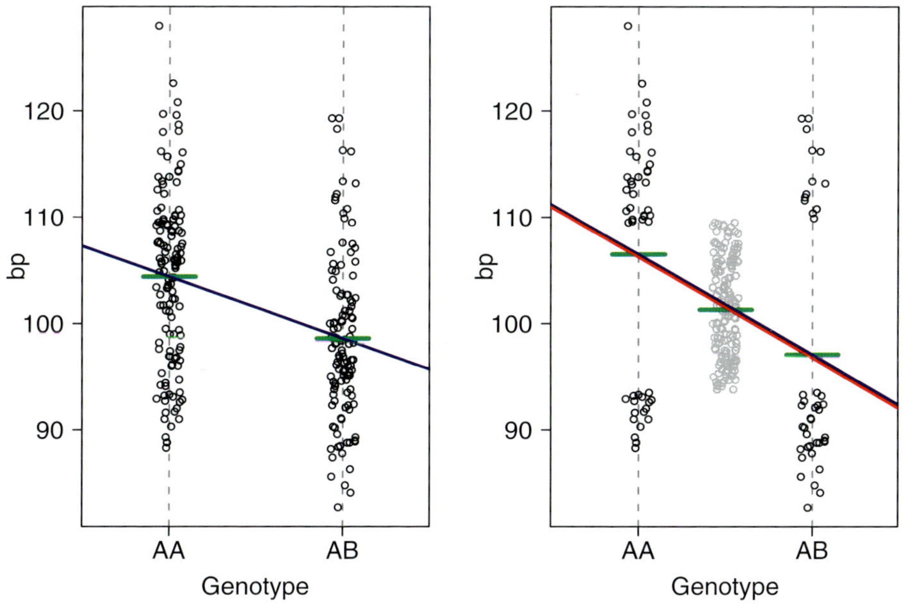

Figure 4.15. Relationship between blood pressure and the genotype at D4Mit214 for the `hyper` data. The horizontal green segments indicate the within-group averages. In the left panel, the data for all individuals is shown with the regression line of phenotype on genotype. In the right panel, the genotype data for all but the 92 individuals with extreme phenotypes is omitted. The blue line is the regression line obtained with only the 92 phenotyped individuals. The red line comes from a regression with all individuals, but with those not genotyped placed intermediate between the two genotype groups, as is done in Haley–Knott regression.

```
> for(i in 1:nchr(hyper))
+   hyper.rev$geno[[i]]$data[ind2drop,] <- NA
```

Now, we perform genome scans using all individuals. We will use each of standard interval mapping, Haley–Knott regression and the extended Haley–Knott method.

```
> hyper.rev <- calc.genoprob(hyper.rev, step=1,
+                            error.prob=0.001)
> out1.em <- scanone(hyper.rev)
> out1.hk <- scanone(hyper.rev, method="hk")
> out1.ehk <- scanone(hyper.rev, method="ehk")
```

The LOD curves from the three methods are shown in Fig. 4.16, which was created as follows.

```
> plot(out1.em, out1.hk, out1.ehk, ylab="LOD score",
+      lty=c(1,1,2))
```

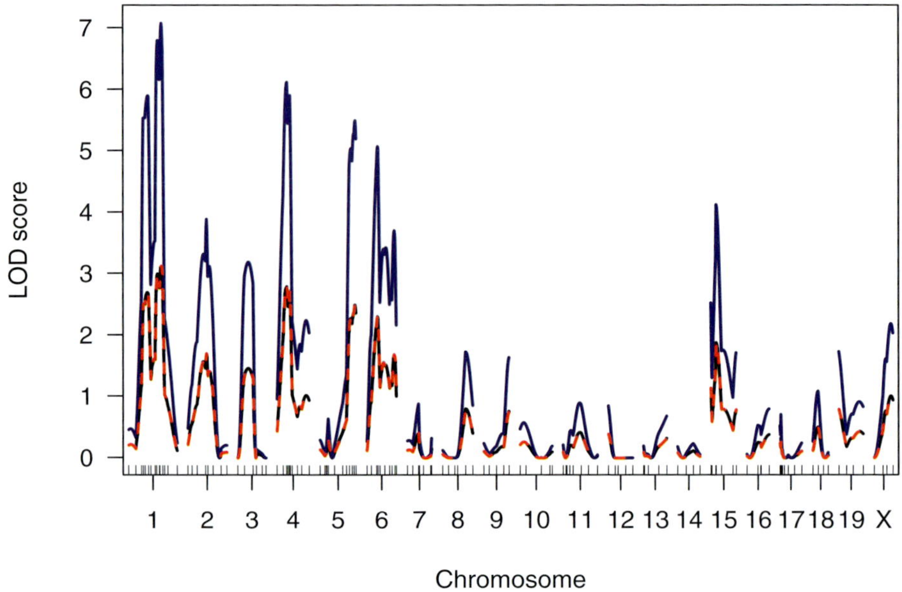

Figure 4.16. LOD curves from standard interval mapping (black), Haley–Knott regression (blue) and the extended Haley–Knott method (red, dashed), for the `hyper` data when all individuals are considered but genotype data for individuals with intermediate phenotype are omitted.

The LOD scores from Haley–Knott regression (in blue) are inflated. The results of standard interval mapping (black) and the extended Haley–Knott method (red) almost completely overlap. Thus, to distinguish among the methods more clearly, we plot the differences between the LOD scores for the Haley–Knott methods and those from standard interval mapping.

```
> plot(out1.hk-out1.em, out1.ehk-out1.em, col=c("blue", "red"),
+       ylim=c(-0.1, 4), ylab=expression(LOD[HK] - LOD[EM]))
> abline(h=0, lty=3)
```

The plot of the differences, in Fig. 4.17, confirm that the extended Haley–Knott method gives results that are virtually identical to standard interval mapping.

We now perform the same analyses, considering only the genotyped individuals. We use `subset.cross` to create a new version of the data with only the genotyped individuals. The vector `ind2drop` was created above to contain the logical values `TRUE` and `FALSE`, with `TRUE` indicating individuals with no genotype data. We use `!ind2drop` to reverse the TRUE/FALSE values (`!` is the logical "not"), to pull out just those individuals with genotype data.

```
> hyper.ex <- subset(hyper.rev, ind = !ind2drop)
> out2.em <- scanone(hyper.ex)
```

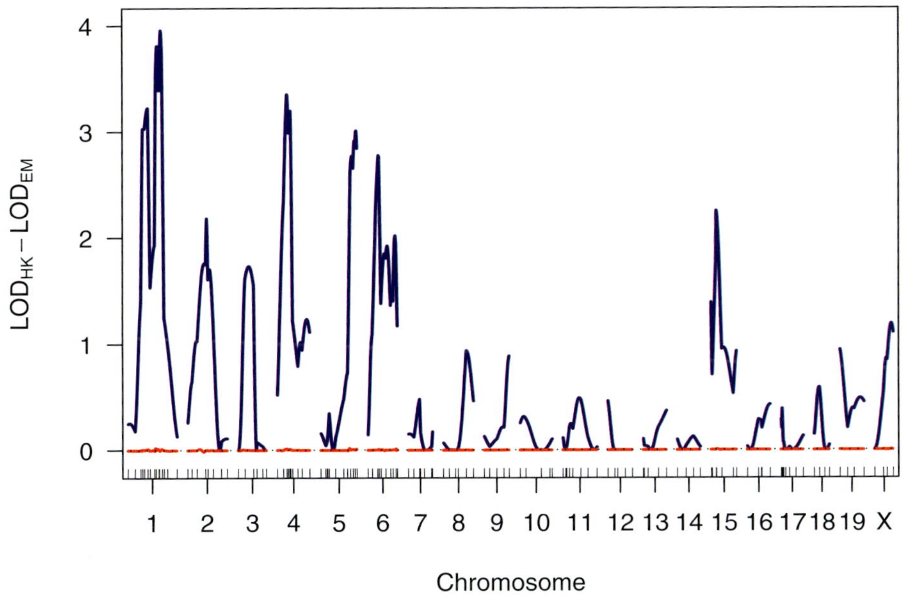

Figure 4.17. Differences in the LOD curves of Haley–Knott regression (blue) and the extended Haley–Knott method (red), from the LOD curves of standard interval mapping, for the hyper data when all individuals are considered but genotype data for individuals with intermediate phenotype are omitted.

```
> out2.hk <- scanone(hyper.ex, method="hk")
> out2.ehk <- scanone(hyper.ex, method="ehk")
```

Let us just look at the differences in the LOD scores for the methods for this case (with only genotyped individuals considered) and the LOD curves from standard interval mapping when all individuals were considered; see Fig. 4.18.

```
> plot(out2.em-out1.em, out2.hk-out1.em, out2.ehk-out1.em,
+      ylim=c(-0.1, 0.2), col=c("black", "green", "red"),
+      lty=c(1,3,2), ylab="Difference in LOD scores")
> abline(h=0, lty=3)
```

The methods all give similar results, and the results are not too different from standard interval mapping when all individuals are considered (but with the genotypes of the individuals with intermediate phenotype omitted).

In summary, standard interval mapping can give spuriously large LOD scores in regions of low genotype information, if the phenotype distribution is better approximated by a normal mixture than by a single normal distribution; the other methods do not have this problem. Haley–Knott regression can provide a poor approximation to interval mapping in the case of low genotype information, and can give inflated LOD scores in the presence of selective

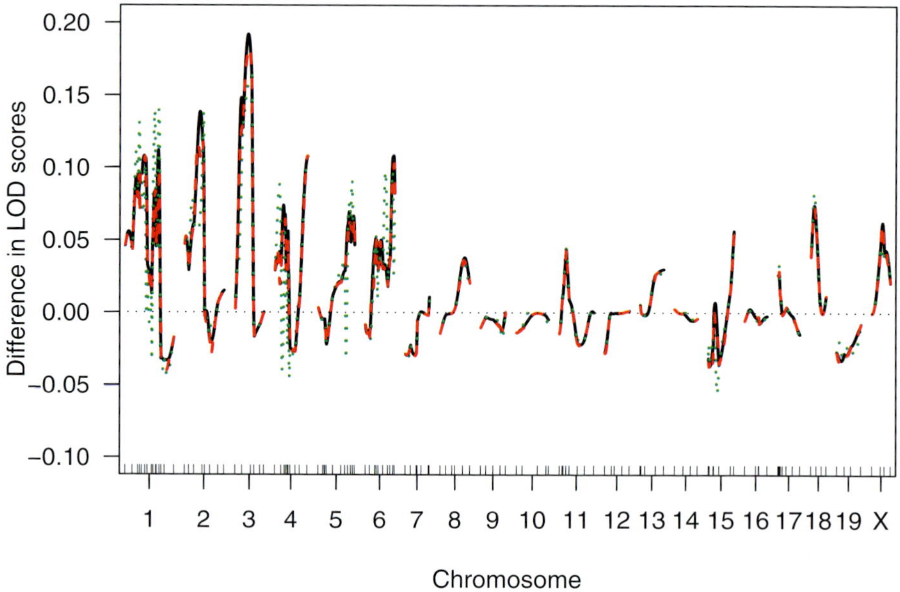

Figure 4.18. Differences in the LOD curves (for the `hyper` data) from standard interval mapping (in black), Haley–Knott regression (in green, dotted) and the extended Haley–Knott method (in red, dashed), all calculated with only the data on the individuals with extreme phenotypes, from the LOD curves of standard interval mapping, with all individuals but with genotype data for individuals with intermediate phenotype omitted.

Table 4.2. Relative advantages and disadvantages of the four interval mapping methods.

	Use of genotype information	Robustness	Selective genotyping	Speed
Standard interval mapping	++	−	+	−
Haley–Knott	−	+	−	+
Extended Haley–Knott	+	+	+	−
Multiple imputation	++	+	+	−−

genotyping. The extended Haley–Knott method may be preferred as it suffers from neither of these problems. These points are summarized in Table 4.2.

Our last point concerns computation time. Haley–Knott regression is especially fast, and the extended Haley–Knott method is intermediate between Haley–Knott regression and standard interval mapping in computation time. The multiple imputation approach is particularly slow for the fit of single-QTL models, and is best reserved for the fit of multiple-QTL models.

We can measure computation time with the `system.time` function. The following times were obtained on a Mac Pro. We will consider the `hyper` data.

First, look at the time to run `calc.genoprob`. We use `step=0.25` so that the later comparison of the methods is most clear.

```
> data(hyper)
> system.time(hyper <- calc.genoprob(hyper, step=0.25,
+                                     error.prob=0.001))

   user  system elapsed
  1.392   0.472   1.882
```

In the output from `system.time`, the first number is the CPU time (in seconds) and the third number is the total time.

Now let us look at the four interval mapping methods.

```
> system.time(test.hk <- scanone(hyper, method="hk"))

   user  system elapsed
  0.238   0.095   0.337

> system.time(test.ehk <- scanone(hyper, method="ehk"))

   user  system elapsed
  0.982   0.093   1.085

> system.time(test.em <- scanone(hyper, method="em"))

   user  system elapsed
  1.105   0.094   1.207
```

The extended Haley–Knott method is intermediate between the other two methods, though here it is not much faster than standard interval mapping.

For the multiple imputation approach, the calculations (in `sim.geno` and `scanone` with `method="imp"`) scale linearly in `n.draws`.

```
> system.time(hyper <- sim.geno(hyper, step=0.25,
+                               error.prob=0.001, n.draws=32))

   user  system elapsed
 11.795   4.253  16.288

> system.time(test.imp <- scanone(hyper, method="imp"))

   user  system elapsed
  3.002   0.608   3.706
```

The imputations themselves are quite time consuming, and the actual analysis is also slower than the EM algorithm. (This will depend on the number of iterations needed for convergence of EM; the time for one iteration of EM is approximately the same as the time analyzing one imputation.)

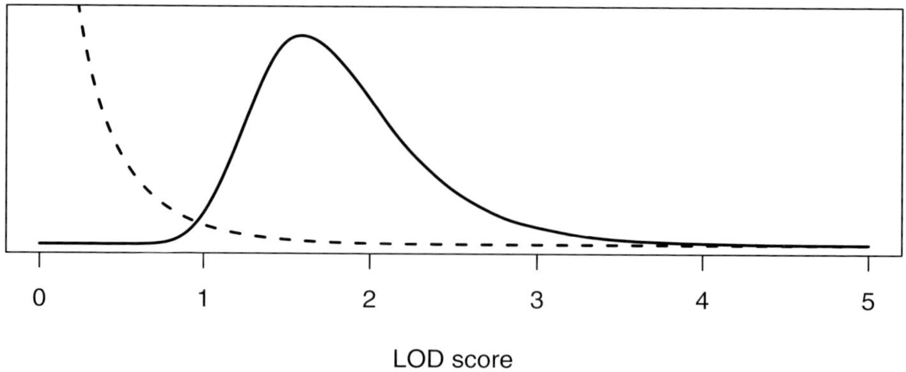

Figure 4.19. Approximate null distribution of the LOD score at a particular position (dashed curve) and of the maximum LOD score genome-wide (solid curve) for a backcross in the mouse.

4.3 Significance thresholds

A LOD score indicates evidence for the presence of a QTL, with larger LOD scores corresponding to greater evidence. The question is, how large is large? In answering this question, we must take account of the genome-wide search for QTL.

We consider the global null hypothesis, that there is no QTL anywhere in the genome. (This is almost always clearly false, as we generally start with inbred lines that show a clear difference in the phenotype, which implies that there must be QTL. Nevertheless, the procedures described here are useful.) Rather than consider the null distribution of the LOD score at a particular position (the dashed curve in Fig. 4.19), we consider the null distribution of the genome-wide maximum LOD score (the solid curve in Fig. 4.19).

As seen in Fig. 4.19, in a backcross with no QTL, at any fixed position in the genome, one will typically see a LOD score of about 0.25, and will seldom see a LOD score greater than 1. However, one will typically see a LOD score > 1.5 *somewhere* in the genome, and will see a LOD score > 2.5 about 10% of the time. We compare our observed LOD scores to the distribution of the genome-wide maximum LOD score, in the case that there were no QTL anywhere. The 95th percentile of this distribution may be used as a genome-wide LOD threshold. Alternatively, one may calculate a genome-scan-adjusted p-value corresponding to an observed LOD score: the chance, under the null hypothesis of no QTL, of obtaining a LOD score that large or larger somewhere in the genome.

The null distribution of the genome-wide maximum LOD depends on a number of factors, including the type of cross (backcross or intercross), the size of the genome (in cM), the number of individuals, the number of typed markers, the pattern of missing genotype data, and the phenotype distribution.

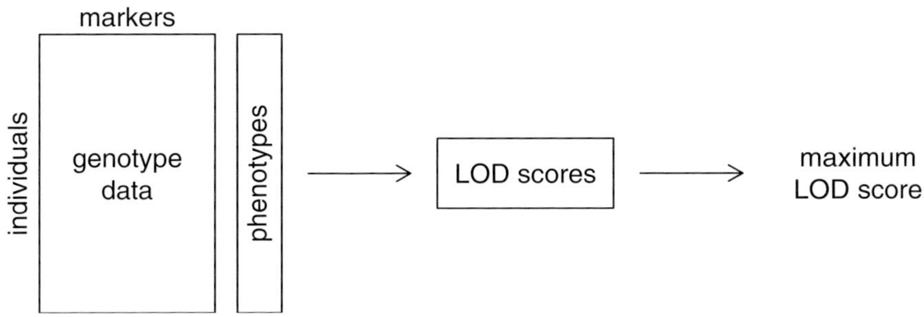

Figure 4.20. Diagram of the interval mapping process.

The type of cross and the genome size have the greatest influence. One may derive this null distribution by several methods: computer simulation, analytic calculations (e.g., for the case of infinitely dense markers), or by a permutation test. We prefer permutation tests, as they best account for the particular features of one's data.

The process involved in interval mapping is illustrated in Fig. 4.20. The data consist of a rectangle of marker genotype data plus a column of phenotypes. QTL analysis results in a set of LOD curves, and one immediately notes the genome-wide maximum LOD curve. The question is, if there were no QTL (so that there was no association between phenotypes and the genotypes), what sort of maximum LOD scores might be obtained?

In a permutation test, we tackle this directly by permuting (i.e., randomizing or shuffling) the phenotypes relative to the genotype data. That is, the genotype data rectangle remains intact, and the observed phenotypes are kept the same, but which phenotype corresponds to which genotypes is randomized. We apply our QTL mapping method to this shuffled version of the data and obtain a set of LOD curves; we then derive the genome-wide maximum LOD score. The process is repeated a number of times, which results in a set of maximum LOD scores, $M_1^*, M_2^*, \ldots, M_r^*$, where r is the number of permutation replicates (generally $r = 1000$ or 10,000). We may use the 95th percentile of the M_i^* as an estimated genome-wide LOD threshold, or we may calculate a genome-scan-adjusted p-value as the proportion of the M_i^* that meet or exceed a particular observed LOD score.

It should be noted that, in the case of selectively genotyped data, the usual permutation test is not appropriate, as the individuals are no longer exchangeable. One should instead use a *stratified* permutation test, shuffling individuals' phenotypes separately within strata of similarly genotyped individuals.

A common question concerns the appropriate number of permutation replicates. A larger number of permutation replicates gives greater precision in the estimated significance thresholds or empirical p-values. While we generally use 1000 permutation replicates initially, we may go up to 10,000 or even 100,000 replicates in order to achieve greater precision. Note that the number of permutation replicates that meet or exceed a given LOD score follows a binomial(r, p) distribution, with r being the number of replicates and p being the true p-value. In the case that the p-value is around 0.05, the standard error of the estimated p-value from the permutation test will be around $\sqrt{0.05 \times 0.95/r}$. Thus, to achieve a standard error of 0.005, one would need $0.05 \times 0.95/(0.005)^2 = 1900$ permutation replicates.

Finally, we wish to emphasize that we strongly oppose the use of strict thresholds for statistical significance; it is better to report genome-scan-adjusted p-values. P-values of 4.6% and 5.4% are essentially the same and so should be treated the same. They shouldn't be called "significant" and "suggestive" according to whether they landed below or above a 5% cutoff. Moreover, the importance accorded to a particular p-value depends upon a number of factors, including the ultimate goals of the experiment.

Example

Let us illustrate the permutation test by considering the `hyper` data. We can perform a permutation test with any of the QTL mapping methods discussed in this chapter through the argument `n.perm` to the `scanone` function. That is, we may use the `scanone` function just as before, but by including `n.perm=1000` in the code, the function performs a permutation test with 1000 replicates rather than doing the genome scan with the data. We must, of course, have previously run `calc.genoprob` to obtain the QTL genotype probabilities; let's reload the `hyper` data and run `calc.genoprob` again, just to be sure.

```
> data(hyper)
> hyper <- calc.genoprob(hyper, step=1, error.prob=0.001)
```

Now we use `scanone` to do the permutation test. We use `verbose=FALSE` to suppress any output.

```
> operm <- scanone(hyper, n.perm=1000, verbose=FALSE)
```

The result is a vector of length 1000, containing the genome-wide maximum LOD score from each of 1000 permutation replicates. We may look at the first 5 results as follows.

```
> operm[1:5]

[1] 1.547 1.434 4.184 2.151 1.060
```

The object `operm` has class `"scanoneperm"`. There is a corresponding `plot` function, for plotting a histogram of the results, and a `summary` function, for calculating LOD thresholds.

A histogram of the permutation results is produced as follows; see Fig. 4.21.

```
> plot(operm)
```

To obtain estimated genome-wide LOD thresholds for significance levels 20% and 5%, we do the following.

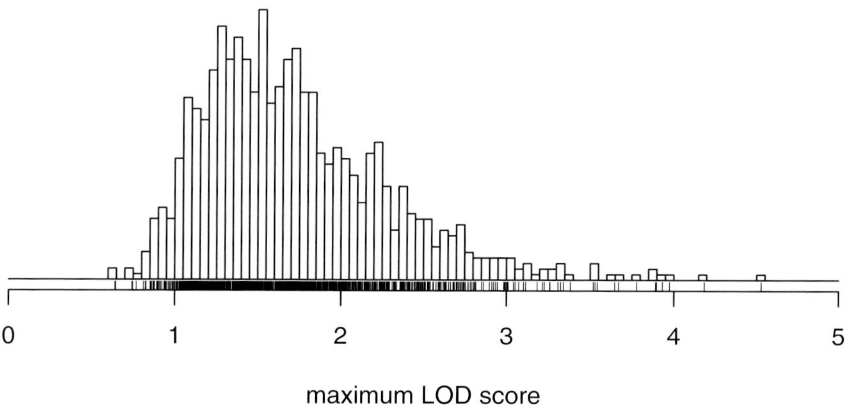

Figure 4.21. Histogram of the genome-wide maximum LOD scores from 1000 permutation replicates with the `hyper` data. The LOD scores were calculated by Haley–Knott regression.

```
> summary(operm, alpha=c(0.20, 0.05))

LOD thresholds (1000 permutations)
     lod
20% 2.16
5%  2.76
```

In the above, we used the traditional permutation test. However, for the `hyper` data, a selective genotyping strategy was used, and so it is best to use a stratified permutation test, permuting individuals' phenotypes separately within strata defined by the extent of genotyping.

We must first define a vector that indicates the strata. This may be done as follows. We place individuals who were genotyped at more than 100 markers in one group and the other individuals in a second group.

```
> strat <- (ntyped(hyper) > 100)
```

We then rerun the permutation test using the argument `perm.strata` to indicate the strata.

```
> operms <- scanone(hyper, n.perm=1000, perm.strata=strat,
+                   verbose=FALSE)
```

In this particular case, we see little difference in the significance threshold when using the stratified permutation test. The 5% threshold is 2.71 (versus 2.76 from the traditional permutation test).

```
> summary(operms, alpha=c(0.20, 0.05))
```

```
LOD thresholds (1000 permutations)
     lod
20% 2.10
5%  2.71
```

Turning now to the use of the permutation results, note that if we provide these results to the `summary.scanone` function, we can have LOD thresholds calculated automatically. For example, the following picks out the LOD peaks (no more than one per chromosome) that meet the 10% significance level.

```
> summary(out.em, perms=operms, alpha=0.1)

         chr  pos  lod
c1.loc45   1 48.3 3.53
D4Mit164   4 29.5 8.09
```

We may further obtain the genome-scan-adjusted p-value for each LOD peak.

```
> summary(out.em, perms=operms, alpha=0.1, pvalues=TRUE)

         chr  pos  lod  pval
c1.loc45   1 48.3 3.53 0.008
D4Mit164   4 29.5 8.09 0.000
```

For the QTL on chromosome 4, our estimated p-value is 0, as no permutations showed a LOD score of 8.1 or greater. Citing a p-value of 0 doesn't seem right, but we can get an upper confidence limit on the true p-value with the `binom.test` function, as follows.

```
> binom.test(0, 1000)$conf.int

[1] 0.000000 0.003682
```

Thus, we might report $p < 0.004$.

4.4 The X chromosome

The X chromosome exhibits special behavior and must be treated differently from the autosomes in QTL mapping. The behavior of the X chromosome depends on the direction of the cross, as well as the sex of the progeny. We enumerate the possibilities in Fig. 4.22 and 4.23.

In Fig. 4.22, the four possible backcrosses to a single strain are presented. For the backcrosses in panels c and d, in which the F_1 parent is male, the X chromosome is not subject to recombination, and so one cannot map QTL on the X chromosome. We omit these crosses from further consideration. In panels a and b, in which the F_1 parent is female, the X chromosome does recombine, and the order of the cross producing the F_1 parent has no impact on the behavior of the X chromosome. Note that the backcross males are hemizygous A or B, while females have genotype AA or AB. Thus, rather than comparing, as for the autosomes, the phenotypic means between the AA and AB genotype groups, the X chromosome requires a comparison of the phenotypic means across four genotypic groups.

In Fig. 4.23, the four possible intercrosses are presented. In all cases, F_2 progeny have a single X chromosome subject to recombination, and male F_2 progeny are, at any given locus, hemizygous A or B. In the case that the F_1 male parent was derived from a cross A $\times$ B (with the A parent being female; panels a and b), the female F_2 progeny are either AA or AB. In the cases that the F_1 male was derived from a cross B $\times$ A (with the B parent being female; panels c and d), the female F_2 progeny are either BB or AB. Note that the direction of the cross giving the female F_1 parent does not affect the behavior of the X chromosome in the F_2 progeny. Thus, when we discuss the direction of the intercross, we consider only the direction of the cross that produced the male F_1 parent. The crosses in Fig. 4.23, panels a and b, are treated the same, and the crosses in Fig. 4.23, panels c and d, are treated the same.

The relevant genotype comparisons are somewhat different for the X chromosome than for autosomes. Further, the null hypothesis of no linkage must be reformulated to avoid spurious linkage to the X chromosome as a result of sex- or cross-direction-differences in the phenotype. (Sex differences are observed in many phenotypes, and systematic phenotypic differences between reciprocal crosses may arise, for example, from parent-of-origin effects. If not taken into account, such systematic differences can lead to large LOD scores on the X chromosome even in the absence of X chromosome linkage.) Finally, to account for the fact that the number of degrees of freedom for the linkage test on the X chromosome may be different from that on the autosomes, an X-chromosome-specific significance threshold is required.

4.4.1 Analysis

The choice of the null and alternative hypotheses requires careful thought. The goal of the R/qtl implementation, which we describe here, is to have a procedure for routine use in QTL mapping. The choice of genotype comparisons was based on the following basic principles. First, a sex- or cross-direction-difference in the phenotype should not lead to spurious linkage to the X chromosome. Second, the set of comparisons should be parsimonious but reasonable. Third, the null hypothesis must be nested within the alternative hypothesis. The choices for all possible cases are presented in Table 4.3; we explain how these choices were made through specific examples, below.

Consider, for example, an intercross performed in one direction and including both sexes. Females then have X chromosome genotype AA or AB, while males are hemizygous A or B. Males that are hemizygous B should be treated separately from the AA or AB females, and so the null hypothesis must then allow for a sex-difference in the phenotype, as otherwise, the presence of such

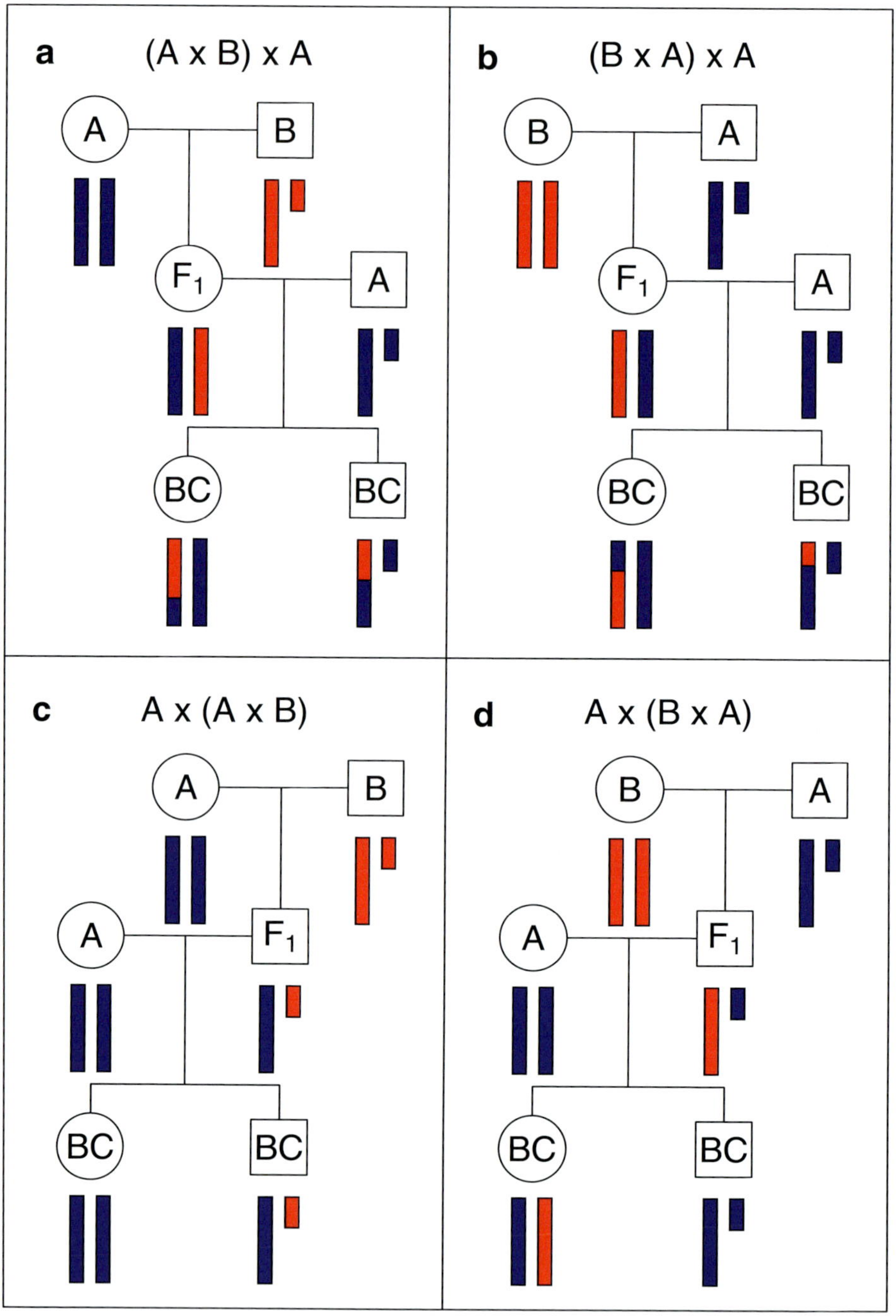

Figure 4.22. The behavior of the X chromosome in a backcross. Circles and squares correspond to females and males, respectively. Blue and red bars correspond to DNA from strains A and B, respectively. The small bar is the Y chromosome.

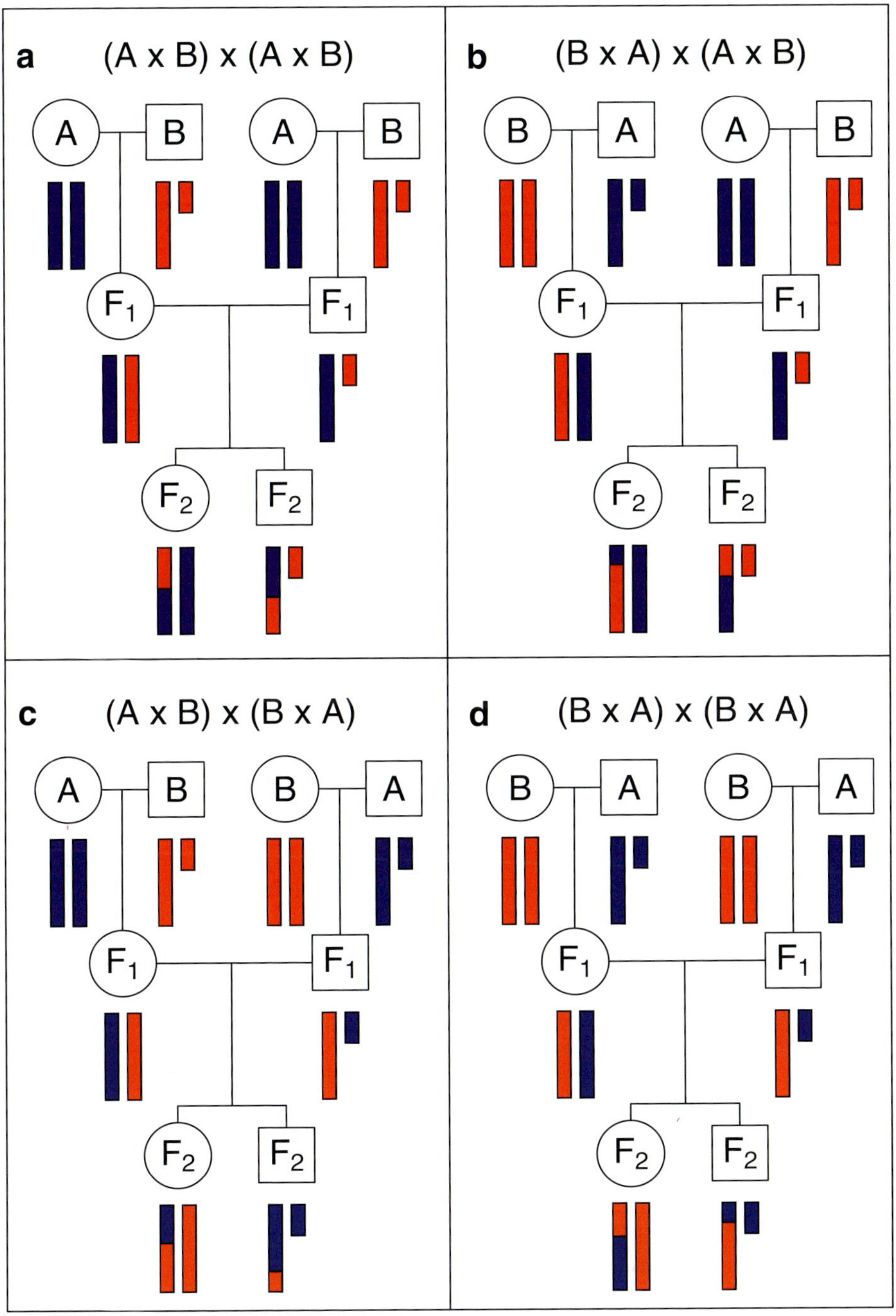

Figure 4.23. The behavior of the X chromosome in an intercross. Circles and squares correspond to females and males, respectively. Blue and red bars correspond to DNA from strains A and B, respectively. The small bar is the Y chromosome.

Table 4.3. Contrasts for analysis of the X chromosome in standard crosses.

Cross	Direction	Sexes	Contrasts	H_0	df
BC		Both	AA:AB:AY:BY	♀:♂	2
BC		♀	AA:AB	Grand mean	1
BC		♂	AY:BY	Grand mean	1
F_2	Both	Both	AA:ABf:ABr:BB:AY:BY	$♀_f$:$♀_r$:♂	3
F_2	Both	♀	AA:ABf:ABr:BB	$♀_f$:$♀_r$	2
F_2	Both	♂	AY:BY	Grand mean	1
F_2	One	Both	AA:AB:AY:BY	♀:♂	2
F_2	One	♀	AA:AB	Grand mean	1
F_2	One	♂	AY:BY	Grand mean	1

a sex difference would cause spurious linkage to the X chromosome. For the null hypothesis to be nested within the alternative, the alternative must allow separate phenotype averages for the genotype groups AA, AB, AY, and BY (see Table 4.3). We will call this the contrast AA:AB:AY:BY. Note that there are two degrees of freedom for this test of linkage, just as for the autosomes, as there are four mean parameters under the alternative and two under the null (the average phenotype for each of females and males).

A somewhat more complex example is for the case that both directions of the intercross were performed, but only females were phenotyped. As the AA individuals come from one direction of the cross (which we will call the "forward" direction) and the BB individuals come from the other direction of the cross (the "reverse" direction), a cross direction effect on the phenotype would cause spurious linkage to the X chromosome if the null hypothesis did not allow for that effect. But then, in order for the null hypothesis to be nested within the alternative, the AB individuals from the two cross directions must be allowed to be different. Thus we arrive at the contrasts AA:ABf:ABr:BB for the alternative and forward:reverse for the null. Here, again, the linkage test has two degrees of freedom.

In the analogous case with males only, since both directions give rise to equal parts hemizygous A and hemizygous B individuals, we need not split the individuals according to the direction of the cross, as a cross direction effect cannot cause spurious linkage to the X chromosome. In this case, the test for linkage has one degree of freedom.

In the most complex case, of an intercross with both directions and both sexes, all four types of females must be allowed to be separate, whereas the males from the two directions may be pooled, and so the simplest comparison includes the contrasts AA:ABf:ABr:BB:AY:BY, with the null hypothesis using the contrasts female forward:female reverse:male. Thus, the linkage test has three degrees of freedom.

For all interval mapping methods, the actual analysis is essentially the same for the X chromosome as for the autosomes. As each backcross or intercross individual has a single X chromosome that was subject to recombination, the calculation of the genotype probabilities given the available multipoint marker genotype data is identical to those for an autosome in a backcross, and so nothing new is needed there. The further analysis is hardly modified; the only changes concern the set of genotype groups and the possible inclusion of sex and/or cross direction as covariates under the null hypothesis. Provided that phenotypes `sex` and `pgm` (if necessary) are included with the data (see Sec. 2.1.1, page 24), the analysis will be performed correctly without any intervention from the user.

4.4.2 Significance thresholds

A further point concerns the need for X-chromosome-specific levels of significance, as the number of degrees of freedom for the X chromosome can differ from that for the autosomes. We can assign a chromosome-specific false positive rate of α_i for chromosome i. We require, however, that the α_i are chosen in order to maintain the desired genome-wide significance level, α. Under the null hypothesis of no QTL and with the assumption of independent assortment of chromosomes, the LOD scores on separate chromosomes are independent, and so we must choose the α_i so that $\alpha = 1 - \prod_i (1 - \alpha_i)$.

Any choice of the α_i satisfying this equation will provide a genome-wide false positive rate that is maintained at the desired level. For example, one could choose $\alpha_1 = \alpha$ and $\alpha_i = 0$ for $i \neq 1$. A key issue, in choosing the α_i, concerns the power to detect a QTL. In the preceding example, one would have high power to detect a QTL on chromosome 1, but no power to detect a QTL on any other chromosome. The usual approach, with a constant LOD threshold across the genome, provides high power to detect a QTL irrespective of its location: in the case of high and uniform marker density and the presence of a single autosomal QTL, the power to detect the QTL would be the same no matter where it resides.

A reasonable approach is to use $\alpha_i = 1-(1-\alpha)^{L_i/L}$, where L_i is the genetic length of chromosome i and $L = \sum_i L_i$. This corresponds approximately to the use of a constant LOD threshold across the autosomes.

Our actual recommendation is slightly different: we use a constant LOD threshold across the autosomes and a separate threshold for the X chromosome. Taking L_A to be the sum of the genetic lengths of the autosomes and L_X to be the length of the X chromosome, so that $L = L_A + L_X$, we use $\alpha_A = 1 - (1 - \alpha)^{L_A/L}$ and $\alpha_X = 1 - (1 - \alpha)^{L_X/L}$. In particular, in a permutation test to determine LOD thresholds, one would calculate, for permutation replicate j, LOD^*_{jA} as the maximum LOD score across all autosomes and LOD^*_{jX} as the maximum LOD score across the X chromosome. The LOD threshold for the autosomes would be the $1 - \alpha_A$ quantile of the LOD^*_{jA}, and the LOD threshold for the X chromosome would be the $1 - \alpha_X$ quantile of the LOD^*_{jX}.

Genome-scan-adjusted p-values can be estimated from the permutation results as follows. For a putative QTL on an autosome, one would first calculate the proportion, call it p, of the LOD^*_{jA} that were greater or equal to the observed LOD score. The adjusted p-value would then be $1-(1-p)^{L/L_A}$. Equations for a locus on the X-chromosome are analogous, replacing the A's with X's.

The precise estimation of the X-chromosome-specific LOD threshold will require considerably more permutation replicates. We have found that one must use roughly L/L_X times more permutation replicates to get the same precision for an adjusted p-value for the X chromosome as one would typically need if a constant LOD threshold were used across the genome.

We have neglected to mention an important detail in the permutations concerning the X chromosome. The rows in the phenotype matrix are shuffled relative to the genotype data. Thus the `sex` (and `pgm`) attached to a particular phenotype is preserved. The X chromosome genotypes, however, are randomized between males and females, but in a special way: the X chromosome genotypes are coded as `1/2` for all individuals, indicating the pattern of recombination and of missing genotype data. These are shuffled across all individuals, but the `1/2` codes are still interpreted as AA/AB for females from one direction of the cross, BB/AB for females from the other direction, and AY/BY for males.

An alternate strategy would be to permute separately within the four strata (males and females in each cross direction). This would be important if there were differences in the pattern of genotype data among the strata.

4.4.3 Example

As an example, we consider the data of Grant *et al.* (2006), which concerns the basal iron levels in the liver and spleen of intercross mice. Both sexes from reciprocal intercrosses with the C57BL/6J/Ola and SWR/Ola strains were used; there are 284 individuals in total. The data are available in the R/qtlbook package as the `iron` data. There are two phenotypes: the level of iron (in μg/g) in the liver and spleen. There are approximately equal proportions of males and females and of mice from each cross direction.

We can get access to the data and make a summary plot as follows; see Fig. 4.24. We use `pheno=1:2` to show histograms of just the liver and spleen phenotypes, suppressing barplots of `sex` and `pgm`.

```
> library(qtlbook)
> data(iron)
> plot(iron, pheno=1:2)
```

Figure 4.25 contains a scatterplot of the $\log_2$(liver) and $\log_2$(spleen) phenotypes, with females as red circles and males as blue $\times$'s. The figure was obtained as follows.

```
> plot(log2(liver) ~ log2(spleen), data=iron$pheno,
+        col=c("red", "blue")[iron$pheno$sex],
+        pch=c(1,4)[iron$pheno$sex])
```

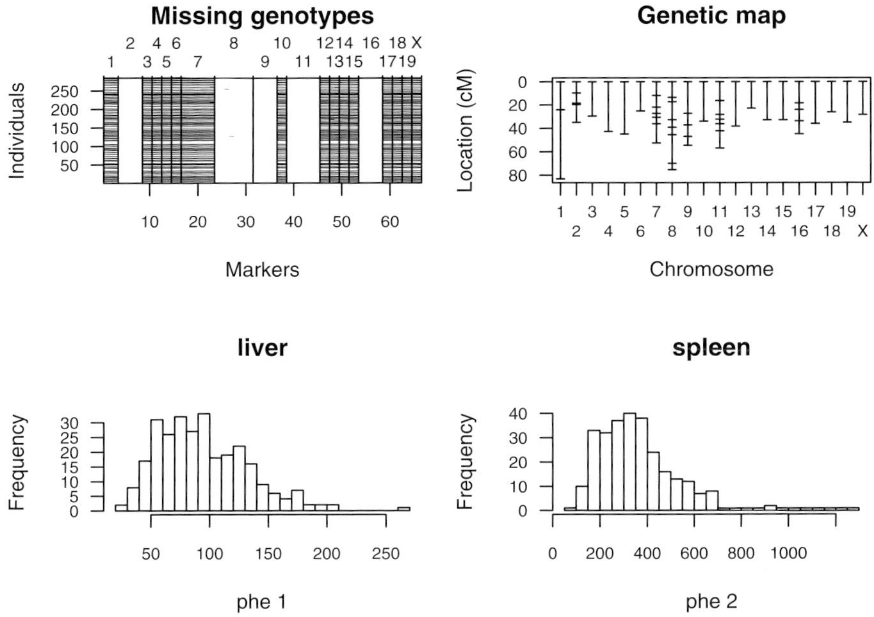

Figure 4.24. The summary plot of the `iron` data.

We use `col` and `pch` to indicate the color and plotting characters to use. For example, `c(1,4)[iron$pheno$sex]` gives a vector of 1's and 4's according to the sexes of the individuals, and with `pch`, 1 corresponds to a circle and 4 to an ×.

Note that both phenotypes show a large sex difference, with females having larger iron levels than males. For example, the average iron levels in the liver of females and males are 112 (SE = 3) and 78 (SE = 3), respectively. If the sex differences were not taken into account in QTL mapping on the X chromosome, the LOD scores for that chromosome would be increased by 12.9 and 4.9 for the liver and spleen phenotypes, respectively.

We will focus on the liver phenotype, but we will consider it on the $\log_2$ scale. (Discussion of the analysis of the spleen phenotype is deferred to Sec. 4.7.) We transform the phenotype, and then use `calc.genoprob` and `scanone` to perform a genome scan by standard interval mapping.

```
> iron$pheno[,1] <- log2(iron$pheno[,1])
> iron <- calc.genoprob(iron, step=1, error.prob=0.001)
> out.liver <- scanone(iron)
```

A plot of the results, obtained with `plot.scanone`, is shown in Fig. 4.26. The peaks with LOD > 3 may be obtained with `summary.scanone`.

```
> summary(out.liver, 3)
```

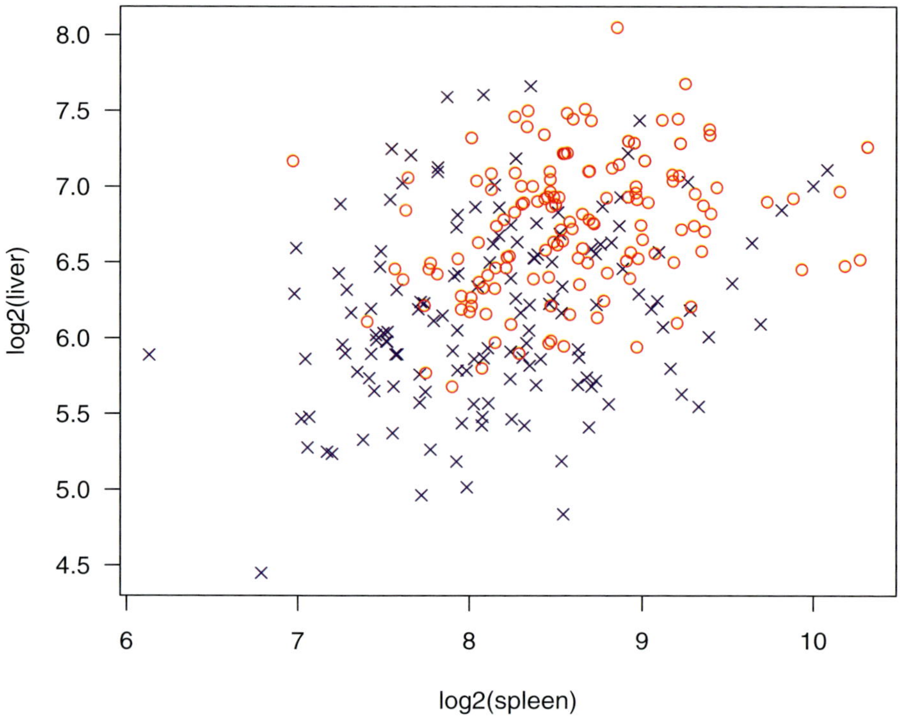

Figure 4.25. Scatterplot of $\log_2$(liver) versus $\log_2$(spleen) for the `iron` data, with females as red circles and males as blue ×'s.

```
           chr  pos  lod
D2Mit17      2 56.8 5.09
c7.loc47     7 48.1 3.41
D8Mit294     8 39.1 3.27
c16.loc22   16 28.6 9.47
```

The maximum LOD score on the X chromosome (0.31) is not so interesting. Nevertheless, it is valuable to show how the permutation test would be done to get autosome- and X-chromosome-specific LOD thresholds. We again use the `scanone` function and use the `n.perm` argument to indicate the number of permutations to perform, but here we also use `perm.Xsp=TRUE` to indicate that we want to perform X-chromosome-specific permutations. In this case, `n.perm` indicates the number of permutations to perform for the autosomes. For the X chromosome, `n.perm` $\times L_A/L_X$ permutations are performed.

```
> operm.liver <- scanone(iron, n.perm=1000, perm.Xsp=TRUE,
+                        verbose=FALSE)
```

The LOD thresholds for a 5% significance level may be obtained as follows.

```
> summary(operm.liver, alpha=0.05)
```

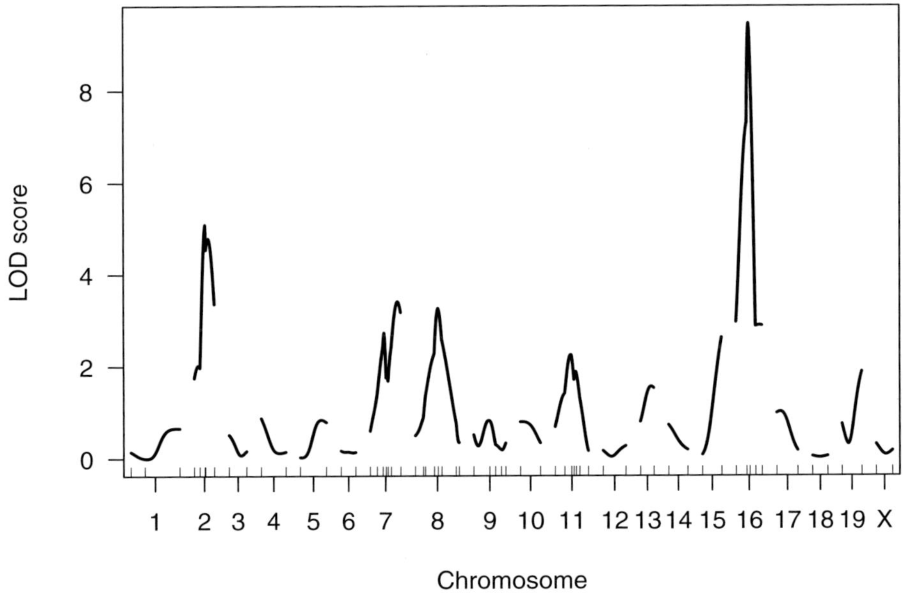

Figure 4.26. LOD curves for the liver phenotype in the `iron` data, calculated by standard interval mapping.

```
Autosome LOD thresholds (1000 permutations)
    lod
5% 3.36

X chromosome LOD thresholds (28243 permutations)
    lod
5% 4.83
```

The threshold for the X chromosome is much larger than that for the autosomes, as the test for linkage on the X chromosome has three degrees of freedom, while for the autosomes there are two degrees of freedom (see Table 4.3 on page 112).

We can again use the permutation results with `summary.scanone` to automatically calculate the relevant LOD thresholds corresponding to a particular significance level and to obtain genome-scan-adjusted p-values.

```
> summary(out.liver, perms=operm.liver, alpha=0.05,
+         pvalues=TRUE)

          chr  pos  lod    pval
D2Mit17     2 56.8 5.09 0.00311
c7.loc47    7 48.1 3.41 0.04449
c16.loc22  16 28.6 9.47 0.00000
```

The alternate strategy, mentioned above, of a stratified permutation test, is performed by first creating a numeric vector that indicates the strata to which the individuals belong.

```
> strat <- as.numeric(iron$pheno$sex) + iron$pheno$pgm*2
> table(strat)

strat
 1  2  3  4
74 75 65 70
```

We then rerun the permutation test with `scanone`, indicating the strata via the `perm.strata` argument.

```
> operm.liver.strat <- scanone(iron, n.perm=1000, perm.Xsp=TRUE,
+                              perm.strata=strat, verbose=FALSE)
```

The LOD thresholds are not too different.

```
> summary(operm.liver.strat, alpha=0.05)

Autosome LOD thresholds (1000 permutations)
    lod
5% 3.41

X chromosome LOD thresholds (28243 permutations)
    lod
5% 4.51
```

4.5 Interval estimates of QTL location

Once one has obtained evidence for a QTL, one may seek an interval estimate of the location of the QTL. The estimated QTL location by interval mapping will deviate somewhat from the true location, and so we seek the range of possible QTL locations that are supported by the data. We assume, in this section, that there is one and only one QTL on the chromosome of interest.

There are two major methods for calculating an interval estimate of QTL location: LOD support intervals and Bayes credible intervals. The 1.5-LOD support interval is the interval in which the LOD score is within 1.5 units of its maximum. See Fig. 4.27 for an illustration. If the LOD score drops down and back up (as in Fig. 4.27), one would obtain a set of disjoint intervals, but we use the conservative approach of taking the wider connected interval. The amount to drop affects the coverage of the LOD support intervals; we prefer to use 1.5-LOD support intervals for a backcross, and 1.8-LOD support intervals for an intercross.

An approximate Bayes credible interval is obtained by viewing 10^{LOD} as a real likelihood function for QTL location. (In fact, it is a profile likelihood,

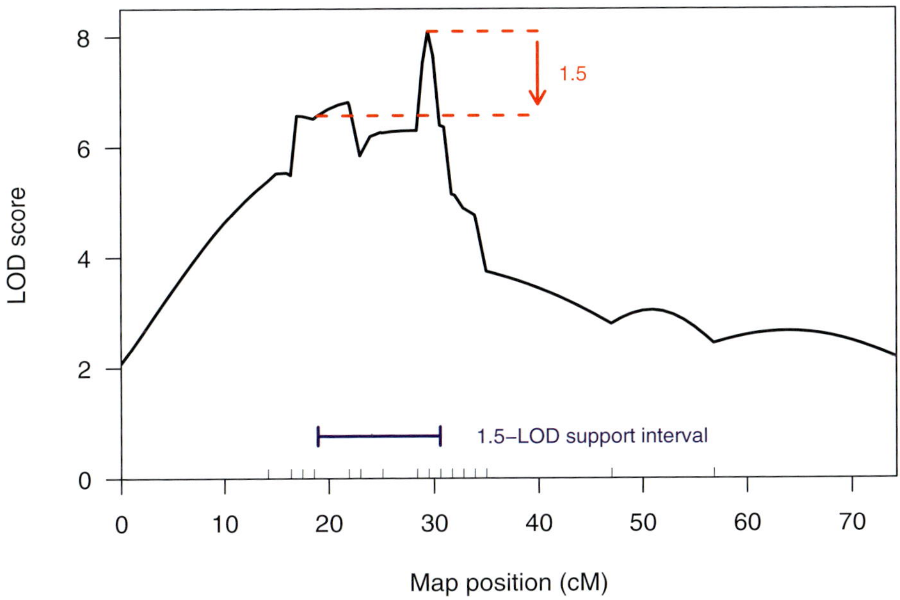

Figure 4.27. Illustration of the 1.5-LOD support interval for chromosome 4 of the `hyper` data. The LOD curve was calculated by standard interval mapping.

as at each point we maximized over the possible QTL effects and the residual SD.) Assuming *a priori* that the QTL is equally likely to be anywhere on the chromosome, the posterior distribution of QTL location is obtained by rescaling 10^{LOD} to be a distribution, $f(\theta \mid \text{data}) = 10^{\text{LOD}(\theta)} / \sum_{\theta} 10^{\text{LOD}(\theta)}$. The 95% Bayes credible interval is defined as the interval I for which $f(\theta \mid \text{data})$ exceeds some threshold and for which $\sum_{\theta \in I} f(\theta \mid \text{data}) \geq 0.95$.

The Bayes interval is illustrated in Fig. 4.28. Plotted is 10^{LOD}, rescaled so that the area underneath the curve is 1. The area shaded in red is approximately 95% of the total area.

It is important to point out that neither the LOD support interval nor the Bayes credible interval behave as true confidence intervals, as interval coverage (the chance of obtaining an interval containing the true QTL location) is not constant, but depends to some extent on the type of cross, marker density, and the size of the QTL effect. Experience has shown, however, that Bayes intervals have remarkably consistent coverage, and so may be preferred.

The use of a nonparametric bootstrap has also been used to create confidence intervals for QTL location. In the nonparametric bootstrap, one samples, with replacement, from the individuals in the cross to create a new data set of the same size as the original, but with some individuals repeated and some omitted. With these new data, interval mapping is performed and the estimated location of the QTL recorded. The process is repeated multiple times, to create a set of QTL location estimates, $\hat{\theta}_1^*, \hat{\theta}_2^*, \ldots, \hat{\theta}_b^*$. A confidence interval

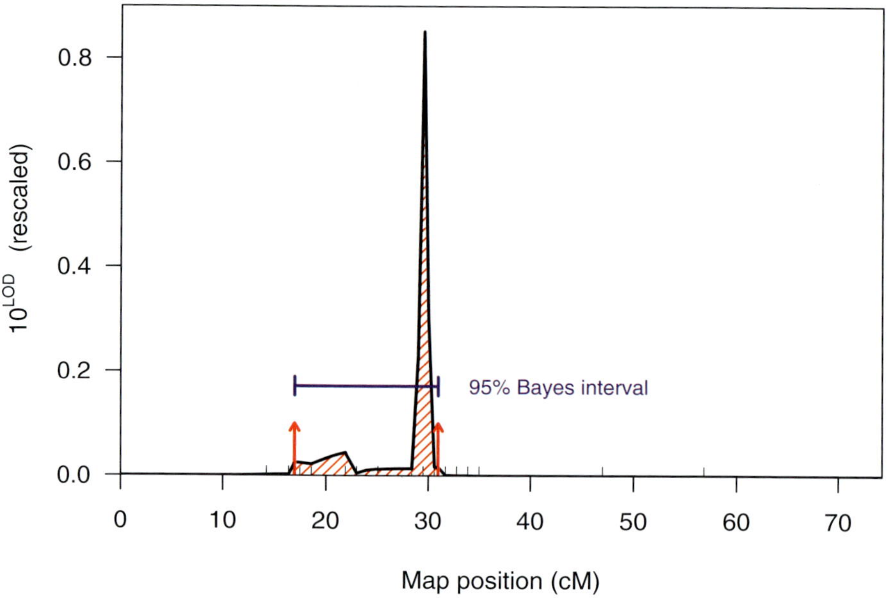

Figure 4.28. Illustration of the approximate 95% Bayes credible interval for chromosome 4 of the `hyper` data. The LOD curve was calculated by standard interval mapping.

is estimated as the region containing 95% of these bootstrap estimates; say the 2.5th to 97.5th percentiles of the $\hat{\theta}_i^*$.

Unfortunately, the bootstrap has been shown to perform poorly in this context, and so it is not recommended. The maximum likelihood estimate of QTL location obtained from interval mapping has a tendency to occur at a marker. Moreover, the standard error of the estimated location depends on the location of the QTL relative to the typed markers. As a result, the coverage of bootstrap confidence intervals for QTL location depends critically on the location of the QTL relative to the markers. In addition, the bootstrap confidence intervals tend to be much wider than the LOD support and Bayes credible intervals.

Example

We can calculate the LOD support interval and Bayes credible interval with the functions `lodint` and `bayesint`, respectively. These take, as input, results from `scanone` and the chromosome to consider, plus the argument `drop` in `lodint`, indicating the amount to drop in LOD (1.5 by default), and `prob` in `bayesint`, indicating the nominal Bayes fraction (95% by default).

We calculate the 1.5-LOD support and 95% Bayes credible intervals for chromosome 4 in the `hyper` data as follows.

```
> lodint(out.em, 4, 1.5)

          chr  pos   lod
c4.loc19    4 19.0 6.566
D4Mit164    4 29.5 8.092
D4Mit178    4 30.6 6.387

> bayesint(out.em, 4, 0.95)

          chr  pos   lod
c4.loc17    4 17.0 6.562
D4Mit164    4 29.5 8.092
c4.loc31    4 31.0 6.359
```

The first and last rows in the results indicate the ends of the intervals; the middle row is the maximum likelihood estimate of QTL location.

Note that the ends of the intervals generally lie between marker locations. One may use the argument `expandtomarkers` to expand the intervals to the nearest flanking markers, as follows.

```
> lodint(out.em, 4, 1.5, expandtomarkers=TRUE)

          chr  pos   lod
D4Mit286    4 18.6 6.507
D4Mit164    4 29.5 8.092
D4Mit178    4 30.6 6.387

> bayesint(out.em, 4, 0.95, expandtomarkers=TRUE)

          chr  pos   lod
D4Mit108    4 16.4 5.490
D4Mit164    4 29.5 8.092
D4Mit80     4 31.7 5.145
```

We can obtain a bootstrap-based confidence interval for the location of a QTL with the function `scanoneboot`, which takes the same set of arguments as `scanone`, though only a single chromosome (indicated with the argument `chr`) is considered, and the number of bootstrap replicates is indicated with the argument `n.boot`. For example, we perform the bootstrap with chromosome 4 of the `hyper` data as follows.

```
> out.boot <- scanoneboot(hyper, chr=4, n.boot=1000)
```

A histogram of the bootstrap results is shown in Fig. 4.29. Note that 80% of the bootstrap locations coincide with genetic markers. The results provide a poor measure of the uncertainty in QTL location.

The output of `scanoneboot` has class `"scanoneboot"`; the actual confidence interval is obtained with the function `summary.scanoneboot`, as follows.

```
> summary(out.boot)
```

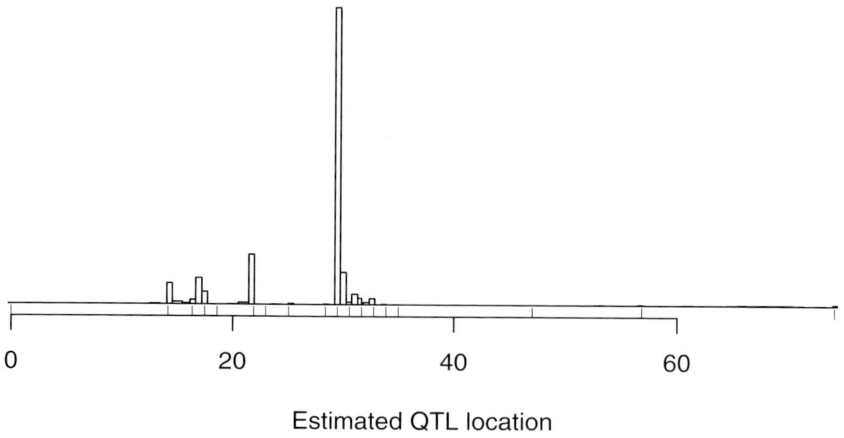

Figure 4.29. Histogram of the estimated QTL locations in 1000 bootstrap replicates with the `hyper` data, chromosome 4, with LOD scores calculated by standard interval mapping. The tick marks beneath the histogram indicate the locations of the genetic markers.

```
          chr  pos   lod
D4Mit41     4 14.2 6.076
D4Mit164    4 29.5 9.151
D4Mit276    4 32.8 3.881
```

This is a bit wider than the LOD support and Bayes credible intervals.

4.6 QTL effects

The effect of a QTL is characterized by the difference in the phenotype averages among the QTL genotype groups. For a backcross, we simply look at the difference between the average phenotypes for the heterozygotes and the homozygotes, $a = \mu_{AB} - \mu_{AA}$. For an intercross, with three possible genotypes, one traditionally considers the additive effect, $a = (\mu_{BB} - \mu_{AA})/2$, and the dominance effect, $d = \mu_{AB} - (\mu_{AA} + \mu_{BB})/2$.

In addition, the QTL effect is often characterized by the proportion of the phenotypic variance explained by the QTL (also called the heritability due to the QTL). Specifically, we consider

$$h^2 = \text{var}\{\text{E}(y|g)\}/\text{var}(y),$$

where y is the phenotype and g is the QTL genotype. For a backcross, we have $h^2 = a^2/(a^2 + 4\sigma^2)$, where σ^2 is the residual variance. For an intercross, the heritability due to the QTL is $h^2 = (2a^2 + d^2)/(2a^2 + d^2 + 4\sigma^2)$.

Knowledge of the QTL effects is especially important in agricultural investigations to develop improved livestock or crops and in studies of evolution. In biomedical research of models of human disease, which is our primary focus, the QTL effects are of lesser importance: while specific genes may carry over from a model organism to humans, the actual alleles are not likely to be the same, and so the QTL effects in the model are not likely to be relevant for humans. Nevertheless, the QTL effects have important influence on one's ability to fine-map the QTL and ultimately identify the causal gene.

In considering the estimated effects of QTL, it is important to recognize that they are often subject to considerable bias. The point is that, the estimated effect of a QTL will vary somewhat from its true effect, but only when the estimated effect is large will the QTL be detected. (We don't estimate the effects of QTL that we have not detected.) Among those experiments in which the QTL is detected, the estimated QTL effect will be, on average, larger than its true effect. This is *selection bias.*

As an illustration, consider Fig. 4.30. We simulated a backcross with 250 individuals, with a single QTL responsible for either 2.5, 5, or 7.5% of the phenotypic variance. The true QTL effect is indicated with a blue vertical line. The distribution of the estimated QTL effect, across 100,000 simulation replicates, is displayed. There is a direct relationship between the estimated percent variance explained by a QTL and the LOD score (see page 77). With a sample size of 250, a LOD score of 3 corresponds to an estimated phenotypic variance explained of 5.4%. Among the cases in which the LOD score is above 3 (the shaded portions of the distributions), the average estimated effect, indicated by the red vertical line, is larger than the true effect.

Note that the selection bias is largest for QTL with small effects (for which we have lower power for detection). QTL with very large effect are always detected, and so the bias in their estimated effects will be minimal.

For a particular inferred QTL with an estimated effect of moderate size, we cannot know whether it is a weak QTL that we were lucky to detect and whose true effect is really rather small, or a QTL of truly large effect that happened to appear to be not so strong in this particular experiment. The estimated effects of QTL will often be overly optimistic, but we cannot be sure about any particular case.

Selection bias in estimated QTL effects has a number of important implications. First, the overall estimated heritability due to a set of identified QTL will almost always be too large and can be markedly so. Second, investigators are often concerned, after repeating a QTL experiment, that an almost completely different set of QTL were identified. One should not conclude, in such a situation, that the unreplicated QTL were false. Instead, it may be that the phenotype is affected by numerous small-effect QTL, for each of which the power of detection is small, and so in any given experiment we identify a random portion of the QTL. Third, in the consideration of a congenic line (in which, for example, the B allele at a QTL is introgressed into the A strain), one may find that the difference between the average phenotypes for the congenic

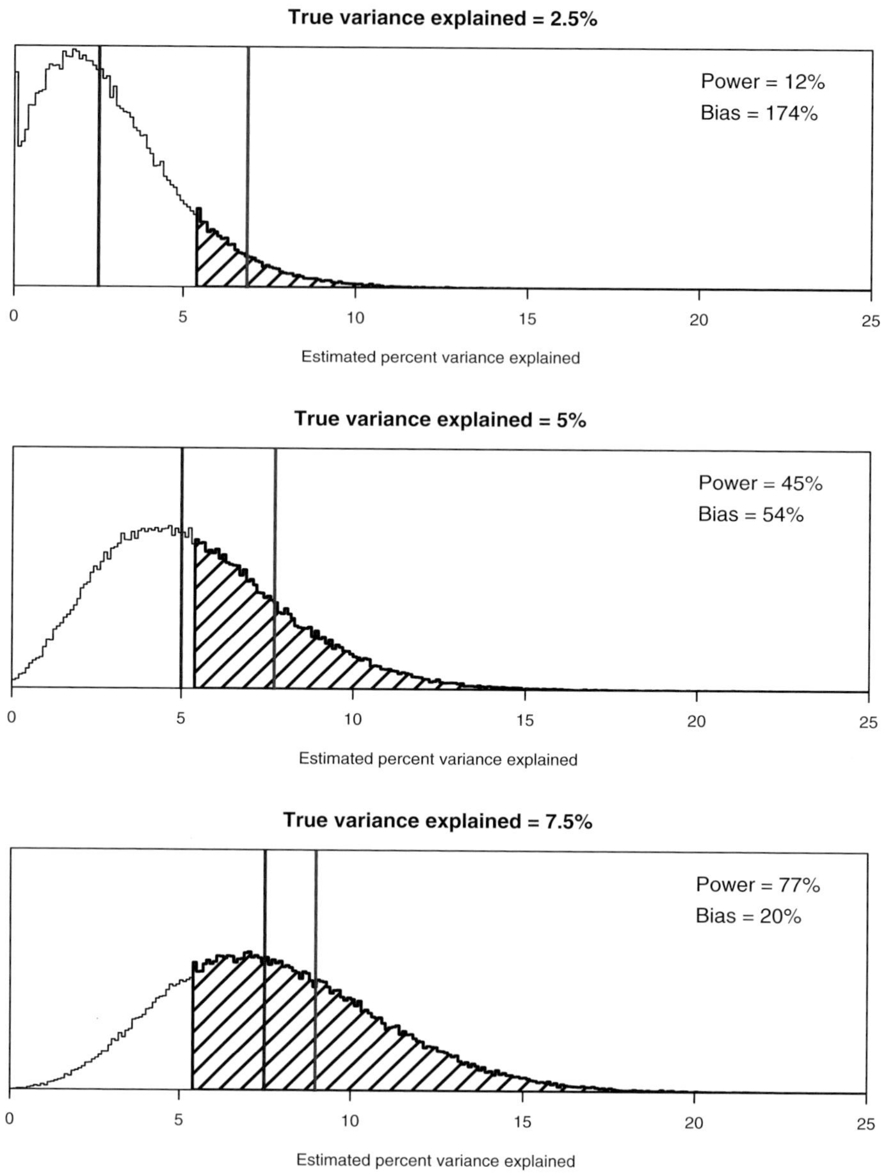

Figure 4.30. Illustration of selection bias in the estimated QTL effect. The curves correspond to the distribution of the estimated percent variance explained by a QTL for different values of the true effect (indicated by the blue vertical lines). The shaded regions correspond to the cases where significant genomewide evidence for the presence of a QTL would be obtained. The red vertical lines indicate the average estimated QTL effect, conditional on the detection of the QTL.

and the recipient strain is not so large as was expected given the QTL mapping results. One might conclude, from this observation, that the QTL consisted of multiple causal genes, and that some have not been included in the congenic line. However, it could just be that the initial estimate of the QTL effect was optimistically large. Finally, in marker-assisted selection efforts, one is likely to find that the progress towards improvement is not so great as had been anticipated, as the true effects of the loci under selection are not so large as their initial estimates.

Example

The function for performing QTL mapping, `scanone`, does not provide estimated QTL effects. Such estimates are best obtained with the function `fitqtl`, particularly for the case of a multiple-QTL model, but we will defer discussion of `fitqtl` to Chap. 9. In the example in this section, we will focus on the use of the function `effectplot`, whose primary purpose is for plotting the phenotype averages for the genotype groups at an inferred QTL.

The function `effectplot` uses the multiple imputation method to obtain estimates of the genotype-specific phenotype averages, taking account of missing genotype data. The estimates are weighted averages of the estimates from the multiple imputations, with the individual imputations being weighted by 10^{LOD}. The standard errors (SEs) include the imputation error. (If imputed genotypes are not available, a call to `sim.geno` with `n.draws=16` is made in order to obtain them.)

The `effectplot` function takes a cross object as input; a marker name is indicated via the `mname1` argument. (The function may be used to plot the estimated effects for two markers, with the second marker indicated via the `mname2` argument; see Chap. 8.) The argument `var.flag` may be used to indicate whether to use a pooled estimate of the residual variance (`var.flag="pooled"`, the default), or to allow different residual variances in the genotype groups (`var.flag="group"`).

One can even use `effectplot` with “pseudomarker” positions (that is, positions on the grid, in between markers). We refer to such pseudomarkers with a chromosome and cM position, in the form `"4@29.5"`, which refers to the pseudomarker closest to 29.5 cM on chromosome 4.

For the `hyper` data, the largest LOD score was obtained at marker D4Mit164 (chr 4, 29.5 cM). If we had known only the position, we could find the name of the nearest marker with the function `find.marker`, which takes as input a cross object, chromosome, and cM position.

```
> find.marker(hyper, 4, 29.5)

[1] "D4Mit164"
```

The `effectplot` for marker D4Mit164 in the hyper data may be obtained as follows; the result appears in the left panel in Fig. 4.31.

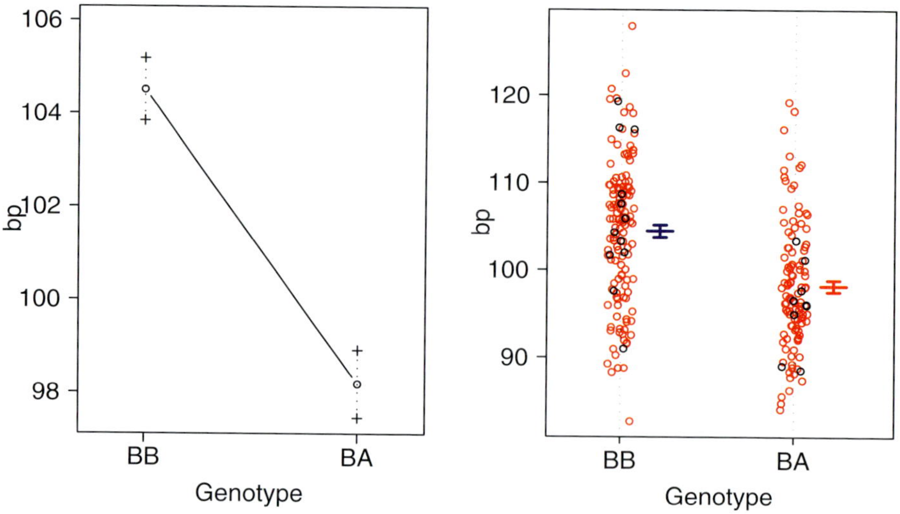

Figure 4.31. Plot of the blood pressure against the genotype at marker D4Mit164 for the `hyper` data. The left panel was produced by `effectplot`. The right panel was produced by `plot.pxg`; red dots correspond to imputed genotypes. Error bars are ± 1 SE.

```
> hyper <- sim.geno(hyper, n.draws=16, error.prob=0.001)
> effectplot(hyper, mname1="D4Mit164")
```

The output of `effectplot` (except for the plot) is generally suppressed, but if one assigns the output to an object, it may be inspected. If one uses `draw=FALSE`, the plot is not created.

```
> eff <- effectplot(hyper, mname1="D4Mit164", draw=FALSE)
> eff

$Means
D4Mit164.BB D4Mit164.BA
     104.51       98.19

$SEs
D4Mit164.BB D4Mit164.BA
     0.6703      0.7298
```

The function `plot.pxg` can be used to create a dot plot of a phenotype against the genotypes at a marker. The function is relatively crude in its treatment of missing genotype data: missing genotypes are filled in by a single random imputation, conditional on the available marker data. Thus, if there are many missing genotypes, one should view the results with some skepticism. Genotypes that were imputed are plotted in red.

As input to `plot.pxg`, we again provide a cross object plus a marker name; the marker name is indicated with the argument `marker`. We can plot the phenotype against the genotypes at D4Mit164 as follows. The plot appears in the right panel of Fig. 4.31. Note that most genotypes are missing.

```
> plot.pxg(hyper, marker="D4Mit164")
```

The argument `marker` can take a vector of marker names, in which case the phenotypes will be plotted against the joint genotypes at the markers; see Chap. 8.

4.7 Multiple phenotypes

In a QTL experiment, one often measures multiple related phenotypes. The joint analysis of multiple phenotypes can increase the power for QTL detection and the precision of QTL localization, and can allow one to test for pleiotropy (that a single QTL influences multiple phenotypes) versus tight linkage of distinct QTL. Unfortunately, R/qtl does not yet include facilities for such joint analysis of multiple phenotypes; multiple phenotypes are only considered individually.

By default, `scanone` performs interval mapping with the first phenotype in a data set. A different phenotype may analyzed via the argument `pheno.col` (for "phenotype column"), which is the numeric index of the phenotype to be analyzed or a character string indicating the phenotype by name. One may also use the `pheno.col` argument to select multiple phenotypes for analysis by providing a vector of numeric indices or phenotype names. For most methods, this is accomplished by a loop over the selected phenotypes. However, for Haley–Knott regression (`method="hk"`) and multiple imputation (`method="imp"`), an improvement in computational efficiency is achieved by application of multivariate regression.

In Haley–Knott regression, one regresses the phenotype, y, on the genotype probability matrix, X, by calculating $(X'X)^{-1}X'y$. One may analyze a set of phenotypes, Y, simultaneously, by calculating $(X'X)^{-1}X'Y$. The construction and inversion of $X'X$ need only be done once.

This trick can also be used for permutation tests: one creates a set of permuted phenotypes, pastes them together into one large matrix, and then performs Haley–Knott regression with the permuted phenotypes as a unit. (Thanks are due to Hao Wu who suggested and implemented this strategy.)

Example

To illustrate the analysis of multiple phenotypes, we will return to the `iron` data discussed in Sec. 4.4. There are two phenotypes: the iron levels in the liver and spleen. We may also want to look at these on the log scale; we can place $\log_2$(liver) and $\log_2$(spleen) in the phenotypes as follows. Note that `data(iron)` reloads the data.

```
> data(iron)
> iron$pheno <- cbind(iron$pheno[,1:2],
+                      log2liver=log2(iron$pheno$liver),
+                      log2spleen=log2(iron$pheno$spleen),
+                      iron$pheno[,3:4])
```

We place them in positions 3 and 4, moving the `sex` and `pgm` phenotypes to the end.

By default, `scanone` would analyze the first phenotype (liver). If we want to consider the $\log_2$(liver) phenotype, we would use `pheno.col=3`, as follows.

```
> iron <- calc.genoprob(iron, step=1, error.prob=0.001)
> out.logliver <- scanone(iron, pheno.col=3)
```

We may also refer to phenotypes by name. For example, in place of `pheno.col=3` we could use `pheno.col="log2liver"`, as follows.

```
> out.logliver <- scanone(iron, pheno.col="log2liver")
```

In addition, one may use `pheno.col` with a numeric vector of phenotypes (with length equal to the number of individuals in the cross). And so, if we were interested solely in the analysis of the $\log_2$(liver) phenotype, we could skip the effort to include the transformed phenotype within the cross object and simply type the following.

```
> out.logliver <- scanone(iron,
+                         pheno.col=log2(iron$pheno$liver))
```

We may analyze all four phenotypes through `pheno.col=1:4`.

```
> out.all <- scanone(iron, pheno.col=1:4)
```

The result has six columns: chromosome, cM position, and then four columns with LOD scores.

```
> out.all[1:5,]

        chr  pos  liver spleen log2liver log2spleen
D1Mit18   1 27.3 0.1947 0.2695    0.1461   0.034035
c1.loc1   1 28.3 0.1869 0.2602    0.1390   0.026292
c1.loc2   1 29.3 0.1789 0.2515    0.1315   0.019184
c1.loc3   1 30.3 0.1705 0.2436    0.1239   0.012920
c1.loc4   1 31.3 0.1619 0.2366    0.1161   0.007736
```

The `summary.scanone` function displays (by default) the peak positions for the first phenotype, but also shows LOD scores at those positions for the other phenotypes. We can look at the peaks for another phenotype with the `lodcolumn` argument. The LOD columns are indexed as 1, 2, 3, 4, and so to get the peaks above 3 for the $\log_2$(spleen) phenotype, we would do the following.

```
> summary(out.all, threshold=3, lodcolumn=4)
```

```
          chr  pos liver spleen log2liver log2spleen
D8Mit4      8 13.6 0.739    4.3     0.887       3.90
c9.loc50    9 56.6 0.183   10.9     0.199      12.64
```

Two other summary formats are available, which display the peak LOD scores for all phenotypes. The method mentioned above corresponds to the use of `format="onepheno"`. The use of `format="allpheno"` gives essentially the same output, but with all of the LOD score columns considered. For each LOD score column, we identify the position of the maximum peak and include that row in the output if the LOD score exceeds its threshold. Thus, the output may display multiple rows for a given chromosome. Note that the `threshold` argument may be a single number (applied to all LOD score columns) or a vector specifying separate thresholds for each LOD score column. For example, the following returns the results for positions at which at least one of the LOD score columns achieved its maximum for a chromosome, provided that maximum LOD score exceeded 3.

```
> summary(out.all, threshold=3, format="allpheno")
```

```
           chr  pos liver spleen log2liver log2spleen
D2Mit17      2 56.8 4.907  1.917     5.086      2.279
c7.loc47     7 48.1 2.935  0.395     3.413      0.596
D8Mit4       8 13.6 0.739  4.303     0.887      3.895
D8Mit294     8 39.1 3.769  1.902     3.268      1.724
c8.loc40     8 40.0 3.786  1.752     3.241      1.581
c9.loc50     9 56.6 0.183 10.897     0.199     12.642
c16.loc21   16 27.6 7.837  0.848     9.338      1.136
c16.loc22   16 28.6 7.829  0.909     9.465      1.214
```

Finally, with `format="allpeaks"`, a single row is given for each chromosome, containing the maximum LOD score for each phenotype column and the position at which it was maximized. Those chromosomes for which at least one of the LOD score columns exceeded its threshold are printed. For example, the following returns the chromosomes at which at least one of the LOD score columns had a peak exceeding 3.

```
> summary(out.all, threshold=3, format="allpeaks")
```

```
   chr  pos liver  pos spleen  pos log2liver  pos log2spleen
2    2 56.8 4.907 58.0  1.994 56.8     5.086 58.0       2.42
7    7 50.1 2.959 53.6  0.818 48.1     3.413 53.6       1.01
8    8 40.0 3.786 13.6  4.303 39.1     3.268 13.6       3.90
9    9 31.6 0.998 56.6 10.897 32.6     0.824 56.6      12.64
16  16 27.6 7.837 30.6  0.981 28.6     9.465 30.6       1.31
```

If `scanone` output that contains multiple LOD score columns is sent to the `plot.scanone` function, the default is again to plot just the first one. A different column may be indicated via the argument `lodcolumn`, which (as

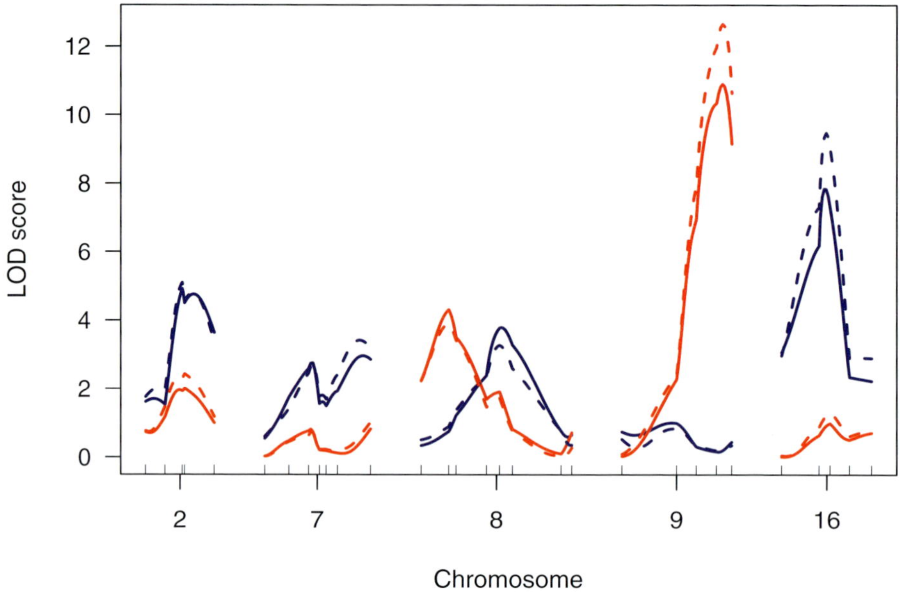

Figure 4.32. LOD curves for selected chromosomes with the `iron` data. Blue and red correspond to the liver and spleen phenotypes, respectively. Solid and dashed curves correspond to the original and log scale, respectively.

in `summary.scanone`) takes values 1, 2, ... Further, `lodcolumn` can take a vector of up to three values. And so, if we want to look at the results for all four phenotypes, we might do the following. The results are in Fig. 4.32.

```
> plot(out.all, lodcolumn=1:2, col=c("blue", "red"),
+       chr=c(2, 7, 8, 9, 16), ylim=c(0,12.7), ylab="LOD score")
> plot(out.all, lodcolumn=3:4, col=c("blue", "red"), lty=2,
+       chr=c(2, 7, 8, 9, 16), add=TRUE)
```

Note the use of `ylim` to set the y-axis limits, to ensure that all of the LOD curves would be contained in the plot.

We can use `scanone` to perform permutation tests on the set of phenotypes simultaneously. As discussed in Sec. 4.4, we should perform separate permutations on the autosomes and X chromosome, which may be accomplished by using `perm.Xsp=TRUE`.

```
> operm.all <- scanone(iron, pheno.col=1:4, n.perm=1000,
+                          perm.Xsp=TRUE)
```

We may again use `summary` to obtain LOD thresholds for each phenotype for the autosomes and the X chromosome.

```
> summary(operm.all, alpha=0.05)
```

```
Autosome LOD thresholds (1000 permutations)
   liver spleen log2liver log2spleen
5%  3.96   7.27       3.4       3.37

X chromosome LOD thresholds (28243 permutations)
   liver spleen log2liver log2spleen
5%  3.72   6.57      4.68       4.63
```

The unusually large LOD thresholds for the spleen phenotype are due to the occurrence of spuriously high LOD scores in regions with large gaps between markers (see Sec. 4.2.5). The log transformed phenotypes are preferred for these data, due to the skewed phenotype distributions (see Fig. 4.24 on page 115).

The permutation results may again be used in `summary.scanone` to automatically calculate thresholds and to obtain genome-scan-adjusted p-values. The significance level for the LOD thresholds is indicated by the argument `alpha`, which may take only one value, applied to all LOD score columns. For example, the following gives, for each chromosome, the maximum LOD score from each LOD score column with the corresponding genome-scan-adjusted p-values. The results for a chromosome are printed if at least one of the LOD score columns exceeded its 5% LOD threshold.

```
> summary(out.all, format="allpeaks", perms=operm.all,
+         alpha=0.05, pvalues=TRUE)

   chr  pos liver   pval  pos spleen  pval  pos log2liver
2    2 56.8 4.907 0.0166 58.0  1.994 0.833 56.8     5.086
7    7 50.1 2.959 0.2104 53.6  0.818 1.000 48.1     3.413
8    8 40.0 3.786 0.0641 13.6  4.303 0.320 39.1     3.268
9    9 31.6 0.998 0.9992 56.6 10.897 0.000 32.6     0.824
16  16 27.6 7.837 0.0000 30.6  0.981 0.999 28.6     9.465
      pval  pos log2spleen   pval
2  0.00104 58.0       2.42 0.2525
7  0.04759 53.6       1.01 0.9967
8  0.06826 13.6       3.90 0.0114
9  1.00000 56.6      12.64 0.0000
16 0.00000 30.6       1.31 0.9513
```

We see evidence for QTL on chromosomes 2, 7, 8 and 16 for $\log_2$(liver) and on chromosomes 8 and 9 for $\log_2$(spleen).

4.8 Summary

In a single-QTL scan, we posit the presence of a single QTL and consider each position, one at a time, as the putative location of that QTL. With dense markers and complete marker genotype data, one may perform analysis of variance at each marker. More typically, however, markers are spaced at 10–20 cM, and some marker genotypes are missing. There are several approaches to interval mapping, for interrogating positions between markers. These approaches differ in the way in which they take account of missing genotype data at a putative QTL.

Standard interval mapping may be viewed as the gold standard, but it is susceptible to spurious linkage peaks in regions of low marker information when the phenotype distribution is not approximately normal. Haley–Knott regression often gives a good approximation to standard interval mapping, at a great improvement in computation time, but it performs poorly in the case of selective genotyping. The extended Haley–Knott regression method gets around these problems. The multiple imputation approach is rather slow for single-QTL analyses, but will show advantages for the fit and exploration of multiple-QTL models.

Statistical significance in a single-QTL scan is generally established through the consideration of the distribution of the genome-wide maximum LOD score, under the global null hypothesis of no QTL. This distribution is best derived via a permutation test.

There are several technical difficulties that arise in the analysis of the X chromosome. Additional covariates need to be incorporated into the null hypothesis, to avoid spurious linkage to the X chromosome, and a separate significance threshold will generally be required for the X chromosome.

Finally, an interval estimate of the location of a QTL may be obtained as the 1.5-LOD support interval: the region in which the LOD score is within 1.5 units of its chromosome-wide maximum. An approximate Bayes credible interval may also be used.

4.9 Further reading

Soller *et al.* (1976) were among the first to clearly describe the use of marker regression. Lander and Botstein (1989) is the seminal paper on interval mapping. The initial paper on the EM algorithm in general (and which coined the term) was Dempster *et al.* (1977).

Haley and Knott (1992) and Martínez and Curnow (1992) independently developed the method we now call Haley–Knott regression. (The discussion in Haley and Knott (1992) is more clear.) Whittaker *et al.* (1996) described a nice trick for Haley–Knott regression: regression of the phenotype on each pair of adjacent markers can give the same information as regression on the conditional QTL genotype probabilities at steps through the interval. This strategy can further reduce computational effort, but requires complete marker genotype data (which is seldom available), and one cannot allow for the presence of genotyping errors.

Feenstra *et al.* (2006) described the extended Haley–Knott method; a similar method was proposed by Xu (1998), though the iteratively reweighted least squares algorithm presented there is not quite right and can give negative LOD scores. The multiple imputation approach was proposed by Sen and Churchill (2001).

Lander and Botstein (1989) estimated significance thresholds for a genome scan by computer simulation as well as analytic means (the latter for the case of an infinitely dense map of markers). Churchill and Doerge (1994) proposed the use of permutation tests in this context. Manichaikul *et al.* (2007) described the use of a stratified permutation test in the presence of selective genotyping.

Broman *et al.* (2006) described the treatment of the X chromosome as presented in this chapter.

Lander and Botstein (1989) proposed the use of 1- and 2-LOD support intervals. Dupuis and Siegmund (1999) provided some support for the use of 1.5-LOD support intervals. Sen and Churchill (2001) described the Bayes intervals. Visscher *et al.* (1996b) described the use of the bootstrap in this context, though Manichaikul *et al.* (2006) showed that it behaves badly, and so the LOD support intervals and especially the Bayes credible intervals are preferred.

Beavis (1994) was the first to raise the issue of selection bias in estimated QTL effects (now often called the Beavis effect); see also Broman (2001).

Jiang and Zeng (1995) provide a good discussion of the joint analysis of multiple phenotypes.

5

Non-normal phenotypes

The methods discussed in Chap. 4 all rely on the assumption that, given QTL genotype, the phenotype follows a normal distribution. This is not the same as to assume that the marginal phenotype distribution is normal—it will follow a mixture of normal distributions. But in the case that no QTL has very large effect, the marginal phenotype distribution would generally be close to normal: unimodal and reasonably symmetric.

While the normality assumption is often reasonable, departures from normality are not uncommon: the phenotype may be dichotomous, highly skewed, or exhibit spikes. (For example, if the phenotype is the mass of gallstones, some individuals may have no gallstones, and so a spike at 0 would be observed.) In practice, application of standard interval mapping will generally give reasonable results, even for a dichotomous trait, provided that statistical significance is established via a permutation test, and except for the problem of spurious LOD scores in regions of low genotype information (see Sec. 4.2.5).

Nevertheless, improved efficiency may be obtained by applying alternate methods. The simplest approach is to transform the phenotype. (For example, for the `iron` data in Sec. 4.4.3, we used a log transformation.) We generally stick to either taking logs, square roots, or no transformation. In this chapter, we describe several alternative interval mapping methods, including nonparametric interval mapping (based on the ranks of the phenotypes), interval mapping specific for binary traits, and a two-part model for the case of a phenotype distribution exhibiting a spike (such as at 0).

We conclude the chapter with a section describing, for the especially computer-savvy reader, how one can implement one's own QTL mapping method in R/qtl. This is illustrated by an implementation of a Cox proportional hazards model for right-censored phenotypes, using the Haley–Knott regression approach.

K.W. Broman, Ś. Sen, *A Guide to QTL Mapping with R/qtl*,
Statistics for Biology and Health, DOI 10.1007/978-0-387-92125-9_5,

5.1 Nonparametric interval mapping

In the case of complete genotype data at a putative QTL, standard interval mapping is equivalent to using a t test (for a backcross) or analysis of variance (for an intercross). The nonparametric analogs of these methods are the Wilcoxon rank-sum test and the Kruskal–Wallis test. Here we describe the extension of these rank-based methods for interval mapping, in which the QTL genotypes are not known but must be inferred on the basis of multipoint marker genotype data. We will focus on the extension of the Kruskal–Wallis test, in which there is an arbitrary number of genotype groups; the Wilcoxon rank-sum test corresponds to the special case of two groups.

Let y_i denote the phenotype of individual i, and rank the phenotypes from $1, \dots, n$, with R_i denoting the rank for individual i. In the case of ties, one may randomize the ranks for any tied phenotypes, though we prefer to assign the average rank within each group of ties.

Consider some fixed position in the genome as the location of a putative QTL, and let $p_{ij} = \Pr(g_i = j | \boldsymbol{M}_i)$, the QTL genotype probabilities given the available multipoint marker data, $\boldsymbol{M}_i$. Whereas in the Kruskal–Wallis test statistic, one considers the sum of the ranks within each group, here the exact assignment of individuals to QTL genotype groups is not known; rather, individual i has prior probability p_{ij} of belonging to group j. Thus we consider the expected rank-sum, $S_j = \sum_i p_{ij} R_i$.

We then form the statistic

$$H = \sum_j \left(\frac{n - \sum_i p_{ij}}{n} \right) \left[\frac{(S_j - E_{0j})^2}{V_{0j}} \right]$$

where E_{0j} and V_{0j} are the mean and variance of S_j under the null hypothesis of no linkage, considering the p_{ij} as fixed. That is, E_{0j} and V_{0j} are the average and variance of the S_j if we take the R_i to be a random permutation of the integers $1, \dots, n$. We seek loci for which the expected rank sums, S_j, deviate from their average under the null hypothesis of no linkage.

After some algebra, we obtain the following formula.

$$H = \frac{12}{n(n+1)} \sum_j \frac{(n - \sum_i p_{ij}) \, (\sum_i p_{ij})^2}{n \, \sum_i p_{ij}^2 - (\sum_i p_{ij})^2} \left[\frac{\sum_i p_{ij} \, R_i}{\sum_i p_{ij}} - \frac{n+1}{2} \right]^2$$

In the case that the putative QTL is at a fully typed genetic marker, the p_{ij} will all be 0 or 1, and the above statistic reduces to the Kruskal–Wallis test statistic.

A standard correction for the case of ties is to use the statistic $H' = H/D$ where $D = 1 - \sum_k (t_k^3 - t_k)/(n^3 - n)$, with t_k being the number of values in the kth group of ties. If there are no ties, $D = 1$ and so $H' = H$. In the presence of ties, the correction results in a slight inflation of the test statistic. Note, however, that if a permutation test is used to establish statistical significance, the correction factor is immaterial, as it will apply uniformly to the observed statistics as well as those from each permutation.

As the nonparametric statistic H' follows, approximately, a χ^2 distribution under the null hypothesis of no linkage, we convert the statistic to the LOD scale by taking LOD $= H'/(2\ln 10)$. The resulting statistic is not a true LOD score, as a LOD score is a $\log_{10}$ likelihood ratio, and the nonparametric interval mapping method does not involve a likelihood. However, this transformation gives statistics whose values are more in line with those from standard interval mapping.

It is important to emphasize that this method for extending rank-based test statistics for the case of missing genotype information is more in the style of Haley–Knott regression than of standard interval mapping. In the case of appreciable missing genotype information, the method may lose efficiency. Thus, an alternate approach deserves mention: one might convert the ranked phenotypes to quantiles of the normal distribution and apply standard interval mapping. In other words, we could use as our phenotypes $z_i = \Phi^{-1}[(R_i - 1/2)/n]$, where Φ^{-1} is the inverse of the cumulative distribution function of the standard normal distribution. (That is, if Z follows a normal distribution with mean 0 and SD 1, $\Phi(z) = \Pr(Z \leq z)$, and Φ^{-1} is the inverse of this function.) This approach will not work well in the case of many ties in the phenotypes, particularly for the case of a spike in the phenotype distribution, but otherwise should give results similar to the nonparametric interval mapping method described above.

Example

To illustrate these nonparametric methods, we consider the `listeria` data, described in detail in Sec. 2.3 and 2.4. This is a mouse intercross; the phenotype concerns survival time following infection with *Listeria monocytogenes*, and exhibits a spike at 264 hours: approximately 30% of the mice survived past the 240-hour time point and were considered to have recovered from the infection; their phenotype was recorded as 264. (For a histogram, see the lower left panel in Fig. 2.7 on page 35.)

We first need to get access to the data (which are distributed with the R/qtl package), and run `calc.genoprob` to calculate the QTL genotype probabilities given the available marker genotype data.

```
> library(qtl)
> data(listeria)
> listeria <- calc.genoprob(listeria, step=1, error.prob=0.001)
```

Nonparametric interval mapping is performed with the `scanone` function, using the argument `model="np"`, as follows. (In `scanone`, the argument `model` refers to the phenotype model, by default taken to be `"normal"`, and the argument `method` refers to the analysis method. The `method` argument is ignored when `model="np"`.)

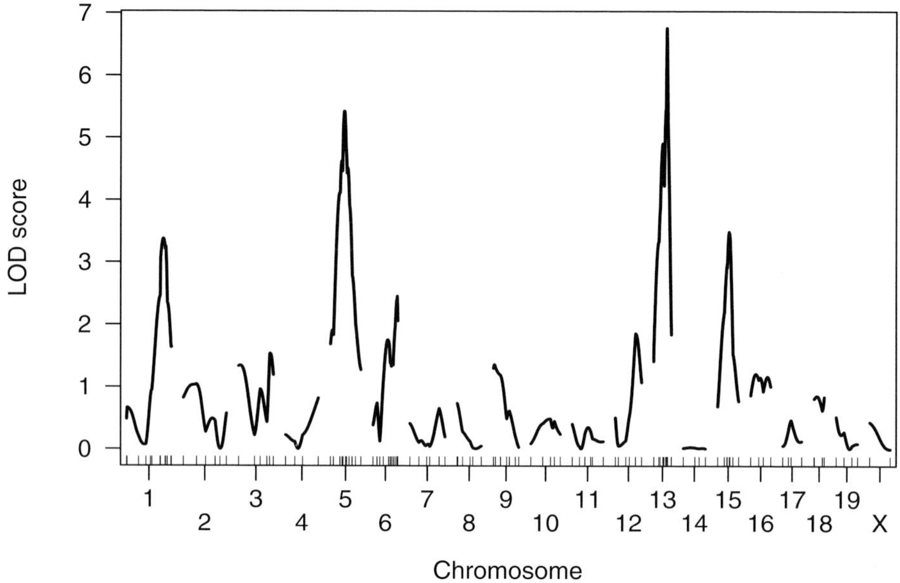

Figure 5.1. LOD scores by nonparametric interval mapping for the `listeria` data.

```
> out.np <- scanone(listeria, model="np")
```

By default, ties in the phenotypes are replaced by the average rank in each group of ties. With the argument `ties.random=TRUE`, ranks for tied phenotypes are randomized.

We can plot the results for all chromosomes as follows; the results appear in Fig. 5.1. We use `alternate.chrid=TRUE` so that the chromosome IDs may be more easily distinguished.

```
> plot(out.np, ylab="LOD score", alternate.chrid=TRUE)
```

As described in Chap. 4, a permutation test may be performed via the `n.perm` argument to `scanone`, as follows. We need to use `perm.Xsp=TRUE` to perform X-chromosome-specific permutations.

```
> operm.np <- scanone(listeria, model="np", n.perm=1000,
+                     perm.Xsp=TRUE)
```

The 5% LOD thresholds are the following.

```
> summary(operm.np, 0.05)

Autosome LOD thresholds (1000 permutations)
    lod
5% 3.20

X chromosome LOD thresholds (25078 permutations)
```

```
   lod
5% 2.33
```

Significant evidence for a QTL is seen on chromosomes 1, 5, 13, and 15.

```
> summary(out.np, perms=operm.np, alpha=0.05, pvalues=TRUE)

           chr  pos  lod   pval
c1.loc76     1 76.0 3.38 0.0343
c5.loc27     5 27.0 5.42 0.0000
D13M147     13 26.2 6.76 0.0000
c15.loc23   15 23.0 3.49 0.0187
```

5.2 Binary traits

In human linkage analysis, much of the focus has been on binary traits, and methods for quantitative traits were developed later. With experimental crosses, on the other hand, researchers have focused almost exclusively on quantitative traits. Nevertheless, interval mapping for binary traits is no more difficult than for quantitative traits.

Let the phenotypes y_i take values either 0 or 1. (For example, assign unaffected individuals the value 0 and affected individuals 1.) Consider some fixed position in the genome as the location of a putative QTL, and let g_i denote the QTL genotype of individual i. Again, let $p_{ij} = \Pr(g_i = j|\boldsymbol{M}_i)$.

Let $\pi_j = \Pr(y_i = 1|g_i = j)$, the penetrance of QTL genotype j. Given the marker data, $\boldsymbol{M}_i$, but not knowing the QTL genotypes g_i, the y_i follow mixtures of Bernoulli distributions (analogous to the mixtures of normal distributions that arise in standard interval mapping).

The likelihood for the parameters $\boldsymbol{\pi} = (\pi_j)$ is then

$$L(\boldsymbol{\pi}) = \prod_i \sum_j p_{ij}\,(\pi_j)^{y_i}\,(1-\pi_j)^{(1-y_i)}$$

We obtain maximum likelihood estimates (MLEs), $\hat{\pi}_j$, using a form of the EM algorithm. At iteration $s+1$, we have estimates of the parameters, $\hat{\boldsymbol{\pi}}^{(s)}$. In the E-step, we calculate weights for each individual and for each genotype:

$$w_{ij}^{(s+1)} = \Pr(g_i = j|y_i, \boldsymbol{M}_i, \hat{\boldsymbol{\pi}}^{(s)}) = \frac{p_{ij}\,(\hat{\pi}_j^{(s)})^{y_i}\,(1-\hat{\pi}_j^{(s)})^{(1-y_i)}}{\sum_k p_{ik}\,(\hat{\pi}_k^{(s)})^{y_i}\,(1-\hat{\pi}_k^{(s)})^{(1-y_i)}}.$$

In the M-step, we reestimate the probabilities π_j as weighted proportions using the weights, $w_{ij}^{(s+1)}$:

$$\hat{\pi}_j^{(s+1)} = \frac{\sum_i y_i w_{ij}^{(s+1)}}{\sum_i w_{ij}^{(s+1)}}.$$

We begin the algorithm by taking $w_{ij}^{(0)} = p_{ij}$, and iterate until the estimates converge, giving the MLE, $\hat{\boldsymbol{\pi}}$.

We next calculate a LOD score for the test of $\mathrm{H}_0 : \pi_j \equiv \pi$. First note that the MLE, under H_0, of the common probability π is the overall proportion, $\hat{\pi}_0 = \sum_i y_i/n$. Letting $\hat{\boldsymbol{\pi}}_0 = (\hat{\pi}_0, \hat{\pi}_0, \hat{\pi}_0)$, the LOD score is LOD $= \log_{10}\{L(\hat{\boldsymbol{\pi}})/L(\hat{\boldsymbol{\pi}}_0)\}$.

As with standard interval mapping, the likelihood under H_0 is calculated once, while the EM algorithm is performed at each position on a grid covering the genome, producing a LOD curve for each chromosome.

Example

Let us apply the binary trait method to the `listeria` data, taking as the binary trait whether the individuals' survived the infection or not. We first create the binary trait and append it to the phenotype data. Note that the function `pull.pheno` can be used to pull out a phenotype column.

```
> binphe <- as.numeric(pull.pheno(listeria, 1) > 250)
> listeria$pheno <- cbind(listeria$pheno, binary=binphe)
```

We need the results of `calc.genoprob`, but these were obtained in the previous section. The binary trait mapping method is performed with `scanone`, using the argument `model="binary"`. The phenotype must have values 0 and 1. We use the argument `pheno.col` to indicate that the new phenotype (named `"binary"`) is to be used; we could also use `pheno.col=3` or even `pheno.col=binphe`.

```
> out.bin <- scanone(listeria, pheno.col="binary",
+                    model="binary")
```

We can plot the results for all chromosomes, together with the results obtained by the nonparametric method, as follows. The results appear in Fig. 5.2.

```
> plot(out.np, out.bin, col=c("blue", "red"), ylab="LOD score",
+      alternate.chrid=TRUE)
```

A permutation test is again performed using the `n.perm` argument to scanone. Again we should use `perm.Xsp=TRUE`.

```
> operm.bin <- scanone(listeria, pheno.col="binary",
+                      model="binary", n.perm=1000,
+                      perm.Xsp=TRUE)
```

The LOD thresholds for the 5% significance level are the following.

```
> summary(operm.bin, alpha=0.05)
```

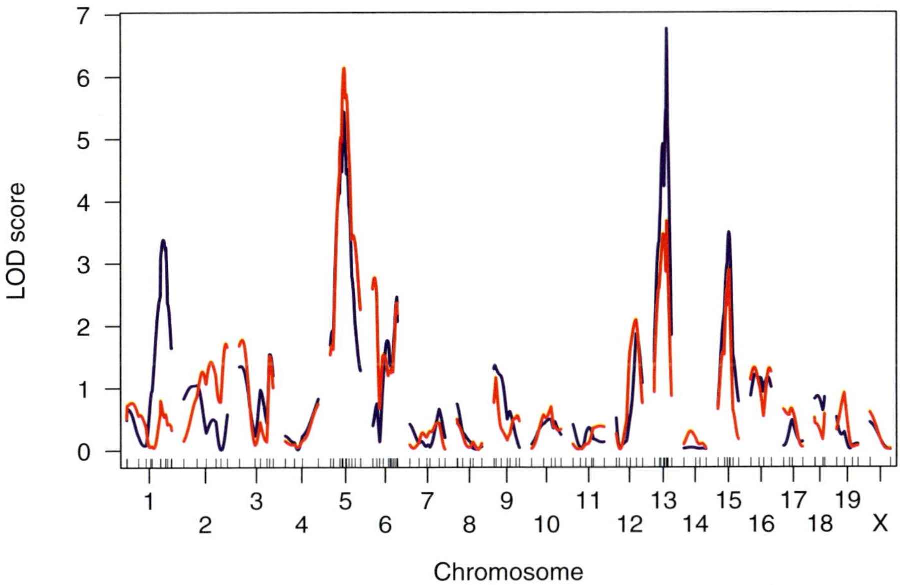

Figure 5.2. LOD scores by nonparametric interval mapping (in blue) and binary trait mapping (in red) for the `listeria` data. The binary trait is defined by survival > 250 hr or not.

```
Autosome LOD thresholds (1000 permutations)
    lod
5% 3.63

X chromosome LOD thresholds (25078 permutations)
    lod
5% 2.53
```

Significant evidence for a QTL is seen on chromosomes 5 and 13.

```
> summary(out.bin, perms=operm.bin, alpha=0.05, pvalues=TRUE)

          chr  pos  lod    pval
c5.loc29    5 29.0 6.13 0.00104
D13M147    13 26.2 3.67 0.04468
```

5.3 Two-part model

One often observes a spike in the phenotype distribution, such as that observed for the survival phenotype in the `listeria` data. Another common example is the case of mass of tumor, with some individuals exhibiting no tumor. In this section, we describe an analysis method particular for this situation.

Assume, without loss of generality, that the spike in the phenotype distribution is at 0. Let y_i denote the quantitative phenotype for individual i. Let $z_i = 0$ if $y_i = 0$, and $z_i = 1$ if $y_i > 0$.

A simple approach for QTL mapping in this situation is to first analyze the quantitative phenotype, y_i, using only the individuals for which $y_i > 0$, by standard interval mapping, and then separately analyze the binary trait z_i. These can be combined in what we call the two-part model.

Assume that the $(\boldsymbol{M}_i, y_i, z_i)$ are mutually independent, that $\Pr(z_i = 1|g_i = j) = \pi_j$, and that $y_i|(g_i = j, z_i = 1) \sim \text{normal}(\mu_j, \sigma^2)$. In other words, the probability that an individual with QTL genotype j has the null phenotype is $1 - \pi_j$; if this individual's phenotype is non-null, it follows a normal distribution with mean μ_j, depending on the QTL genotype, and with SD σ, independent of genotype.

In an intercross, this model contains seven parameters, $\boldsymbol{\theta} = (\pi_1, \pi_2, \pi_3, \mu_1, \mu_2, \mu_3, \sigma)$. The likelihood function is the following:

$$L(\boldsymbol{\theta}) = \prod_i \sum_j p_{ij} \; (1 - \pi_j)^{1-z_i} \; \{\pi_j \; \phi(y_i; \mu_j, \sigma)\}^{z_i}$$

where $\phi(y; \mu, \sigma)$ is the density function for a normal distribution with mean μ and SD σ.

We may again obtain MLEs with a form of the EM algorithm. Assume at iteration $s+1$ we have estimates $\hat{\boldsymbol{\theta}}^{(s)}$. In the E-step, we calculate weights for each individual and each genotype:

$$w_{ij}^{(s+1)} = \Pr(g_i = j | y_i, z_i, \boldsymbol{M}_i, \hat{\boldsymbol{\theta}}^{(s)}) = \begin{cases} \dfrac{p_{ij} \; (1-\hat{\pi}_j^{(s)})}{\sum_k p_{ik} \; (1-\hat{\pi}_k^{(s)})} & \text{if } z_i = 0 \\[2ex] \dfrac{p_{ij} \; \hat{\pi}_j^{(s)} \; \phi(y_i; \hat{\mu}_j^{(s)}, \hat{\sigma}^{(s)})}{\sum_k p_{ik} \; \hat{\pi}_k^{(s)} \; \phi(y_i; \hat{\mu}_k^{(s)}, \hat{\sigma}^{(s)})} & \text{if } z_i = 1 \end{cases}$$

In the M-step, we obtain revised estimates of the parameters according to the following equations:

$$\hat{\pi}_j^{(s+1)} = \frac{\sum_i w_{ij}^{(s+1)} \; z_i}{\sum_i w_{ij}^{(s+1)}}$$

$$\hat{\mu}_j^{(s+1)} = \frac{\sum_i y_i \; w_{ij}^{(s+1)} \; z_i}{\sum_i w_{ij}^{(s+1)} \; z_i}$$

$$\hat{\sigma}^{(s+1)} = \sqrt{\frac{\sum_i \sum_j (y_i - \hat{\mu}_j^{(s+1)})^2 \; w_{ij}^{(s+1)} \; z_i}{\sum_i z_i}}$$

We again start the algorithm by taking $w_{ij}^{(0)} = p_{ij}$, and iterate until the estimates converge, producing the MLEs, $\hat{\boldsymbol{\theta}}$.

We may calculate a LOD score for the test of $\mathrm{H}_0 : \pi_j \equiv \pi, \mu_j \equiv \mu$. We first note that, under H_0, the MLEs of the three parameters, π, μ, and σ, are

$$\hat{\pi}_0 = \frac{\sum_i z_i}{n}$$

$$\hat{\mu}_0 = \frac{\sum_i z_i\, y_i}{\sum_i z_i}$$

$$\hat{\sigma}_0 = \sqrt{\frac{\sum_i (y_i - \hat{\mu}_0)^2\, z_i}{\sum_i z_i}}.$$

In other words, $\hat{\pi}_0$ is the proportion of individuals with a positive phenotype, and $\hat{\mu}_0$ and $\hat{\sigma}_0$ are the sample mean and SD, among individuals with positive phenotypes. Letting $\hat{\boldsymbol{\theta}}_0 = (\hat{\pi}_0, \hat{\pi}_0, \hat{\pi}_0, \hat{\mu}_0, \hat{\mu}_0, \hat{\mu}_0, \hat{\sigma}_0)$, the LOD score is LOD $= \log_{10}\{L(\hat{\boldsymbol{\theta}})/L(\hat{\boldsymbol{\theta}}_0)\}$.

We calculate two additional sets of LOD scores. First, we consider the hypothesis $\mathrm{H}_0' : \pi_j \equiv \pi$, but allowing the μ_j to vary; the corresponding LOD scores assess evidence for QTL that specifically influence the chance that an individual has the null phenotype. Second, we consider the hypothesis $\mathrm{H}_0'' : \mu_j \equiv \mu$, but allowing the π_j to vary; the corresponding LOD scores assess evidence for QTL that influence the average phenotype, among individuals with a non-null phenotype. We could say that H_0' concerns the penetrance of the disease and H_0'' concerns the severity of the disease.

Note that in the case of complete QTL genotype information (i.e., when the putative QTL is at a marker that has been fully typed), the p_{ij} are all either 1 or 0, and the two parts of the model fully separate. In this case, the MLEs under the two-part model are exactly those obtained by the two separate analyses (the analysis of the binary trait and the conditional analysis of the quantitative trait, for those individuals with nonzero phenotype). Further, the LOD score for the two-part model is simply the sum of the LOD scores from the two separate analyses.

Example

As an illustration, we again consider the `listeria` data. Analysis with the two-part model is performed with `scanone` using `model="2part"`. The spike in the phenotype is assumed to be either the largest or the smallest observed phenotype. By default, the smallest phenotype is assumed; for the `listeria` data we must use the argument `upper=TRUE` to indicate that it is the largest observed phenotype (264 hr) that is to be treated as the spike. We will consider log survival time as the phenotype, and so we first append log survival to the phenotype data.

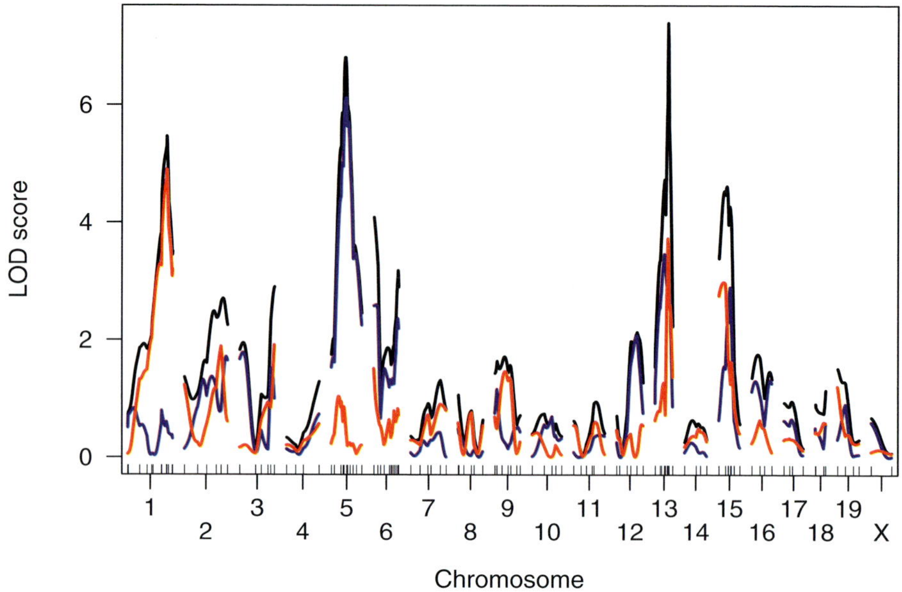

Figure 5.3. LOD scores from the two-part model, with LOD(π, μ) in black, LOD(π) in blue and LOD(μ) in red, for the `listeria` data.

```
> y <- log(pull.pheno(listeria, 1))
> listeria$pheno <- cbind(listeria$pheno, logsurv=y)
```

We now use `scanone` to calculate the LOD curves.

```
> out.2p <- scanone(listeria, model="2part", upper=TRUE,
+                   pheno.col="logsurv")
```

The results (see Fig. 5.3) contain three LOD score columns: LOD(π, μ), for the overall test of $\pi_j \equiv \pi$ and $\mu_j \equiv \mu$; LOD(π), for the test of $\pi_j \equiv \pi$; and LOD(μ), for the test of $\mu_j \equiv \mu$. We plot all three together as follows. The argument `lodcolumn` is used to plot all three LOD score columns.

```
> plot(out.2p, lodcolumn=1:3, ylab="LOD score",
+      alternate.chrid=TRUE)
```

The results indicate that the locus on chromosome 1 largely affects time-to-death, given that an individual has died (LOD(μ), in red, is large while LOD(π), in blue, is small). The locus on chromosome 5 largely affects the chance of survival (LOD(π), in blue, is large while LOD(μ), in red, is small). The loci on chromosomes 13 and 15 affect both aspects of the phenotype (both LOD(μ) and LOD(π) are large).

A permutation test is performed as follows.

```
> operm.2p <- scanone(listeria, model="2part", upper=TRUE,
+                     pheno.col="logsurv", n.perm=1000,
+                     perm.Xsp=TRUE)
```

The results, for each of the autosomes and X chromosome, contain 3 columns: the genome-wide maxima of $\text{LOD}(\pi, \mu)$, $\text{LOD}(\pi)$, and $\text{LOD}(\mu)$, for each permutation replicate. We obtain LOD thresholds as follows. (Note that π is denoted `p` in the output.)

```
> summary(operm.2p, alpha=0.05)

Autosome LOD thresholds (1000 permutations)
   lod.p.mu lod.p lod.mu
5%      4.8  3.63   3.88

X chromosome LOD thresholds (25078 permutations)
   lod.p.mu lod.p lod.mu
5%     3.18  2.54   2.58
```

If we consider just the overall LOD score, $\text{LOD}(\pi, \mu)$, significant evidence for a QTL is seen on chromosomes 1, 5, and 13.

```
> summary(out.2p, perms=operm.2p, alpha=0.05, pvalues=TRUE)

         chr  pos lod.p.mu    pval lod.p    pval lod.mu
c1.loc81   1 81.0     5.46 0.02183 0.594 1.00000  4.890
c5.loc27   5 27.0     6.80 0.00416 6.030 0.00104  0.779
D13M147   13 26.2     7.39 0.00312 3.667 0.04571  3.726
            pval
c1.loc81 0.00832
c5.loc27 1.00000
D13M147  0.06750
```

As discussed in Sec. 4.7, we may use the `format` argument of `summary.scanone` to get the maximum LOD scores for each of the three LOD score columns in `out.2p`.

```
> summary(out.2p, perms=operm.2p, alpha=0.05, pvalues=TRUE,
+         format="allpeaks")

   chr  pos lod.p.mu    pval  pos lod.p    pval  pos lod.mu
1    1 81.0     5.46 0.02183 12.0 0.825 1.00000 81.0   4.89
5    5 27.0     6.80 0.00416 29.0 6.117 0.00104 15.0   1.03
13  13 26.2     7.39 0.00312 26.2 3.667 0.04571 26.2   3.73
      pval
1  0.00832
5  1.00000
13 0.06750
```

Note that the locus on chromosome 15 did not achieve the 5% significance level here, but was seen to have a significant effect by nonparametric interval mapping (see Sec. 5.1). The linkage test in nonparametric interval mapping concerns two degrees of freedom, while with the two-part model, the test concerns four degrees of freedom. Thus, the two-part model has a higher significance threshold and so lower power for QTL detection. On the other hand, the separation of penetrance and severity can give a more detailed understanding of the QTL effects.

5.4 Other extensions

Essentially any phenotype model that is of interest in linear regression would also find application for QTL mapping. Only a limited number have been implemented in R/qtl. Thus, we conclude this chapter with a description, for the more computationally-savvy reader, of how such alternative mapping methods may be implemented with R/qtl.

We consider, as an illustration, the case of right-censored phenotypes. A phenotype is right-censored if it is only known to be greater than some value. For example, in the `listeria` data, a large proportion of individuals had not died of the *Listeria monocytogenes* infection at 264 hr. In the two-part model described in the previous section, these were viewed as having recovered from the infection, but we could also view the outcome as a survival time, and that the survival time for these individuals was right-censored.

A common approach for the analysis of such data is the use of a Cox proportional hazards model. Consider a random survival time, Y, and let $f(y)$ denote its density and $S(y) = \Pr(Y > y)$ denote its survival function. The *hazard function* is $h(y) = f(y)/S(y)$, which is essentially the chance that an individual will die immediately at time y given that it has survived to that point.

In the Cox proportional hazards model, we assume that the hazard function for individuals with QTL genotype g is $h_g(y) = h_0(y)e^{\beta_g}$, where h_0 is some completely unspecified baseline hazard function, and the β_g are the effects of the QTL genotypes. While the QTL genotypes will generally not be known, we may use an approach analogous to Haley–Knott regression. Let $p_{ij} = \Pr(g_i = j|\boldsymbol{M}_i)$ denote the QTL genotype probabilities given the available marker data, and assume that the hazard function for individual i is $h_0(y)\exp[\sum_j \beta_j p_{ij}]$.

To fit the Cox proportional hazards model, we use the `survival` package, which is distributed with R.

A function to perform the analysis appears in Fig. 5.4. As described in Sec. 4.4, the X chromosome requires special treatment, but we will just omit it from consideration here.

In line 4, we use the `require` function to ensure that the `survival` package is loaded. In lines 6–10 we omit individuals with missing phenotypes and make

```
scanone.cph <-
function(cross, pheno.col=1)
{
  require(survival)

  pheno <- pull.pheno(cross, pheno.col)
  cross <- subset(cross, ind=!is.na(pheno))
  pheno <- pheno[!is.na(pheno)]
  if(class(pheno) != "Surv")
    stop("Need the phenotype to be of class \"Surv\".")

  chrtype <- sapply(cross$geno, class)
  if(any(chrtype=="X")) {
    warning("Dropping X chromosome.")
    cross <- subset(cross, chr=(chrtype != "X"))
  }
  chr <- names(cross$geno)

  result <- NULL
  for(i in 1:nchr(cross)) {
    if(!("prob" %in% names(cross$geno[[i]]))) {
      warning("First running calc.genoprob.")
      cross <- calc.genoprob(cross)
    }
    p <- cross$geno[[i]]$prob

    # pull out map; drop last column of probabilities
    map <- attr(p, "map")
    p <- p[,,-dim(p)[3],drop=FALSE]

    lod <- apply(p, 2, function(a,b)
                 diff(coxph(b ~ a)$loglik)/log(10), pheno)

    z <- data.frame(chr=chr[i], pos=map, lod=lod)

    # special names for rows
    w <- names(map)
    o <- grep("^loc-*[0-9]+", w)
    if(length(o) > 0) # locations cited as "c*.loc*"
      w[o] <- paste("c",names(cross$geno)[i],".",w[o],sep="")
    rownames(z) <- w

    result <- rbind(result, z)
  }
  class(result) <- c("scanone", "data.frame")
  result
}
```

Figure 5.4. The scanone.cph function.

sure that the phenotype has been appropriately converted to be a survival time (see below). In lines 12–17, we omit the X chromosome and store the names of the chromosomes.

In line 19, we create a dummy object that will contain all of the results. In line 20, we begin a loop over the chromosomes. In lines 21–25 we check that the results of `calc.genoprob` are available, and pull out the probabilities for the current chromosome.

In line 28, we pull out the genetic map for the grid on which the QTL genotype probabilities were calculated. In line 29, we drop the last column from the QTL genotype probabilities, since the analysis will include an intercept term.

In lines 31–32, we perform the actual analysis. We use the `apply` function to send the QTL genotypes, one column at a time, to the `coxph` function. The `coxph` function is part of the `survival` package, and performs the Cox proportional hazards regression. Part of the output of `coxph` is the log (base e) likelihood, for the null model (with just the intercept) and for the alternative model (including the covariates: here, the QTL genotype probabilities). We take the difference of these log likelihoods and then divide by $\ln(10)$ to convert the result to the LOD scale.

In line 34, we paste together the chromosome IDs, map positions, and LOD scores into a data frame. In lines 36–41 we create special row names, used to ensure clarity regarding which positions are markers and which are between markers. In line 43, we append the results for this chromosome to the end of our growing set of results.

Finally, in line 45, we change the "class" of the result to include `"scanone"`, so that we may use `plot` and `summary` and have the data sent to `plot.scanone` and `summary.scanone`. The final line contains the return value for the function.

Example

Let us now apply the method to the `listeria` data. We first need to load the `survival` package and the code for the `scanone.cph` function; the latter may be done with the `source` function.

```
> library(survival)
> source("scanone_cph.R")
```

We need to convert the phenotype into a censored survival time, using the function `Surv` in the `survival` package. This is done to indicate which phenotypes are to be viewed as censoring times rather than actual survival times. `Surv` takes two arguments: the survival/censoring times and an indicator of whether the values were observed or were censored. We append this revised phenotype to the end of the phenotype data. (This will now be the fifth phenotype in the data set.)

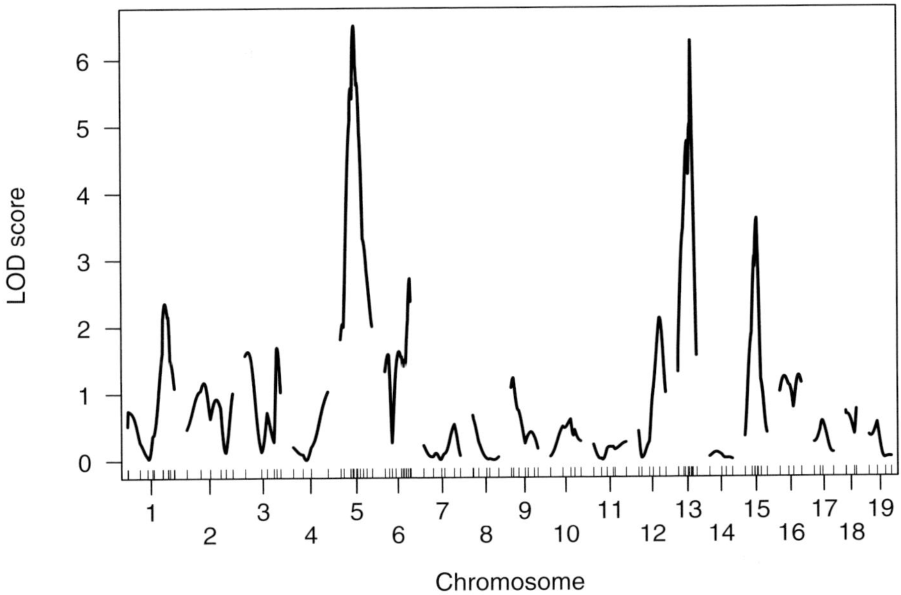

Figure 5.5. LOD scores from the Cox proportional hazards model, using an approach analogous to Haley–Knott regression, for the `listeria` data.

```
> y <- pull.pheno(listeria, 1)
> y <- Surv(y, y<250)
> listeria$pheno <- cbind(listeria$pheno, surv=y)
```

We may now send the data to `scanone.cph` to perform the analysis. We may refer to the phenotype by its name, `"surv"`.

```
> out.cph <- scanone.cph(listeria, pheno.col="surv")
```

We may plot the resulting LOD scores as follows. (See Fig. 5.5.)

```
> plot(out.cph, ylab="LOD score", alternate.chrid=TRUE)
```

We have not written code to do a permutation test, and so we will perform the permutation test by brute force, using a `for` loop.

```
> n.perm <- 1000
> operm.cph <- cbind(lod=1:n.perm)
> chr <- names(listeria$geno)
> temp <- subset(listeria, chr=(chr != "X"))
> n.ind <- nind(listeria)
> for(i in 1:n.perm) {
+   temp$pheno <- temp$pheno[sample(n.ind),]
+   out <- scanone.cph(temp, pheno.col="surv")
+   operm.cph[i] <- max(out[,3], na.rm=TRUE)
```

```
+ }
> class(operm.cph) <- "scanoneperm"
```

The LOD threshold for the 5% significance level is the following.

```
> summary(operm.cph, 0.05)

LOD thresholds (1000 permutations)
    lod
5% 3.51
```

Significant evidence for a QTL is seen on chromosomes 5, 13 and 15.

```
> summary(out.cph, perms=operm.cph, alpha=0.05, pvalues=TRUE)

          chr  pos  lod  pval
c5.loc28    5 28.0 6.50 0.000
D13M147    13 26.2 6.26 0.000
c15.loc24  15 24.0 3.60 0.039
```

5.5 Summary

With the interval mapping methods of Chap. 4, one assumes that the residual variation in the phenotype follows a normal distribution. While the normality assumption is often appropriate, many phenotypes show clear departures from normality. Application of interval mapping often performs reasonably anyway, particularly if the phenotype is transformed to give approximate normality. However, there are alternatives. Nonparametric interval mapping considers the ranks of the phenotype values. An interval mapping method specific for binary traits is simple to develop. In general, one may generate an interval mapping method tailored to any sort of phenotype model.

5.6 Further reading

The first three methods discussed in this chapter were considered in Broman (2003), in which the two-part model was proposed. Kruglyak and Lander (1995) proposed the nonparametric interval mapping approach for backcrosses; they described a somewhat different method for dealing with intercrosses, but we prefer the extension of the Kruskal–Wallis test statistic. Xu and Atchley (1996) described the approach for binary traits, though in a somewhat more general context that included marker covariates. Visscher *et al.* (1996a) and McIntyre *et al.* (2001) both described approximate methods for interval mapping with binary traits, but these are not recommended.

Several authors have described QTL mapping methods for survival times. Symons *et al.* (2002) used a Monte Carlo method to fit a semiparametric

Cox proportional hazards model (Cox, 1972), making more precise use of the genotype data than the method described in Sec. 5.4. Diao and Lin (2005) considered the same model, but used a different, and less computationally demanding method for estimation. Diao *et al.* (2004) described maximum likelihood for a fully parametric Weibull model. Moreno *et al.* (2005) studied the performance of the Weibull and Cox proportional hazards model relative to standard interval mapping.

Several other methods deserve mention, but have not yet been implemented in R/qtl. Hackett and Weller (1995) described a method for the analysis of ordinal traits. Jansen (1992, 1993b) described the use of generalized linear models for QTL mapping, which could be applied for a variety of types of phenotypes, including count data.

6

Experimental design and power

Sound experimental design is essential to good science. Scientific review committees reviewing research proposals usually look for evidence of careful experimental design. They may also desire, and it may be in the researcher's self-interest, that experiments be economical. Much of good experimental design is a mixture of common sense, pragmatism, and careful forethought, which are difficult to codify. Nonetheless, some general principles for experimental design can be laid out. In this chapter, we discuss issues special to QTL experiments with the help of the R package, R/qtlDesign.

In the context of QTL experiments this would include adjusting for covariates potentially influencing the phenotype, performing reciprocal crosses (if maternal effects are suspected or if the QTL is X-linked), deciding whether to consider one sex or both, adjusting for litter or foster mom effects, etc.

Before a QTL experiment can be conducted, the experimenter has to make many choices. What strains should be crossed, and what type of cross should be used? What phenotypes should be measured? Should they be replicated? What covariates should be collected? What markers should be used, and how dense should the genotyping be? Can selective genotyping be used to save money? How many progeny should be collected?

The calculations performed with R/qtlDesign provide quick answers to the above questions but rely on a number of approximations that may not always be accurate. A more cumbersome but potentially more accurate approach is to use computer simulation. We conclude the chapter with a brief discussion of the use of computer simulation to estimate the power to detect a QTL and the precision of localization of QTL.

6.1 Phenotypes and covariates

The most important choice is the phenotype of interest to the experimenter. Sometimes, the choice is natural. Someone interested in hypertension may measure blood pressure, while someone interested in cancer may count tumors.

K.W. Broman, Ś. Sen, *A Guide to QTL Mapping with R/qtl*,
Statistics for Biology and Health, DOI 10.1007/978-0-387-92125-9_6,

Researchers studying human diseases in model systems, such as mice or rats, have to consider what phenotype best approximates the human analog. For example, a researcher studying anxiety in mice may use the open field test.

In addition to the primary phenotype of interest, it is useful to record all covariates that may affect the phenotype. Sex and cross direction should always be recorded. Whenever possible date/time of experiment, experiment batch, technician name, cage number, litter number, and parent IDs should be recorded. These are useful for diagnosing and correcting oddities in the data. In addition, factors that may affect the primary phenotype of interest should be recorded. For example, an obesity study might want to record average room temperatures because lower temperatures might lead to greater energy expenditure and lower body weight.

6.2 Strains and strain surveys

After the primary phenotype of interest has been chosen, the experimenter has to decide what strains to cross. We would generally want to cross two strains that exhibit a consistent difference in the phenotype of interest (although a cross between strains of similar phenotype may segregate QTL). Preliminary data in the experimenter's laboratory might suggest natural choices for strains to cross. In the absence of such data, an experimenter may perform a strain survey by comparing the phenotype of interest in a number of available strains. This may be done in one of two ways. The strain survey may be done *in silico*, that is, in the comfort of one's office with a computer, by simply surveying published data on the phenotypes of many strains (e.g., the Mouse Phenome Database, `http://www.jax.org/phenome`). Alternatively, it may be done the hard way, by actually phenotyping all of the strains in one's own laboratory.

The advantage of the first approach is convenience, and the ability to survey a large number of strains relatively easily. The disadvantage is that some phenotypes vary from lab to lab, and with time (as equipment and the strains may slowly drift with time). A recent study showed that the stability of phenotyping over time and space varies by phenotype, and there were no discernible differences in stability between strains (Wahlsten *et al.*, 2003). The advantage of the second approach is that the experimenter has complete control over the phenotyping protocols, and therefore more reliable data is obtained. The disadvantage is cost, and time.

Once there is reliable data on the phenotypes of strains, if we see differences in a phenotype between two strains, then we can be confident that the difference is genetic. Our next step would be to cross the two strains creating genetic variation, so that we can study the association between genetic markers and the phenotype.

6.3 Theory

Having decided which strains to cross, an investigator will have to decide what type of cross to use (e.g., backcross, intercross, or recombinant inbred lines), how many progeny to raise and phenotype, and the genotyping and phenotyping strategies. The ability to detect QTL and the precision of estimated QTL effects depend on design choices through three key quantities: the variance attributable to a given locus, the residual error variance, and the information content of the cross. We will now review the theory behind each of these quantities. This will help us understand how we can leverage cross type, number of progeny, and genotyping strategies to our advantage.

For simplicity we assume that we wish to detect a single locus contributing to variation in the phenotype. The phenotypic variance is composed of genetic and nongenetic components. A part of the genetic variance may be attributable to a locus, the rest being background genetic variance which may be due to multiple loci.

The variance attributable to a locus depends on the cross type and the mode of inheritance. Specifically, a locus with an additive mode of action (i.e., with the average phenotype for heterozygotes at the midpoint between the phenotype averages for the two homozygote groups) explains twice as much variance in an intercross, and four times as much variance in recombinant inbred lines (RILs), compared to a backcross population. On the other hand, the genetic similarity of a backcross population is greater than an intercross population, which, in turn, is more similar than a RIL population. Thus the background genetic variance is greatest in RILs, followed by intercross populations, and least in backcross populations.

Sample size calculations for QTL experiments are usually presented in terms of the proportion of variance explained by the locus of interest we wish to detect (i.e., the heritability due to the QTL). This approach provides a partial picture for comparing different cross types because the variance attributable to a locus and the background variance depend on the cross (for an illustration see Sec. 6.4).

Therefore, we present a complementary approach which directly considers the effect of a locus on the phenotype, as well as the background genetic variance in the population. To develop this approach, we begin by considering the variance attributable to a locus, followed by a discussion of the residual error variance which includes the background genetic variance. Finally, we consider manipulation of information content (or the "effective sample size") using selective genotyping and variable marker spacing.

6.3.1 Variance attributable to a locus

Let the two strains under consideration be A and B, and let AA and BB be the parental genotypes at a locus of interest. The possible genotypes at each autosomal locus are AA, AB, and BB. Let μ be the overall mean, α be the

Table 6.1. Genotype probabilities, genotype means, and the variance attributable to a segregating locus, as a function of genetic model and cross type. BC_A denotes backcross to the A strain, and BC_B denotes backcross to the B strain.

		Probability			
Genotype	Mean	Intercross	BC_A	BC_B	RILs
AA	$\mu+\alpha-\delta/2$	1/4	1/2	0	1/2
AB	$\mu+\delta/2$	1/2	1/2	1/2	0
BB	$\mu-\alpha-\delta/2$	1/4	0	1/2	1/2
		Variance attributable to locus			
Model	Parameter	Intercross	BC_A	BC_B	RILs
General		$\frac{1}{4}\delta^2+\frac{1}{2}\alpha^2$	$\frac{1}{4}(\alpha-\delta)^2$	$\frac{1}{4}(\alpha+\delta)^2$	α^2
No dominance	$\delta=0$	$\frac{1}{2}\alpha^2$	$\frac{1}{4}\alpha^2$	$\frac{1}{4}\alpha^2$	α^2
A dominant	$\delta=\alpha$	$\frac{3}{4}\alpha^2$	0	α^2	α^2
B dominant	$\delta=-\alpha$	$\frac{3}{4}\alpha^2$	α^2	0	α^2
No additive	$\alpha=0$	$\frac{1}{4}\delta^2$	$\frac{1}{4}\delta^2$	$\frac{1}{4}\delta^2$	0

additive effect of the locus (half of the difference between the means of the homozygotes), and δ be the dominance effect (the difference between the mean of the heterozygotes and the average of the homozygote means). The variance attributable to a segregating locus depends on the type of cross as well as the genetic model (see Table 6.1).

A purely additive locus (no dominance, $\delta=0$) segregating in a set of RILs accounts for twice the variance compared to an intercross, and four times compared to the two possible backcrosses. A dominant locus ($\delta=\pm\alpha$) segregating in an intercross explains 75% of the variance compared to RILs. If we happen to perform a backcross to the correct parental strain, the locus explains as much variance as in RILs, otherwise it does not contribute to the phenotypic variation. Thus, for a suspected dominant locus, it is safer to perform an intercross, barring specific knowledge about the dominant allele. A segregating locus would explain the most variance in RILs unless the locus is overdominant ($|\delta| > \alpha$). When a locus has no additive effect, it will explain no variance in RILs, but both an intercross and a backcross would explain the same amount of variance. Thus, an intercross, which segregates all three genotypes, offers the opportunity to detect the widest variety of genetic models.

6.3.2 Residual error variance

The residual error variance is composed of background genetic variance due to all QTL not linked to the locus of interest and the nongenetic residual variance. If measurement error is negligible, then the nongenetic residual variance is due to environmental effects specific to each individual. By using biological replication (more than one individual with the same genotype), we can reduce nongenetic residual variance. This is usually only possible for RILs, where we can create multiple individuals with the same genotype. On a similar note, we can reduce the measurement error contribution to the nongenetic residual variance by replicating measurements.

Nongenetic error variance

Let the measurement error variance be σ_M^2 (the variance of phenotype measurements in the same individual), and the environmental variance be σ_E^2 (the variance of phenotypes in individuals with the same genotype, assuming no measurement error). Assume we have m individuals per unique genotype. For backcross and intercross populations, m=1. For RILs, we can choose m, subject to cost constraints. Assume that we have k replicate measurements per individual (e.g., the number of times blood pressure is measured on the same mouse). Then the nongenetic residual error variance is $(\sigma_E^2+\sigma_M^2/k)/m$. It is in the interest of the investigator to choose an instrument with negligible measurement error (σ_M^2), or to replicate the measurement enough times so that σ_M^2/k is small. In that case, the nongenetic residual error variance is approximately σ_E^2/m. Estimates of the measurement error variance (σ_M^2) and the environmental variance (σ_E^2) may be obtained from pilot studies.

Genetic variance

As mentioned earlier, the background genetic variance depends on the cross type. For simplicity, assume that the variance attributable to any single locus is small, so that the background genetic variance is approximately equal to the genetic variance. Let c be a constant that depends on the cross type, equal to 4 for backcrosses, 2 for intercrosses, and 1 for RILs, and let $\sigma_G^2(c)$ be the corresponding genetic variance. Then, the variance attributable to an additive locus is α^2/c (see Table 6.1). If we assume that all loci are additive, then the genetic variance in a cross is $\sigma_G^2(c) = \sigma_G^2/c$, where σ_G^2 is the genetic variance in RILs. Then the residual error variance would be $\sigma_G^2/c + (\sigma_E^2 + \sigma_M^2/k)/m$. Thus the ratio of the variance attributable to an additive locus to the residual error variance (the "signal-to-noise ratio"), would be

$$\begin{aligned}\frac{\alpha^2/c}{\sigma_G^2/c + (\sigma_E^2 + \sigma_M^2/k)/m} &= \frac{\alpha^2}{\sigma_G^2}\left[1 + \frac{c}{m}\left(\frac{\sigma_E^2}{\sigma_G^2} + \frac{\sigma_M^2}{k\sigma_G^2}\right)\right]^{-1} \\ &\simeq \frac{\alpha^2}{\sigma_G^2}\left(1 + \frac{c}{m}\frac{\sigma_E^2}{\sigma_G^2}\right)^{-1}.\end{aligned}$$

Table 6.2. Effect size multiplier by cross type, number of environmental replicates per genotype (m), and the ratio of environmental relative to genetic variance, measured by σ_E^2/σ_G^2, where σ_E^2 is the within-genotype variance and σ_G^2 is the between-genotype variance in RILs. We assume that all loci are additive, so that the between-genotype variance (the genetic variance) is $\sigma_G^2/2$ and $\sigma_G^2/4$ in intercross and backcross populations, respectively.

	Backcross	Intercross	RILs	
σ_E^2/σ_G^2	m=1	m=1	m=1	m=4
1/16	0.80	0.89	0.94	0.98
1/4	0.50	0.67	0.80	0.94
1/2	0.33	0.50	0.67	0.89
1	0.20	0.33	0.50	0.80
2	0.11	0.20	0.33	0.67
4	0.06	0.11	0.20	0.50
16	0.015	0.030	0.059	0.200

The approximation holds when σ_M^2/k is negligible (i.e., when technical measurement error is small or we have sufficient technical replicates). It is easier to detect a QTL when the signal-to-noise ratio is higher.

The signal-to-noise ratio depends on α^2, σ_G^2, σ_E^2, c, and m. Of these, the first three are determined by nature, over which the experimenter has no control. However, by choosing the cross type (which determines c), and the number of environmental replicates per genotype (m), the experimenter can manipulate the signal-to-noise ratio. We focus on the effect size multiplier, $\left[1+c\,\sigma_E^2/(m\,\sigma_G^2)\right]^{-1}$, displayed in Table 6.2 as a function of cross, number of replicates per genotype, and the ratio of the environmental variance to the genetic variance. As one would expect, the signal-to-noise ratio is highest when the environmental variance is small; in this setting, there is little difference between the different crosses. However, when the environmental variance is high, the signal-to-noise ratio is low for all crossing designs. In this setting, RILs are most advantageous. This advantage can be magnified by replication.

6.3.3 Information content

We have greater control over the information content of a cross compared to our control over the effect size and the error variance. The information content of an experiment may be interpreted as the effective sample size. It is a fundamental quantity which affects the power to detect a QTL, the expected LOD score, and the precision of the estimated QTL effects. We define the information content in the sense of Fisher information (see Cox and Hinkley,

1974): the reciprocal of the expected variance of the genetic model parameters for unit residual variance.

The information content depends on the number of progeny, genotyping strategies, and phenotyping strategies. Genotyping strategies that can affect information content include marker spacing and selective genotyping (where a fraction of the individuals with extreme phenotypes are genotyped). Phenotyping strategies include choosing a subset of individuals based on their genotype for expensive phenotyping, and replication of noisy measurements. By making choices that maximize information content, based on the cost structure of the experiment, the experimenter can most efficiently allocate resources.

6.4 Examples with R/qtlDesign

R/qtlDesign is an R package for facilitating experimental design choices for QTL mapping. It may be obtained from the Comprehensive R Archive Network (CRAN, `http://cran.r-project.org`) and installed in the same way that the R/qtl package is installed (see App. A).

We assume that the phenotypes follow a normal distribution, that the effect size of a QTL is small relative to the residual variance, and that the sample size is large. (A warning is printed if the effective sample size is smaller than 30.) The normality assumption facilitates analytical tractability, and the sample size assumption is needed to use χ^2 approximations in power calculations. The small effect size assumption simplifies information content calculations; the resulting approximation is accurate for most practical situations. We also assume that measurement error is negligible. If that is not the case, it may be advisable to consider replicating measurements.

The package assumes that cost functions are linear; this ignores economies of scale, but provides a useful guide. The optimal marker spacing and the selection fraction (the proportion of extreme phenotypic individuals that are genotyped) should therefore be seen as approximations; they are not necessarily optimal. The function approximating the residual error variance assumes that all loci are additive. This is intended as an approximation; in practice one would consider a range of possibilities (see the examples below).

6.4.1 Functions

The main functions in the R/qtlDesign package are the following:

`powercalc` Calculates the power to detect a QTL given the effect size, the residual error variance, the genome-wide LOD threshold, the width of the marker interval containing the QTL, the sample size, and the proportion of extreme individuals genotyped (the selection fraction).

`detectable`	Calculates the minimum detectable QTL effect as a function of the target power and other `powercalc` inputs.
`samplesize`	Calculates the minimum sample size needed to detect a QTL effect given the desired power and other `powercalc` inputs.
`info`	Calculates the approximate expected information as a function of the selection fraction and the marker interval width.
`optspacing`	Calculates the optimal marker spacing and selection fraction given the genotyping cost and the genome size.
`optselection`	Calculates the optimal selection fraction given the genotyping cost, marker spacing, and genome size.
`error.var`	Calculates the residual error variance given the cross type, environmental variance, background genetic variance, and the number of environmental replicates per unique genotype.
`thresh`	Calculates the genome-wide LOD threshold required for QTL detection given the cross type, genome length, and marker density, using the approximations of Dupuis and Siegmund (1999).

6.4.2 Choosing a cross

Barring specific knowledge about the mode of action of the QTL, the intercross is often the best cross choice. It segregates all possible genotypes, and therefore permits detection of QTL with any mode of action (dominant, recessive, additive, or overdominant). If the phenotype is noisy, with a lot of environmental variation, then RILs (provided that they are available) are the best choice, as we can use replicate individuals to decrease noise. If we are confident of the nature of the effect, or suspect substantial genetic variance due to epistasis, then performing a backcross would be a good choice. Sometimes investigators will perform more than one type of cross, and then combine the evidence from multiple populations.

Power calculations in research grant applications traditionally are presented in terms of the proportion of variance detectable with high power given a population of a certain size. Let us explore what effects we can detect using a backcross or intercross population with 100 individuals. We assume that we desire 80% power in a mouse cross.

We must first load the R/qtlDesign package.

```
> library(qtlDesign)
```

We first estimate the 5% genome-wide LOD threshold that we will use for the mouse genome (of size 1440 cM), assuming infinitely dense markers. We use the `thresh` function.

```
> thresh(G=1440, cross="bc", p=0.05)

[1] 3.190
```

We get the analogous threshold for an intercross as follows.

```
> thresh(G=1440, cross="f2", p=0.05)

[1] 4.183
```

Note that the LOD threshold for an intercross population is a bit higher. This is because we have two degrees of freedom in an intercross (versus one in a backcross), and because the recombination density in an intercross is twice that of a backcross.

We can now calculate the minimum detectable effect sizes using the function `detectable`.

```
> detectable(cross="bc", n=100, sigma2=1, thresh=3.2)

     effect percent.var.explained
[1,]  0.936                 17.97

> detectable(cross="f2", n=100, sigma2=1, thresh=4.2)

     additive.effect dominance.effect percent.var.explained
[1,]           0.726                0                 20.86
```

Thus, we can detect loci explaining a similar percentage of variance in backcrosses and intercrosses. However, the effect size that can be detected in an intercross is smaller if the environmental variance is high. We will see this in examples below.

Table 6.3 displays the approximate detectable effects (measured as the percent variance explained by a QTL) for a range of sample sizes, in each of a backcross and an intercross.

Suppose we are planning to map blood pressure QTL in mice by crossing the A and B strains, whose blood pressure means are 85 mm of Hg and 105 mm of Hg, respectively. The within-strain standard deviations are 8 mm of Hg. Let us also assume that we are interested in detecting a locus with an additive effect of 5 mm of Hg. We will compare crossing designs assuming that we want to detect the locus with 80% power. (For brevity, we suppress measurement units below.) How do the choices of backcross, intercross, and recombinant inbred lines shape up?

We estimate σ_E^2, the environmental variance, by the within-strain variance, 8^2=64. We estimate the background genetic variance with some assumptions, and therefore consider a range of possibilities, guiding our choices using the

Table 6.3. The minimum percent variance attributable to a QTL for it to be detectable with 80% power, as a function of sample size, marker spacing (in cM), and the selection fraction (the proportion of extreme phenotypic individuals genotyped). We use a significance threshold of 3.2 for a backcross and 4.2 for an intercross. These are the estimated thresholds corresponding to dense markers for the mouse genome (1440 cM). The power is calculated for a locus at the center of a marker interval with the given spacing.

	Selection fraction							
	100%				50%			
	Marker spacing in cM				Marker spacing in cM			
Sample size (n)	0	5	10	20	0	5	10	20
Backcross								
100	17.9	18.7	19.5	21.4	19.1	19.9	20.7	22.7
200	9.9	10.3	10.8	12.0	10.5	11.0	11.6	12.8
400	5.2	5.4	5.7	6.4	5.6	5.8	6.1	6.8
Intercross								
100	20.8	21.7	22.6	24.7	22.1	23.0	23.9	26.1
200	11.6	12.2	12.7	14.1	12.4	13.0	13.6	15.0
400	6.2	6.5	6.8	7.6	6.6	6.9	7.3	8.1

between-strain variance, $(105-85)^2/4 = 100$. This would be the genetic variance in RILs if a single QTL accounted for the strain difference. If there are two or more QTL, the genetic variance would be smaller, assuming no epistasis and that all QTL have effects of the same sign. Thus we consider three values of σ_G^2, 25, 50, and 100 (these correspond to percent variance attributable to the QTL equal to 14.0%, 12.3% and 9.9%, respectively). We may use the `samplesize` function to get an approximate sample size. For σ_G^2=25, we use:

```
> samplesize(cross="f2", effect=c(5,0), env.var=64, gen.var=25,
+             thresh=4.2)

     sample.size percent.var.explained
[1,]         162                 14.04
```

This gives us a sample size of 162. The additive and dominance effects we wish to detect are given in the `effect` argument. The residual error variance is approximated using our estimates of the environmental and genetic variances (alternatively one can specify the residual error variance directly). If the genetic variance is 50 or 100, we get sample size estimates of 188 and 241, respectively.

```
> samplesize(cross="f2", effect=c(5,0), env.var=64, gen.var=50,
+             thresh=4.2)

     sample.size percent.var.explained
[1,]         188                 12.32

> samplesize(cross="f2", effect=c(5,0), env.var=64,
+             gen.var=100, thresh=4.2)

     sample.size percent.var.explained
[1,]         241                 9.881
```

By default, the software assumes that our target LOD threshold is 3, that our desired power is 80%, and that all individuals are typed densely.

For a backcross population, we would need more individuals, as the following results show.

```
> samplesize(cross="bc", effect=5, env.var=64, gen.var=25,
+             thresh=3.2)

     sample.size percent.var.explained
[1,]         247                  8.17

> samplesize(cross="bc", effect=5, env.var=64, gen.var=50,
+             thresh=3.2)

     sample.size percent.var.explained
[1,]         269                 7.553

> samplesize(cross="bc", effect=5, env.var=64, gen.var=100,
+             thresh=3.2)

     sample.size percent.var.explained
[1,]         312                 6.562
```

To estimate how many RILs we would need (if such a resource existed), we will first have to estimate the LOD threshold required. Assuming that the RILs were created by sibling mating, we can get the threshold by using the `thresh` function for a backcross population, multiplying the genome length by 4 (because of the four-fold map expansion in RILs by sibling mating).

```
> thresh(G=1440*4, cross="bc", p=0.05)

[1] 3.834
```

RILs are most advantageous when the genetic variance is small relative to the environmental variance. We therefore consider the setting when σ_G^2 is 25.

```
> samplesize(cross="ri", effect=5, env.var=64, gen.var=25,
+             thresh=3.8)
```

```
     sample.size percent.var.explained
[1,]          90                 21.93
```

We would need about 90 animals, which is favorable in terms of the number of animals needed, but might be infeasible because RIL populations of such size are expensive to create and maintain. This assumed that we used just one animal per line. With replication, the number of unique RILs decreases, but to a point. We see this by using the `bio.rep` argument.

```
> samplesize(cross="ri", effect=5, env.var=64, gen.var=25,
+            thresh=3.8, bio.rep=2)

     sample.size percent.var.explained
[1,]          58                 30.49

> samplesize(cross="ri", effect=5, env.var=64, gen.var=25,
+            thresh=3.8, bio.rep=4)

     sample.size percent.var.explained
[1,]          42                 37.88

> samplesize(cross="ri", effect=5, env.var=64, gen.var=25,
+            thresh=3.8, bio.rep=16)

     sample.size percent.var.explained
[1,]          30                  46.3

> samplesize(cross="ri", effect=5, env.var=64, gen.var=25,
+            thresh=3.8, bio.rep=100)

     sample.size percent.var.explained
[1,]          26                 49.37
```

This indicates that, with 4–20 replicate animals per line, we can detect the desired effects in a RIL population of modest size (30–40 lines). If we used 4 replicate animals, we would use about 168 animals. The number of intercross animals needed is about 162, and the number of backcross animals needed is about 247. Since the cost of breeding replicate animals is smaller for RILs, one would conclude that using a RIL population with about 40 lines would be a good choice if the genetic variance is small. However, the intercross is quite competitive.

6.4.3 Genotyping strategies

Once a cross has been performed, genotyping strategies can be used to reduce experimental cost. Because of linkage between adjacent markers on the same chromosome, there are diminishing returns as genotyping density increases. An investigator might want to know what genotyping density provides a good

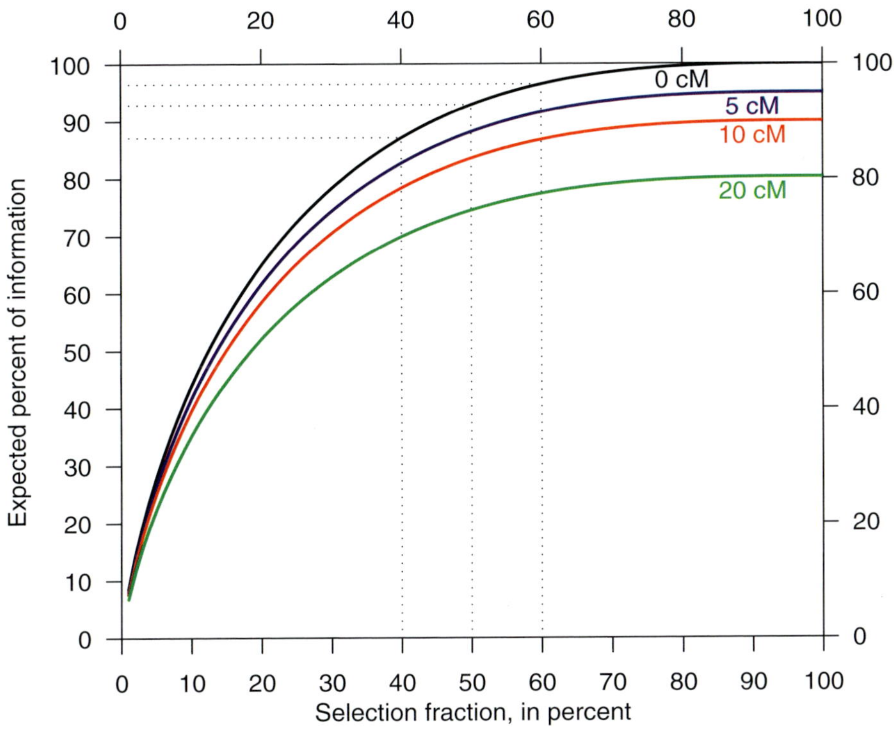

Figure 6.1. Expected information from a selectively genotyped backcross as a function of the selection fraction (the proportion of extreme phenotypic individuals genotyped), in the middle of a marker interval of length 0, 5, 10, and 20 cM. The information is plotted relative to a fully genotyped cross, where all individuals are genotyped at a dense set of markers.

return on investment. Further, if there is a single phenotype of primary interest, selective genotyping, where only extreme phenotypic individuals are genotyped, may be used. This strategy is most effective when the cost of phenotyping and raising an individual is low relative to the cost of genotyping. We will consider these issues with our design tools. First, let us take a look at how information content ("effective sample size") varies with marker density and the selection fraction (the proportion of extreme phenotypic individuals genotyped).

The expected information from a selectively genotyped backcross is displayed in Fig 6.1. The information is plotted relative to a fully genotyped cross where all individuals are genotyped at a dense set of markers. We can see that if the cross is densely genotyped, genotyping between 40% to 60% of the cross gives us 85% to 95% of the information in the cross. The gains diminish with wider marker spacings. The figure was created using the function `info`.

Suppose we are performing an intercross, and suppose a genotyping facility charges 10 cents per genotype, and that the animal facility per diem rate for mice is $1.20. The total cost of housing a mouse for 25 weeks is about $30; therefore the genotyping cost in the units of raising the mouse is about 1/300. To find the genotyping density that gives the best information-to-cost ratio, we use the function `optspacing`.

```
> optspacing(cost=1/300, G=1440, sel.frac=1, cross="f2")
```

```
Marker spacing (cM)  Selection fraction
              17.93                1.00
```

This suggests that relatively sparse genotyping (18 cM density) would be adequate for detecting QTL.

If we had a single phenotype and could perform selective genotyping, we could find the selection fraction and genotyping density combination that give the best information to cost ratio. We do this by setting the `sel.frac` argument to `NULL`.

```
> optspacing(cost=10/3000, G=1440, sel.frac=NULL, cross="f2")
```

```
Marker spacing (cM)  Selection fraction
            14.5573              0.6063
```

This suggests that the most economical option would be to genotype approximately 60% of the cross at a 15 cM marker density.

6.4.4 Phenotyping strategies

For many investigators, phenotyping costs, rather than genotyping costs, are high relative to the cost of raising an individual. Examples include behavioral phenotyping or microarrays. In these settings it may be more useful to perform *selective phenotyping*. Selective phenotyping involves phenotyping a subset of individuals, chosen based on their genotype or by based on another (inexpensive, but related) phenotype.

The idea of selective phenotyping based on observed genotypes is to select the most genotypically diverse subset of given size. For example, we may want to select 40 out of 200 intercross individuals for microarray phenotyping. One can select a set of dissimilar individuals using the MMA (minimum moment aberration) method. This method is implemented in the `mma` function.

We illustrate the use of the former strategy by simulating data on five chromosomes of length 100 cM. Then we use the `mma` function to select 40 individuals based on chromosome 1 genotypes (where the QTL is). We compare the LOD scores obtained by selective phenotyping to that obtained from a random subset of 40 individuals.

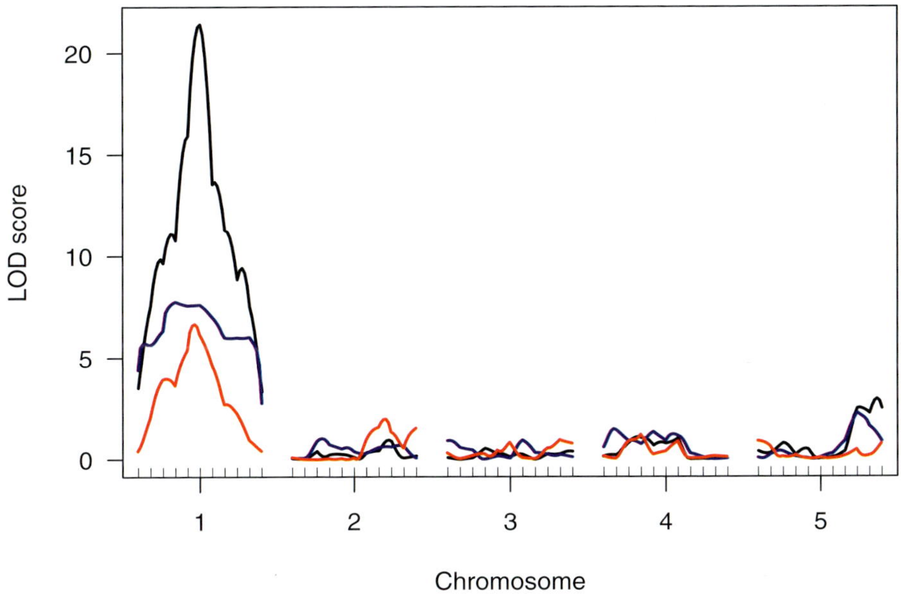

Figure 6.2. Illustration of selective phenotyping using simulated data. We compare the LOD scores obtained using three different strategies: the full cross of 200 individuals (black), 40 individuals selectively phenotyped based on chromosome 1 genotypes (blue), and a random set of 40 individuals (red).

```
> library(qtl)
> mp <- sim.map(len=rep(100,5), n.mar=11, include.x=FALSE,
+                  eq.spacing=TRUE)
> cr <- sim.cross(mp, model=c(1,50,1,0), n.ind=200, type="f2")
> idx40 <- mma(pull.geno(cr, chr=1), p=40)
> cr <- calc.genoprob(cr, step=2)
> out1 <- scanone(cr)
> out2 <- scanone(subset(cr, ind=idx40$cList))
> out3 <- scanone(subset(cr, ind=sample(1:200, 40)))
> plot(out1, out2, out3, ylab="LOD score")
```

This type of selective phenotyping is most effective when we have credible knowledge about the region where the putative QTL might be. The better the knowledge, the more effective it is. Otherwise, selectively phenotyping is about as good as phenotyping a random subset of individuals.

6.4.5 Fine mapping

After a QTL has been detected, one will usually want to narrow the location of the QTL. Planning for fine mapping involves different considerations than in the detection stage discussed earlier in the chapter. It is helpful to have access to genotyping resources to effectively narrow the location of every crossover in every individual. With dense genotyping, the average width of confidence intervals is proportional to the *inverse of the sample size*. By contrast, for most statistical problems, the lengths of confidence intervals are approximately proportional to the inverse of the *square root* of the sample size. This happy circumstance is tempered by the fact that the width of confidence intervals varies from cross to cross depending on the configuration of marker genotypes in the neighborhood of the true QTL. For this reason, the R/qtlDesign function, `ci.length`, for calculating confidence interval widths, reports the *median* confidence interval width.

The following commands give us the median width of the 95% confidence interval for QTL location for a backcross or intercross with 250 individuals, assuming dense genotyping. In practice, confidence intervals are likely to be slightly wider than these calculations indicate.

```
> ci.length(cross="bc", n=250, effect=5, p=0.95,
+            gen.var=25, env.var=64)

[1] 13.47

> ci.length(cross="f2", n=250, effect=c(5,0), p=0.95,
+            gen.var=25, env.var=64)

[1] 7.334
```

We see that with an intercross we will get confidence intervals that are expected to be about half as wide as those obtained from a backcross. As with the power for detecting QTL, the expected widths also depend on the background genetic variance, as well as the environmental variance.

6.5 Other experimental populations

Although backcrosses, intercrosses, and recombinant inbred lines are the staples of experimental geneticists, a number of other populations are in widespread use for QTL mapping. All of them are based on the same fundamental idea. We create (or assemble) a genetically diverse set of individuals, genotype and phenotype them, and look for associations between genotype and phenotype. Statistical methods to examine these associations depend on the population. In the following, we briefly describe a number of other populations that might be considered for QTL mapping.

Advanced intercross lines (AIL) are constructed by intercrossing an intercross population for multiple generations. At any given locus, they have approximately the same diversity as an intercross; however the span of linkage disequilibrium (LD) is much shorter due to the multiple generations of recombination. They are useful for fine-mapping, but require a greater breeding effort and very dense genotyping.

Heterogeneous stock (HS) are similar in spirit to advanced intercross lines, but are derived from multiple strains. Typically, an outbreeding mating scheme is used to maintain heterozygosity. The genotypic diversity of HS is greater than AIL, and the span of LD is smaller as well. In the analysis of both HS and AIL, one generally will need to take account of familial relations among individuals.

Introgression lines are a set of congenic strains spanning a locus, a chromosome, or the whole genome. They consist of a series of introgressions from a donor strain onto a recipient strain (i.e., a genomic segment from the donor strain is inserted into the recipient strain by a series of crosses). Thus, they are a collection of single-factor perturbations of a genetic system. They may be used for either genome scanning or fine mapping. Consomic strains (or chromosome substitution strains, CSS) are a special case in which the introgressed segment consists of a whole chromosome.

Recombinant congenic strains (RCS) are like introgression lines, but each strain may contain multiple small introgressed segments, rather than just one.

Inbred–outbred crosses between an inbred strain and an outbred populations may offer benefits of both linkage and association mapping. By tracking identity by descent (IBD) from the parental strains, one can perform linkage mapping (as in a backcross or intercross). By examining association between alleles at any given locus and the phenotype, one can use historical recombinations within the outbred stock to narrow down the location of a QTL.

Strain collections (association mapping) The prospect of using historical recombination for QTL detection or fine mapping also underlies association mapping using a collection of strains. The advantage of this method is that one can tap the great genetic diversity of strain collections. A disadvantage is that the complex relationships among strains may lead to spurious associations.

Natural populations also offer some of the same advantages of strain collections, especially the prospect of using historical recombination for fine mapping. However, one may have to contend with hidden population structure in the collection, and the trait of interest may not be segregating in the population.

Selection experiments apply a selective pressure on a population, and let the population evolve for a small number of generations before genotyping. By observing which genotypes survive the selective pressure, one can identify loci that underly fitness to survive selection.

An affecteds-only design may be seen as a variant of the above where one only genotypes the affected individuals (or individuals exhibiting a particular extreme of a continuous trait). By observing which genotypes are over-represented, one can map loci for the trait of interest.

The Collaborative Cross (CC) is a proposed collection of a large number of recombinant inbred lines derived from eight parental mouse strains. It seeks to establish an immortal, genetically diverse set of experimental genetic factors for mouse biology. Because it is genetically more diverse than RILs derived from two strains and has a smaller span of linkage disequilibrium, it is expected to be more efficient for mapping loci than RILs. The CC lines can be used as reference strains for complex phenotyping that generate data comparable across laboratories or years.

6.6 Estimating power and precision by simulation

The R/qtlDesign package provides quick answers to numerous design questions. However, the calculations in R/qtlDesign rely on a number of approximations that may not always be appropriate, and so the results are not always accurate.

An alternate approach to assessing the power to detect QTL and the precision of localization of QTL is to use computer simulation. Computer simulations can be cumbersome and time consuming, but they are extremely flexible and so allow one to address more complex questions.

In this section, we illustrate the use of computer simulation to assess the power to detect a QTL and the precision of localization of a QTL. We focus on the simple case of an intercross with a single QTL.

The simulation of QTL mapping data was described in Sec. 2.5. We must start with a genetic map of marker locations. It is convenient to use the `map10` object that is distributed with R/qtl. This is a genetic map modeled after the mouse genome, with evenly spaced markers at approximately a 10 cM spacing. (The marker spacing is slightly different on the different chromosomes so that the markers will be evenly spaced on each chromosome but with chromosome lengths matching those of the mouse genome.)

We first load R/qtl and get access to the `map10` object.

```
> library(qtl)
> data(map10)
```

To assess the power to map a QTL, we must first obtain a significance threshold. While one might use a permutation test for each simulated QTL mapping data set, that would be extremely time consuming. Instead, we perform some initial simulations under the null model (of no QTL) to estimate the null distribution of the genome-wide maximum LOD score.

We consider the case of an intercross with 250 individuals and assume no crossover interference and complete marker genotype data with no genotyping errors. We focus solely on the autosomes (chromosomes 1–19) and perform 10,000 simulation replicates.

The code to accomplish this is not too complicated, but requires a bit of knowledge of R.

```
> n.sim <- 10000
> res0 <- rep(NA, n.sim)
> for(i in 1:n.sim) {
+   x <- sim.cross(map10[1:19], n.ind=250, type="f2")
```

```
+   x <- calc.genoprob(x, step=1)
+   out <- scanone(x, method="hk")
+   res0[i] <- max(out[,3])
+ }
```

We first create an empty vector to contain the genome-wide maximum LOD scores. We use a `for` loop to do the 10,000 simulations. We call `sim.cross` to simulate the data (under the null hypothesis of no QTL). We use `calc.genoprob` to calculate the QTL genotype probabilities and then `scanone` to perform a genome scan. (We use Haley–Knott regression, for the sake of speed.) We pull out the maximum LOD score with `max(out[,3])`, since the LOD scores form the third column in the output.

The 95th percentile of the results serves as our estimate of the 5% genome-wide LOD threshold. We use `print` in order to simultaneously assign the 95th percentile to `thr` and print the value.

```
> print(thr <- quantile(res0, 0.95))

  95%
3.582
```

It is interesting to compare this result to that obtained with the function `thresh` in R/qtlDesign. To use `thresh`, we first need to calculate the length of the genome. This may done by a call to `summary.map`. We add up the values in the first 19 rows (corresponding to the autosomes) in the column labeled "length."

```
> print(G <- sum(summary(map10)[1:19,"length"]))

[1] 1568
```

We now use `thresh` to estimate the significance threshold. We use `d=10` (the marker spacing) and `p=0.05` (the significance level). (Note that we would need to use `library(qtlDesign)` to load the R/qtlDesign package, if it had not already been loaded.)

```
> thresh(G, "f2", d=10, p=0.05)

[1] 3.424
```

This is slightly smaller than that estimated by our simulations.

We can also make a histogram of the genome-wide maximum LOD scores. See Fig. 6.3. We use the function `rug` to create, underneath the histogram, line segments at the individual data points.

```
> hist(res0, breaks=100, xlab="Genome-wide maximum LOD score")
> rug(res0)
```

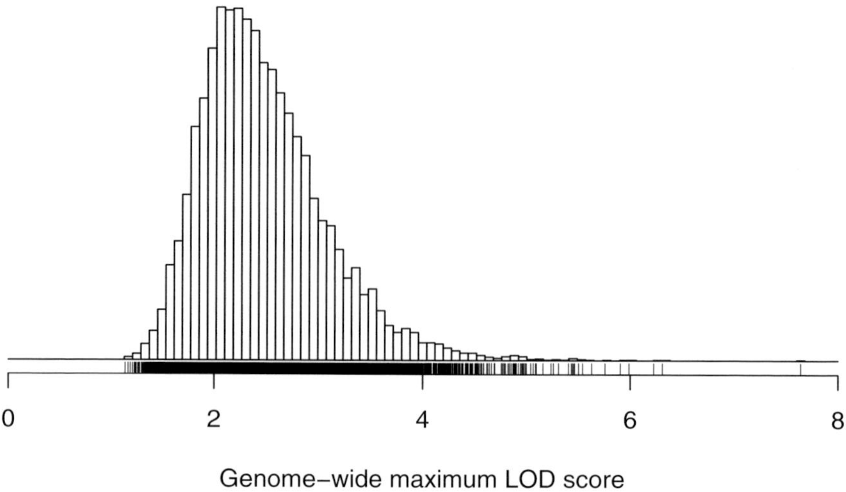

Figure 6.3. Distribution of the genome-wide maximum LOD scores under the null hypothesis of no QTL, for the case of an intercross with 250 individuals and with a genome modeled after that of the mouse and with equally spaced markers at a 10 cM spacing.

With our LOD threshold in hand, we may now turn to simulations in the presence of QTL. We will consider the simplest possible case, of a single QTL. We will assume that the alleles act additively (that is, the average phenotype for the heterozygote is halfway between the averages for the two homozygotes). We will simulate a QTL responsible for 8% of the phenotypic variance. If the average phenotypes for the two homozygotes are $-\alpha$ and α and the residual variance is 1 (as assumed in `sim.cross`), then the heritability due to the QTL is $\alpha^2/2/(\alpha^2/2+1)$. (See Table 6.1 on page 156.) Thus, we need $\alpha = \sqrt{2 \times 0.08/(1-0.08)} \approx 0.417$.

We will place the QTL at 54 cM on chromosome 1 (halfway between two markers). We may assess power and precision at the same time, and we will also study the width and coverage of the 1.5-LOD support interval (see Sec. 4.5). We need only simulate data for the chromosome containing the QTL. This will save a great deal of computation time.

The code to perform the simulations is a bit more complicated than before.

```
> alpha <- sqrt(2*0.08/(1-0.08))
> n.sim <- 10000
> loda <- est <- lo <- hi <- rep(NA, n.sim)
> for(i in 1:n.sim) {
+   x <- sim.cross(map10[1], n.ind=250, type="f2",
+                  model=c(1, 54, alpha, 0))
+   x <- calc.genoprob(x, step=1)
```

```
+    out <- scanone(x, method="hk")
+    loda[i] <- max(out[,3])
+    temp <- out[out[,3]==loda[i],2]
+    if(length(temp) > 1) temp <- sample(temp, 1)
+    est[i] <- temp
+    li <- lodint(out)
+    lo[i] <- li[1,2]
+    hi[i] <- li[nrow(li),2]
+ }
```

We first calculate the effect of the QTL that corresponds to heritability of 8%. We create empty vectors that will contain the results: the maximum LOD scores, the estimated QTL positions, and the lower and upper endpoints of the 1.5-LOD support intervals. We must be careful about the case that multiple positions give exactly the same LOD score; in such cases, we pick a random location (among those with the maximum LOD) as the estimated location of the QTL. We use the `lodint` function to calculate the 1.5-LOD support interval, and we again must be careful of the case that multiple positions share the maximum LOD score. The first and last elements in the second column of the output from `lodint` are the endpoints of the interval; while there are generally three rows in the output, there can be more.

The estimated power is the proportion of the simulation replicates with LOD score exceeding our threshold.

```
> mean(loda >= thr)
```

```
[1] 0.7383
```

This is slightly larger than the estimate provided by `powercalc` in R/qtlDesign. Note that we use `theta=0.09`, the approximate recombination fraction between two markers (from the Haldane map function, for a genetic distance of 10 cM).

```
> powercalc("f2", 250, sigma2=1, effect=c(alpha,0), thresh=thr,
+           theta=0.09)
```

```
      power percent.var.explained
[1,] 0.6855                     8
```

We might also wish to look at the distribution of the maximum LOD scores. In 10,000 simulation replicates, we obtained LOD scores as large as 14.7 (see Fig. 6.4).

```
> hist(loda, breaks=100, xlab="Maximum LOD score")
```

Turning to the precision of localization of the QTL, we first consider a histogram of the estimated QTL locations.

```
> hist(est, breaks=100, xlab="Estimated QTL location (cM)")
> rug(map10[[1]])
```

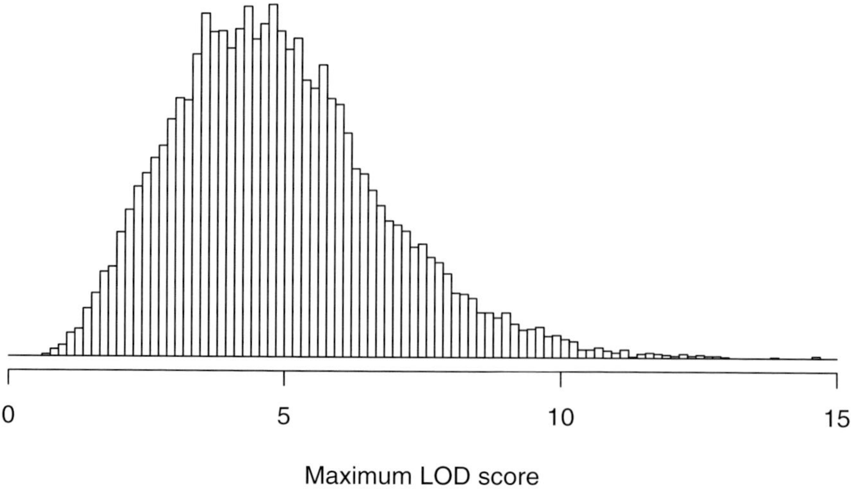

Figure 6.4. Distribution of the chromosome-wide maximum LOD scores in the presence of a single QTL responsible for 8% of the phenotypic variance, for the case of an intercross with 250 individuals and with equally spaced markers at a 10 cM spacing.

We use `rug(map10[[1]])` to place tick marks at the marker locations. (In the results, in Fig. 6.5, we used a slightly fancier method for defining the breakpoints in the histogram; see the detailed code for this and all figures in the book in the online complements at `http://www.rqtl.org/book`.)

Note the large spike in the distribution at the two markers flanking the QTL. The QTL is estimated to be at one or the other of these markers approximately 12% of the time.

The estimate of QTL location is approximately unbiased. (Recall that the QTL was located at 54 cM.)

```
> mean(est)
[1] 54.37
```

The estimated standard error of the estimated QTL location is approximately 11.6.

```
> sd(est)
[1] 11.62
```

It is interesting to consider the precision of the estimated QTL location among those cases in which there was significant evidence for a QTL (i.e., for which the LOD score exceeded our threshold, 3.58). We first create an indicator of the cases in which the LOD score exceeded the threshold, and then calculate the SD of the estimated QTL location among those cases.

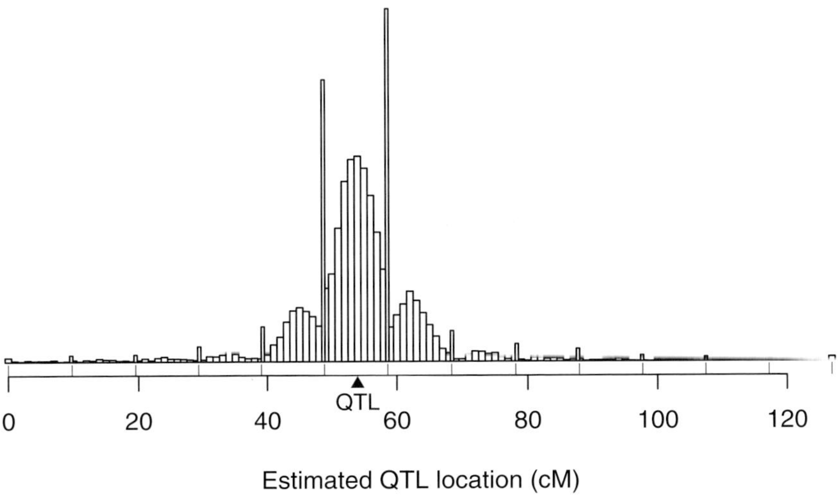

Figure 6.5. Estimated QTL location in 10,000 simulation replicates of an intercross with 250 individuals, with a QTL located at 54 cM (indicated by the blue triangle) and responsible for 8% of the phenotypic variance. The tick marks at the bottom indicate the marker locations (~10 cM spacing).

```
> sig <- (loda >= thr)
> sd(est[sig])
```

```
[1] 9.07
```

We see that the QTL location is more precisely estimated in the cases in which we had significant evidence for a QTL.

Let us turn to the 1.5-LOD support intervals. We are particularly interested in the estimated coverage: the proportion of simulation replicates in which the left endpoint was ≤ 54 and the right endpoint was ≥ 54.

```
> mean(lo <= 54 & hi >= 54)
```

```
[1] 0.9795
```

Also interesting is coverage conditional on having significant evidence for a QTL.

```
> mean(lo[sig] <= 54 & hi[sig] >= 54)
```

```
[1] 0.9743
```

In either case, coverage is a bit higher than 95%.

Finally, let us look at the distribution of the width of the 1.5-LOD support interval. We will focus on the cases in which there was significant evidence for a QTL.

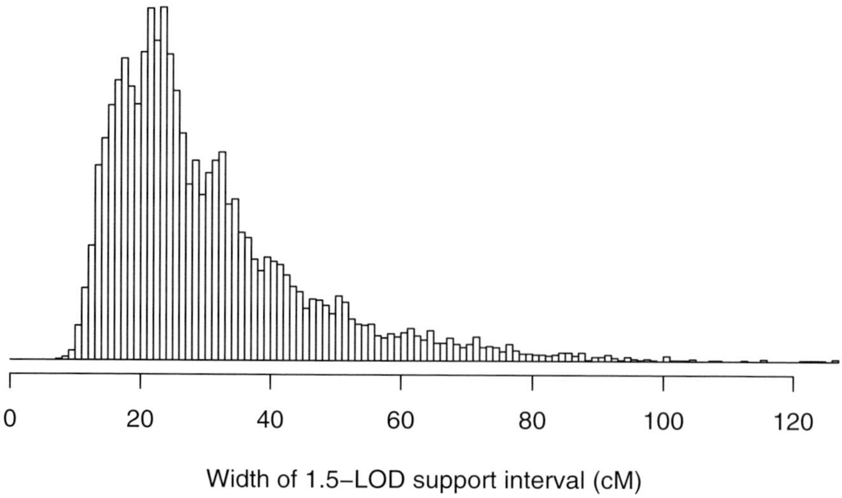

Figure 6.6. Distribution of the width of the 1.5-LOD support interval, conditional on having significant evidence for a QTL, for the case of a single QTL responsible for 8% of the phenotypic variance, in an intercross with 250 individuals and markers at a 10 cM spacing.

```
> hist(hi[sig]-lo[sig], breaks=100,
+        xlab="Width of 1.5-LOD support interval (cM)")
```

The median width of the 1.5-LOD support interval was 26 cM, but it typically varied from 14 to 63 cM long (see Fig. 6.6).

In summary, while the calculations in R/qtlDesign are often accurate, they do rely on a number of approximations. An alternative is to use computer simulations. Simulations can be time consuming and require more detailed knowledge of R, but are quite flexible and so may be used to address more complex design questions.

6.7 Summary

Sound experimental design is the bedrock of good science. A QTL experimenter should follow the general principles of good experimental design. The special structure of QTL experiments offer some additional choices, including the type of cross, the number of progeny to raise, and genotyping and phenotyping strategies. Using the package R/qtlDesign, the experimenter can make choices based on the cost structure of the experiment and the nature of the QTL effects one seeks to identify. The calculations in R/qtlDesign are efficient and convenient but rely on a number of approximations and are not always accurate. Computer simulations, while more cumbersome, can give more accurate estimates of the power to detect a QTL and the precision of localization of QTL.

6.8 Further reading

The planning of experimental crosses is discussed in Silver (1995) and Lynch and Walsh (1998). Belknap (1998) compared the sample size requirements for recombinant inbred lines (RILs) relative to intercrosses and backcrosses by quantifying the error variance. The advantages of selective genotyping were analyzed by Lander and Botstein (1989) and Darvasi and Soller (1992). Darvasi (1998) gave a comprehensive account of design options for model organisms, including selective genotyping. Selective phenotyping was proposed and analyzed by Jin *et al.* (2004). For a review reflecting recent developments, see Flint *et al.* (2005). Dupuis and Siegmund (1999) discussed genome-wide thresholds for QTL detection and confidence interval construction. Sen *et al.* (2005) framed experimental design through its information content. Also see Sen *et al.* (2009). The web-based program of Purcell *et al.* (2003) performs power calculations for complex trait analysis, although it is not designed for inbred line crosses. Sen *et al.* (2007) is the paper introducing R/qtlDesign.

7

Working with covariates

It is often of interest to take account of a covariate (such as sex or an environmental factor, such as diet) in QTL mapping. If such a covariate has a large effect on the phenotype, its inclusion in the analysis will result in reduced residual variation and so will enhance our ability to detect QTL. It is also of interest to assess possible QTL × covariate interactions. For example, does a QTL have different effects in the two sexes?

When there is evidence for a QTL with large effect, one may wish to include a nearby typed marker as a covariate in further analysis, in order to reduce the residual variation and so improve our ability to detect further QTL. This is related to the method of composite interval mapping (CIM), and is a step towards the multiple-QTL models that will be described in detail in Chap. 9.

In this chapter, we describe the use of covariates in interval mapping (i.e., in a single-QTL model), and of tests for QTL × covariate interaction. We conclude the chapter with a discussion of composite interval mapping and the use of genetic markers as covariates in interval mapping.

7.1 Additive covariates

The usual model for interval mapping is that $y_i|g_i \sim N(\mu_{g_i}, \sigma^2)$, where y_i is the phenotype and g_i is the QTL genotype for individual i. This is the sort of model that one sees in analysis of variance (ANOVA): that the different genotype groups have possibly different phenotypic means, and that the residual variation is normally distributed with constant variance.

Just as ANOVA may be viewed as a special case of linear regression, the above model may be equivalently expressed as a linear model. In a backcross, take $z_i = -1/2$ if $g_i =$ AA and $z_i = +1/2$ if $g_i =$ AB. We then have

$$y_i = \mu + \alpha z_i + \epsilon_i$$

where we assume that the ϵ_i are independent and are normally distributed with mean 0 and constant variance, σ^2.

K.W. Broman, Ś. Sen, *A Guide to QTL Mapping with R/qtl*,
Statistics for Biology and Health, DOI 10.1007/978-0-387-92125-9_7,

In an intercross, take $z_{i1} = -1, 0, +1$ according to whether g_i is AA, AB, BB, and take $z_{i2} = +1$ if $g_i =$ AB and $z_{i2} = 0$ otherwise. We then have

$$y_i = \mu + \alpha z_{i1} + \delta z_{i2} + \epsilon_i$$

The coding of the QTL genotypes is an annoyance, as is the need to treat the backcross and intercross separately, and so we will generally use the following as short hand.

$$y_i = \mu + \beta g_i + \epsilon_i$$

It is to be understood that β may have two components and g_i must be recoded.

Now consider a covariate, such as sex or weight, denoted x. (We generally code sex as $x = 0$ for females and $x = 1$ for males.) The above models could be expanded to include the covariate as follows.

$$y_i = \mu + \beta_x x_i + \beta_g g_i + \epsilon_i$$

In this case, we call x an *additive covariate*. Note that the average phenotype is linear in x, and the QTL is assumed to have constant effect, independent of x. That is, there is no QTL $\times$ covariate interaction.

For example, consider a backcross with $g = 0$ for the AA genotype and $g = 1$ for the AB genotype, and with sex as the covariate (coded as 0 for females and 1 for males). This is illustrated in Fig. 7.1A. The average phenotype for females with genotype AA is μ, and the average phenotype for females with genotype AB is $\mu + \beta_g$. The average phenotype for males with genotype AA is $\mu + \beta_x$ and the average phenotype for males with genotype AB is $\mu + \beta_x + \beta_g$. For both sexes, the effect of the QTL is β_g, but the average phenotype is allowed to be different in the two sexes. The coefficient β_x is the difference between the sexes, constant for the two QTL genotype groups. Note that we also assume that the residual variation is the same in both sexes.

In the case of a quantitative covariate, we have two regression lines that describe the average phenotype as a function of the covariate for individuals with QTL genotype AA and AB, respectively (see Fig. 7.1B). With an additive covariate, the two lines are parallel, and β_x is the slope while β_g is the distance between the two lines at any fixed value of the covariate.

Covariates can often be assumed to be independent of QTL genotype. This is true for sex (except with regard to genotypes on the X chromosome) or if the covariate is some external environmental effect (such as dietary differences imposed on the individuals). However, if a phenotype (such as body weight) is to be used as a covariate, there may be loci that affect the covariate. Thus, one should be cautious of the use of secondary phenotypes as covariates in QTL mapping. The key issue is that the meaning of the analysis changes; we are looking at the residual effect of QTL after accounting for the covariate. This may be useful for evaluating a pathway: does the QTL have a direct effect on the primary phenotype or only an indirect effect, acting through the

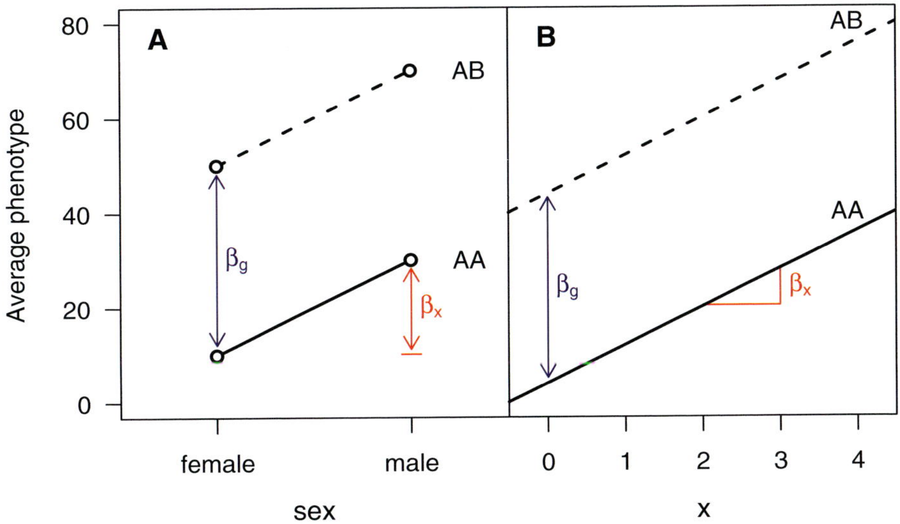

Figure 7.1. Illustration of the effects of a QTL and an additive covariate in a backcross in the case of (A) sex as the covariate and (B) a quantitative covariate.

secondary phenotype? In teasing apart such pathways, measurement error in the phenotypes can confuse things.

For a phenotype like mass of tumor, one might consider the phenotype relative to body weight: using y_i/w_i as the phenotype, where y_i is tumor mass and w_i is body weight. We would consider the model

$$(y_i/w_i) = \mu + \beta_g g_i + \epsilon_i$$

Note that this is quite different from considering body weight, w_i, as an additive covariate. In a backcross, the use of y/w as the phenotype implies the model

$$y_i = \begin{cases} \mu w_i + \epsilon_i' & \text{if } g_i = 0 \\ (\mu + \beta_g) w_i + \epsilon_i' & \text{if } g_i = 1 \end{cases}$$

where the ϵ_i' have SD increasing linearly with w_i. We thus assume that the effect of the QTL on y_i is increasing linearly with w_i. This is illustrated in Fig. 7.2.

Either of the two models (that with weight as an additive covariate, as in Fig. 7.1B, or that based on y/w, as in Fig. 7.2) may be reasonable. Most important is that one understands the assumptions underlying one's choice. A scatterplot of y versus w, with points colored by the genotype at an inferred QTL, may be useful in assessing the appropriateness of the assumptions.

We now turn to the task of obtaining LOD scores for evidence of QTL. In standard interval mapping, in the absence of a covariate, we obtain a LOD score, indicating support for the presence of a QTL, as the $\log_{10}$ likelihood ratio comparing the following two models.

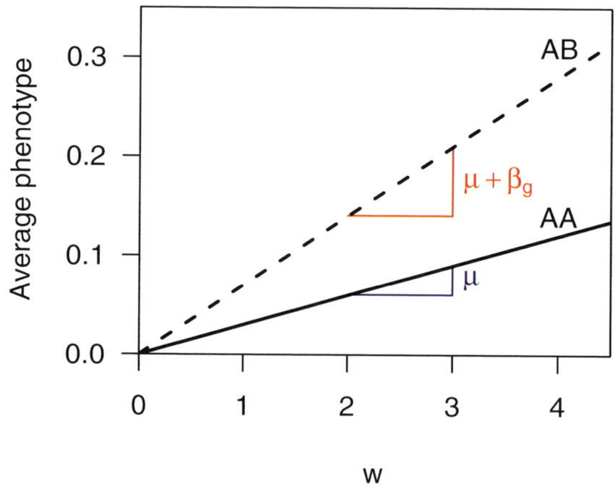

Figure 7.2. Illustration of the effect of a QTL as a function of w, in the model implied by the use of y/w as the phenotype in QTL mapping.

$$y_i = \mu + \beta_g g_i + \epsilon_i$$
$$y_i = \mu + \epsilon_i$$

If a covariate is considered, evidence for the QTL is obtained by comparing the model with both the QTL and the covariate to the model with the covariate alone.

$$y_i = \mu + \beta_x x_i + \beta_g g_i + \epsilon_i$$
$$y_i = \mu + \beta_x x_i + \epsilon_i$$

As with standard interval mapping, this analysis would be performed at a grid of putative QTL locations across the genome. The model with only the covariate must be fit once. The model containing both the covariate and the QTL is fit at each position on the grid.

Statistical significance, adjusting for the genome scan, may be established as before. We prefer the use of a permutation test, which may be performed essentially unchanged, though we must ensure that the relationship between the phenotype and the covariate is preserved, just as the association among marker genotypes should be preserved. This is accomplished by maintaining the correspondence between the covariate data and the phenotype, but shuffling the individuals' phenotype and covariate data relative to their genotype data. Consider Fig. 7.3. We maintain the structure of the genotype data matrix and the structure of the phenotype/covariate matrix, but we shuffle the rows in the genotype data relative to the rows in the phenotype/covariate data.

The fit of the model with both the QTL and the covariate requires some explanation. The QTL genotypes will generally not be known; they must be inferred from the available marker genotype data. We discussed (in Chap. 4) four

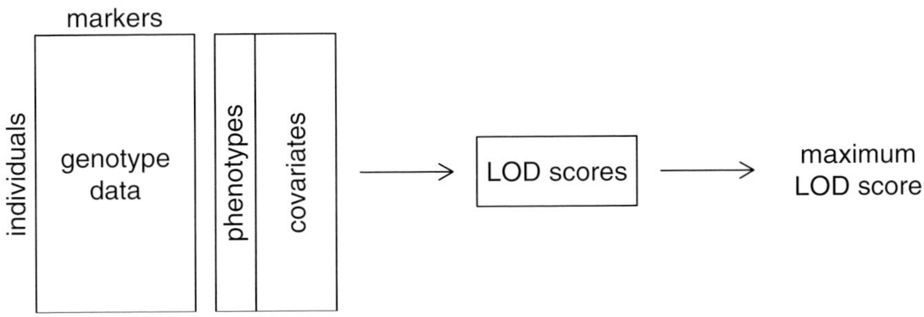

Figure 7.3. Diagram of the interval mapping process in the presence of additive covariates.

methods for fitting the single-QTL model in the absence of a covariate: standard interval mapping, Haley–Knott regression, the extended Haley–Knott method, and multiple imputation. These four methods may all be extended for the case that covariates are to be included. The Haley–Knott regression and multiple imputation methods are easily extended, as both are based on simple linear regression.

Let us briefly describe how model fit is accomplished in the extension of standard interval mapping to include a covariate. It is best to use matrices. Let x continue to denote the additive covariate, and let X be a matrix containing both the additive covariate (which will be known) and the genotypes at the putative QTL (which will not be known). Our model is $y = X\beta + \epsilon$.

If the QTL genotypes were known, we would estimate β as the solution of the normal equations, $(X'X)\hat{\beta} = X'y$, where $'$ denotes transpose. But the QTL genotype data are generally not known, and so we again use an EM algorithm to estimate β.

At iteration s of the EM algorithm, we have estimates $\hat{\beta}^{(s-1)}$ and $\hat{\sigma}^{(s-1)}$. While X and $X'X$ are not known, we may calculate their expected values (element-wise), given the available marker genotype data (denoted $\boldsymbol{M}$), the phenotypes, the covariate, and the current parameter estimates. This is the E-step.

$$Z^{(s)} = E(X|y, x, \boldsymbol{M}, \hat{\beta}^{(s-1)}, \hat{\sigma}^{(s-1)})$$
$$W^{(s)} = E(X'X|y, x, \boldsymbol{M}, \hat{\beta}^{(s-1)}, \hat{\sigma}^{(s-1)})$$

In the M-step, we obtain updated estimates of the parameters, β, as the solution of the normal equations with Z used in place of X and W used in place of $X'X$.

$$W^{(s)}\hat{\beta}^{(s)} = \left[Z^{(s)}\right]' y$$

The updated estimate of the residual SD is obtained as follows.

$$\hat{\sigma}^{(s)} = \sqrt{\left(y'y - y'Z^{(s)}\hat{\beta}^{(s)}\right)/n}$$

We have discussed the fit of a model that contains both the covariates and the QTL. An alternate approach is to first regress the phenotype on the covariates and then use the residuals in standard interval mapping. If the covariates are not correlated with the genotypes at a putative QTL, the two approaches will provide similar results, but the simultaneous fit is preferred.

One final point, before turning to an example: when should covariates be included in the analysis? If sex or an environmental covariate has an appreciable effect on the phenotype, it should definitely be included in the QTL analysis, as its inclusion will reduce the residual variation and so we will have greater power to detect QTL. If the covariate has little or no effect on the phenotype, its inclusion will not improve power, and the estimation of its effect will add noise, and so may reduce our power. If the sample size is large and only a handful of such extraneous covariates are included, there is little worry. But if the sample size is small and a large number of useless covariates are included, we may seriously erode our ability to detect QTL.

Example

As an example, we consider data on gut length in a large mouse intercross. The cross was reported in Owens *et al.* (2005), and the gut length phenotype was discussed in Broman *et al.* (2006). These data are available in the R/qtlbook package as the data set `gutlength`.

Reciprocal intercrosses were performed using the C3HeBFeJ (C3) and C57BL/6J (B6) strains, though one of the B6 parents carried the $Sox10^{Dom}$ mutation, a mouse model for Hirschsprung disease. Over 2000 intercross mice were generated, but only the 1068 mice carrying the $Sox10^{Dom}$ mutation were genotyped and are included in the data. A selective genotyping strategy was used with these data: 323 individuals with extreme aganglionosis phenotype (which is not the phenotype we are considering here) were genotyped at more than 100 markers; the remaining 745 individuals were typed at fewer than 15 markers.

First, we load the necessary packages and get access to the data.

```
> library(qtl)
> library(qtlbook)
> data(gutlength)
```

All individuals are heterozygous at *Sox10*, located on chromosome 15, which results in an unusual segregation pattern on that chromosome. For simplicity, we will omit chromosome 15 from our analysis.

```
> gutlength <- subset(gutlength, chr = -15)
```

We will consider using sex and cross as additive covariates in the QTL analysis, and so we first inspect the relationship between these covariates and the phenotype. We can use `boxplot` to create boxplots of the phenotypes, split by sex and cross. A box plot shows the median, 25th and 75th percentiles,

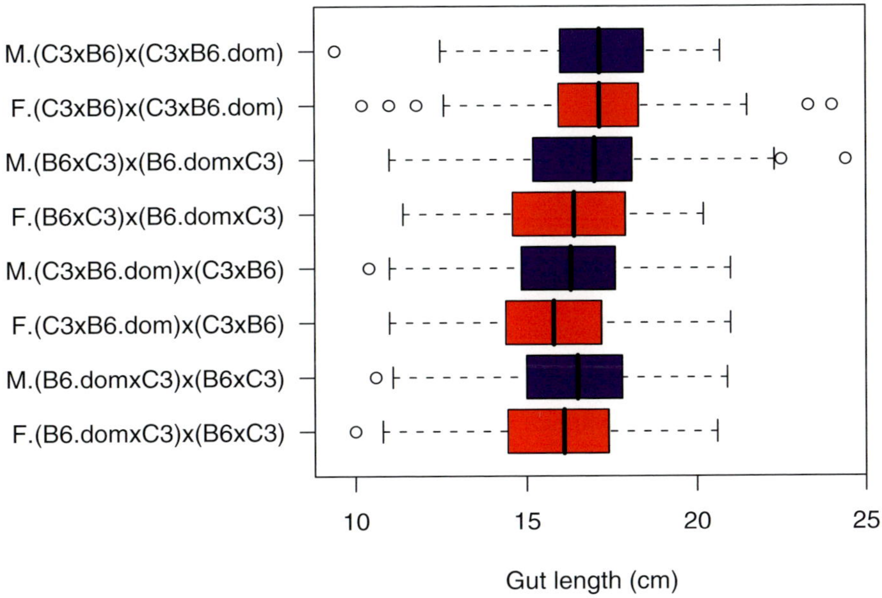

Figure 7.4. Box plots of the gut length phenotype by sex and cross in the `gutlength` data.

and the range of the phenotypes. We use the following code. The argument `col` is used to highlight the males in blue and the females in red. The results are shown in Fig. 7.4.

```
> boxplot(gutlength ~ sex*cross, data=gutlength$pheno,
+          horizontal=TRUE, xlab="Gut length (cm)",
+          col=c("red","blue"))
```

Note that the crosses are written as female × male. Thus, for example, individuals from the (B6.dom×C3)×(B6×C3) cross received the *Sox10*^{Dom} mutation from their maternal grandmother.

Males generally have somewhat longer guts than females (though the individual with the shortest gut was male), and individuals receiving the mutation from their father (the top four groups) generally had longer guts than those receiving the mutation from their mother (the bottom four groups).

We can confirm these features by performing an analysis of variance. The function `aov` is used to perform the ANOVA, and `anova` is used to create the ANOVA table.

```
> anova(aov(gutlength ~ sex*cross, data=gutlength$pheno))

Analysis of Variance Table

Response: gutlength
```

```
              Df Sum Sq Mean Sq F value      Pr(>F)
sex            1     36      36    8.12      0.0045
cross          3    167      56   12.56 0.000000045
sex:cross      3      8       3    0.63      0.5986
Residuals 1060   4711       4
```

Sex and cross both show clear effects on gut length, but there is no apparent sex × cross interaction. Note that this was done leaving the four cross groups completely unstructured, and so we cannot tell whether the cross differences are due to an effect of the parent-of-origin of the mutation or some other difference. To study the cross differences more carefully, let us us separate the four-level cross factor into two parts: whether the mutation was received from the mother or the father, and whether the F_1 individuals were created by the cross B6×C3 (which we will call the forward direction) or C3×B6.

We create indicators of whether the mutation came from the mother or father and whether the F_1 was done in the forward direction, and we paste these back into the phenotype data.

```
> cross <- as.numeric(pull.pheno(gutlength, "cross"))
> frommom <- as.numeric(cross < 3)
> forw <- as.numeric(cross == 1 | cross == 3)
> gutlength$pheno$frommom <- frommom
> gutlength$pheno$forw <- forw
```

We now perform the ANOVA again, using `frommom` and `forw` to get more detail about the relationship between cross and gut length.

```
> anova(aov(gutlength ~ sex*frommom*forw, data=gutlength$pheno))
```

```
Analysis of Variance Table

Response: gutlength
                    Df Sum Sq Mean Sq F value      Pr(>F)
sex                  1     36      36    8.12      0.0045
frommom              1    134     134   30.06 0.000000052
forw                 1      1       1    0.27      0.6062
sex:frommom          1      1       1    0.23      0.6306
sex:forw             1      2       2    0.39      0.5334
frommom:forw         1     33      33    7.49      0.0063
sex:frommom:forw     1      5       5    1.10      0.2936
Residuals         1060   4711       4
```

The parent-of-origin of the mutation has a large effect on gut length, and while the cross direction has little marginal effect, it shows a strong interaction with the parent-of-origin of the mutation (that is, the parent-of-origin effect appears to be different in the two cross directions). There is no interaction with sex.

The large effects of cross and sex suggest that they should be included as additive covariates in QTL mapping. This may be performed using the scanone function; the only tricky part is that the covariates must be strictly numeric, while we have factors. We first convert the sex factor to a quantitative covariate, coding females and males as 0 and 1, respectively.

```
> sex <- as.numeric(pull.pheno(gutlength, "sex") == "M")
```

We also wish to use cross as a covariate. This is a factor with four levels, and so we need to form a matrix with three columns. It is easiest to use the frommom and forw indicators, created above, and their product.

```
> crossX <- cbind(frommom, forw, frommom*forw)
```

Finally, we paste the two together to create a matrix with four columns.

```
> x <- cbind(sex, crossX)
```

Now we are set for interval mapping with these additive covariates. We use the scanone function, indicating the covariates using the argument addcovar. In the following, we perform the QTL analysis with and without the covariates. Recall that we must first use calc.genoprob to calculate the QTL genotype probabilities, given the available marker genotype data.

```
> gutlength <- calc.genoprob(gutlength, step=1,
+                            error.prob=0.001)
> out.0 <- scanone(gutlength)
> out.a <- scanone(gutlength, addcovar=x)
```

Note that we are using standard interval mapping; we could have also used Haley–Knott regression, the extended Haley–Knott method, or multiple imputation.

A plot of the results is obtained as follows. The results appear in Fig. 7.5. Note the use of alternate.chrid=TRUE, which allows the chromosome IDs to be more easily distinguished.

```
> plot(out.0, out.a, col=c("blue", "red"), lty=1:2,
+      ylab="LOD score", alternate.chrid=TRUE)
```

There is a clear QTL for gut length on chromosome 5, and a possible further QTL on the X chromosome, but the inclusion of the covariates in the analysis makes little difference. To better see the effect of the inclusion of covariates, we can plot the differences in the LOD scores; see Fig. 7.6. We include a horizontal dashed line at 0.

```
> plot(out.a - out.0, ylab="LOD w/ covar - LOD w/o covar",
+      ylim=c(-1, 1), alternate.chrid=TRUE)
> abline(h=0, lty=2)
```

We now should perform permutation tests, so that we may assess the statistical significance of the putative QTL. We again must treat the autosomes and X chromosome separately. Further, selective genotyping was used with

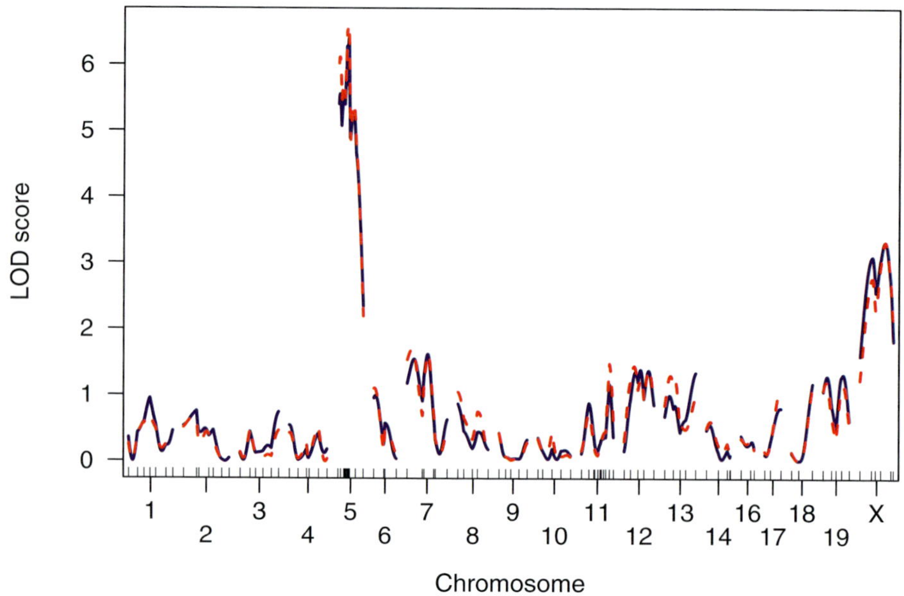

Figure 7.5. Plot of LOD scores for gut length with no covariates (in blue) and with inclusion of sex and cross as additive covariates (in red, dashed) for the `gutlength` data.

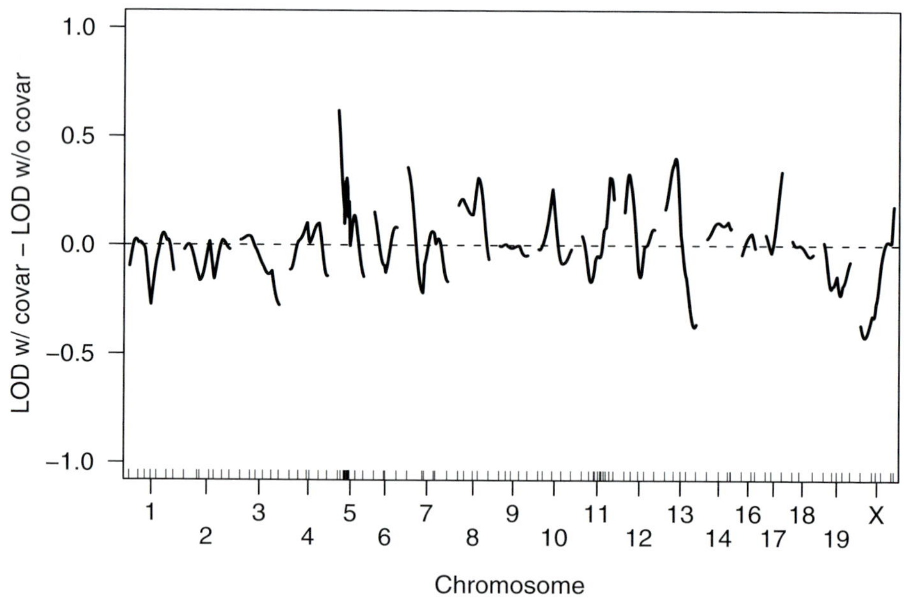

Figure 7.6. Plot of differences between the LOD scores for gut length with sex and cross as additive covariates versus without the covariates for the `gutlength` data.

these data: about 300 individuals were typed at nearly all of the 117 markers, while the remainder were genotyped at only about 10 markers. Thus it would be best to perform a stratified permutation test: permute the genotypes separately within the highly genotyped group and the group with very little genotyping. (The selective genotyping was not based on the phenotype under consideration, and so the *stratified* permutation test may not be necessary.)

We first create a numeric vector that indicates the two strata.

```
> strat <- (nmissing(gutlength) < 50)
```

Now we perform the permutation tests, using `perm.Xsp=TRUE` to indicate that we wish to treat the autosomes and X chromosome separately and `perm.strata=strat` to indicate that we wish to perform a stratified permutation test.

```
> operm.0 <- scanone(gutlength, n.perm=1000, perm.Xsp=TRUE,
+                    perm.strata=strat)
> operm.a <- scanone(gutlength, addcovar=x, n.perm=1000,
+                    perm.Xsp=TRUE, perm.strata=strat)
```

Summaries of the results are somewhat easier to study if the results with and without covariates are combined. This may be done using `c.scanone` for the `scanone` results and `cbind.scanoneperm` for the permutation results, as follows. Note the use of the `labels` argument to attach meaningful labels to the results.

```
> out.both <- c(out.0, out.a, labels=c("nocovar", "covar"))
> operm.both <- cbind(operm.0, operm.a,
+                     labels=c("nocovar", "covar"))
```

The 5% LOD thresholds from the permutation tests with and without the use of the additive covariates are then the following.

```
> summary(operm.both, 0.05)

Autosome LOD thresholds (1000 permutations)
   lod.nocovar lod.covar
5%        3.51      3.51

X chromosome LOD thresholds (16118 permutations)
   lod.nocovar lod.covar
5%        3.71      3.67
```

The following gives a summary of the main results; we display chromosomes with LOD score exceeding the 20% genome-wide significance level. We use `format="allpeaks"` to get the peaks for each of the two LOD scores. Only the chromosome 5 locus meets the 5% significance level. The X chromosome has p-value $\approx$ 10%. The use of the additive covariates had little effect on the results.

```
> summary(out.both, perms=operm.both, format="allpeaks",
+            alpha=0.2, pvalues=TRUE)

  chr pos lod.nocovar  pval pos lod.covar   pval
5   5  22        6.40 0.000  20      6.59 0.0000
X   X  57        3.33 0.105  58      3.33 0.0933
```

7.2 QTL × covariate interactions

In the previous section, we considered additive covariates, in which case the effect of the QTL was constant for all possible values of the covariate. The chief advantage of the inclusion of additive covariates in the QTL analysis is to reduce the residual variation in the case that the covariate has a strong effect on the phenotype, which will enhance our ability to detect QTL.

A covariate may interact with a QTL, meaning the effect of the QTL may vary with the covariate. If x is an *interactive covariate*, we have the following model.

$$y_i = \mu + \beta_x x_i + \beta_g g_i + \gamma x_i g_i + \epsilon_i$$

Note that this is again short-hand notation. In an intercross, there will be two degrees of freedom for g_i, and so also two degrees of freedom for $x_i g_i$.

For example, consider a backcross with $g = 0$ for the AA genotype and $g = 1$ for the AB genotype, and with sex (coded as 0 for females and 1 for males) as an interactive covariate. This is illustrated in Fig. 7.7A. Then females with genotype AA have average phenotype μ, and females with genotype AB have average phenotype $\mu + \beta_g$, and so β_g is the effect of the QTL in females. Males with genotype AA have average phenotype $\mu + \beta_x$ and males with genotype AB have average phenotype $\mu + \beta_x + \beta_g + \gamma$, and so the effect of the QTL in males is $\beta_g + \gamma$. Thus the coefficient for the QTL × sex interaction, γ, is the difference in the QTL effect between males and females. The coefficient β_x is the effect of sex in the AA genotype group. The existence of a QTL × covariate interaction may depend on the scale at which the phenotype was measured. For example, if there is no interaction on the ordinary scale, there will be an interaction if the square-root of the phenotype is considered (unless either sex or genotype has no effect).

If the interactive covariate, x, is quantitative (see Fig. 7.7B), then in the model above we assume that the effect of the QTL changes linearly in x. For each unit increase in x, the effect of the QTL changes by γ, and β_g is the effect of the QTL when $x = 0$.

Note that we always include the main effect for any interactive covariate. If the $x_i g_i$ term is included in the model but x_i is not, the coding of the QTL genotypes, g_i, becomes important. We prefer to maintain a hierarchy in such models: whenever an interaction is included, all relevant main effects are also included.

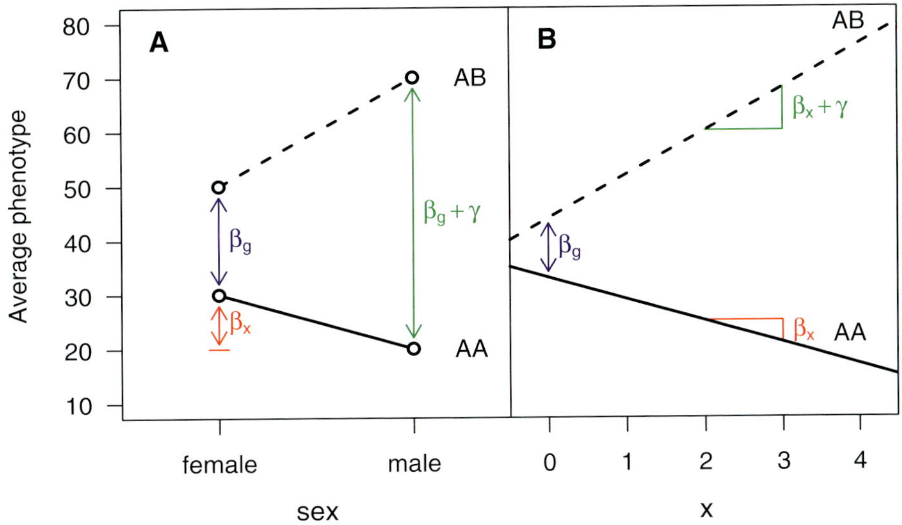

Figure 7.7. Illustration of the effects of a QTL and an interactive covariate in a backcross in the case of (A) sex as the covariate and (B) a quantitative covariate.

Regarding evidence for the presence of a QTL in the context of interactive covariates, and, perhaps most interesting, evidence for QTL × covariate interaction, there are three models that we must consider.

$$(\mathrm{H}_f) \quad y_i = \mu + \beta_x x_i + \beta_g g_i + \gamma x_i g_i + \epsilon_i$$
$$(\mathrm{H}_a) \quad y_i = \mu + \beta_x x_i + \beta_g g_i + \epsilon_i$$
$$(\mathrm{H}_0) \quad y_i = \mu + \beta_x x_i + \epsilon_i$$

In the previous section, we considered the LOD score ($\log_{10}$ likelihood ratio) comparing models H_a and H_0. This indicates evidence for a QTL, allowing for the effect of the additive covariate. We will call this LOD_a.

The LOD score comparing models H_f and H_0 indicates the combined evidence for the QTL and its possible interaction with the covariate. We will call this LOD_f.

To assess evidence for the QTL × covariate interaction, we compare models H_f and H_a. A LOD score for this comparison may be obtained as the difference between the above LOD scores, $\mathrm{LOD}_i = \mathrm{LOD}_f - \mathrm{LOD}_a$, since the log likelihood for the model H_0 cancels out.

There are several possible approaches for testing for QTL × covariate interactions. First, one may look for loci with clear marginal effects (LOD_a is large, adjusting for the genome scan) and test for the QTL × covariate interaction at those positions. Second, we may look for loci for which the combined effect of the QTL and its possible interaction with the covariate is clear (LOD_f is large, adjusting for the genome scan) and again test for the QTL × covariate interaction at those positions, with no further adjustment for multiple testing. Finally, we may look for positions for which LOD_i, the LOD score for the QTL × covariate interaction, is large, adjusting for the genome scan. We prefer the second strategy, though it may be overly conservative, and a large value of LOD_i, in isolation, may still be interesting. See the example below, as well as the case study in Chap. 11.

A permutation test may again be used to establish LOD thresholds or calculate p-values for LOD_f, adjusting for the genome scan. We use the same strategy as described in the previous section: the connection between the phenotype and the covariates is preserved, and the rows in the phenotype/covariate data are shuffled relative to the rows in the genotype data.

The same permutation test might be used to determine statistical significance for the LOD_i scores, indicating evidence for the QTL × covariate interaction. However, the permutations eliminate the effect of the QTL, and so we must assume that the distribution of LOD_i (and its association along the chromosomes) in the absence of a QTL × covariate interaction is the same, whether or not there is a QTL with marginal effect.

The model with sex as an interactive covariate is similar to splitting on sex: performing the QTL analysis separately in males and females. In both cases, the phenotype averages for each QTL genotypes is allowed to vary completely in the two sexes. The only difference is that, in the combined analysis, with sex as an interactive covariate, the residual variance is constrained to be the same in males and females, whereas when the sexes are analyzed separately, the residual variances are estimated separately in the two sexes. If the two sexes show similar residual variation, the sum of the LOD scores from the two sexes, analyzed separately, should be very similar to the LOD score from the combined analysis, LOD_f.

The combined analysis, with sex as an interactive covariate, is generally preferred, but the separate analysis of the two sexes is perhaps more easy to understand, and is probably more often used. The key advantage of the combined analysis is that it allows one to test for the QTL × sex interaction. For example, suppose that separate analyses are performed and there is significant evidence for a QTL in females but that there is little evidence for a QTL in the corresponding region in males. One should not conclude, from such a result, that there is a female-specific QTL (that is, a QTL having effect only in females). Indeed, one cannot even conclude, from these results alone, that the effect of the locus is different in males and females, as the lack of evidence of a QTL in males is not sufficient to conclude that the locus has no effect in males. *Absence of evidence is not the same as evidence of absence.*

It is difficult (and perhaps impossible) to assess whether a locus is truly female-specific, as the effect in males may be simply too small to detect. We can, however, demonstrate that the effect of the locus is different in the two sexes. To do so, we must use the combined analysis of both sexes, with sex included as an interactive covariate, and inspect LOD_i, which, in the case of appreciable QTL × sex interaction (that the effect of the QTL is different in the two sexes), should be large.

Example

We again consider the `gutlength` data. We will continue to use sex and cross as additive covariates. These were placed in the matrix `x`, which we continue to use here. Let us now consider sex as an interactive covariate (coded as 0 for females and 1 for males and placed in the object `sex`). We again use `scanone`, and indicate the additive covariates with the `addcovar` argument and the interactive covariates with the `intcovar` argument. Just as with the additive covariates, the interactive covariates must be numeric.

```
> out.i <- scanone(gutlength, addcovar=x, intcovar=sex)
```

The LOD scores in the output are the LOD_f scores described above, concerning the combined evidence for a QTL or its interaction with the covariate. That is, LOD_f being large indicates that the locus has effect in at least one of the sexes. The LOD scores for the QTL × sex interaction are obtained as $\text{LOD}_i = \text{LOD}_f - \text{LOD}_a$, where LOD_a comes from the analysis in Sec. 7.1, with only the additive covariates. If LOD_i is large, the locus is indicated to have different effects in the two sexes.

A plot of LOD_f and LOD_i may be obtained as follows. The result appears in Fig. 7.8.

```
> plot(out.i, out.i - out.a, ylab="LOD score",
+       col=c("blue", "red"), alternate.chrid=TRUE)
```

Note that $\text{LOD}_i = 0$ for the X chromosome, as sex is implicitly used as an interactive covariate for the X chromosome (see Sec. 4.4). LOD_f is large for the chromosome 5 locus, but LOD_i is small: there is strong evidence for a QTL on chromosome 5, but there is no evidence for QTL × sex interaction. Of particular interest are chromosomes 4 and 18, which show reasonably large values for LOD_f and also large values for LOD_i. At these loci, there is an indication of sex differences in the QTL effects.

To assess the statistical significance of these findings, we again perform permutation tests. LOD thresholds for LOD_f may be obtained as before, but permutation results for LOD_i require us to calculate the differences, $\text{LOD}_f - \text{LOD}_a$, for each permutation replicate. This requires some care. We must perform permutations with sex as an interactive covariate and then again with sex as solely an additive covariate, and we must ensure that the permutations are perfectly matched. This may be accomplished by setting the "seed" for the random number generator, using the function `set.seed`, prior to each set of permutations. Thus, we must rerun the permutations with sex as a solely additive covariate.

```
> set.seed(54955149)
> operm.a <- scanone(gutlength, addcovar=x, n.perm=1000,
+                    perm.Xsp=TRUE, perm.strata=strat)
> set.seed(54955149)
> operm.i <- scanone(gutlength, addcovar=x, intcovar=sex,
```

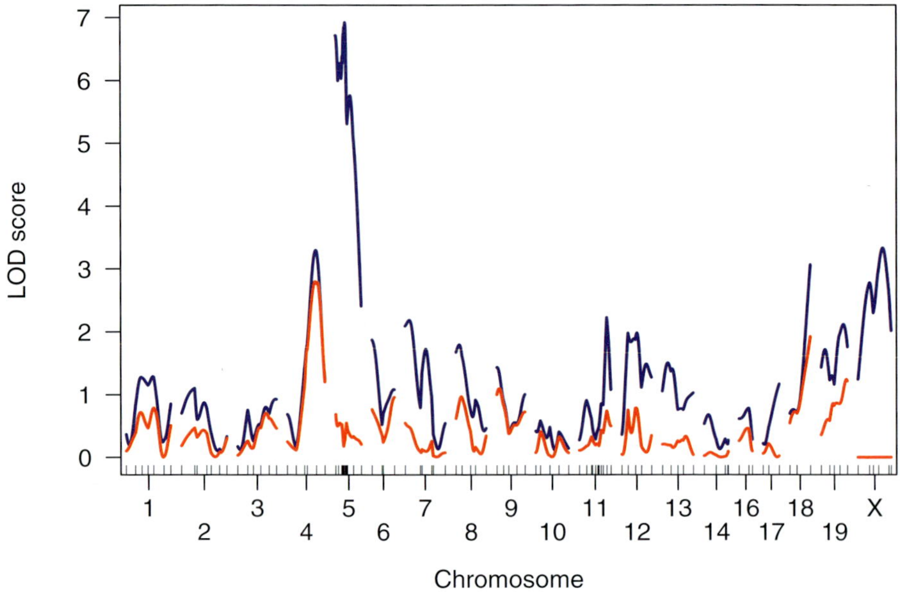

Figure 7.8. Plot of LOD_f (in blue), with cross as an additive covariate and sex as an interactive covariate, and LOD_i (in red), for the QTL × sex interaction, for the `gutlength` data.

```
+                         n.perm=1000, perm.Xsp=TRUE,
+                         perm.strata=strat)
```

It is helpful to combine LOD_f and LOD_i, and their respective permutation results, for later analysis.

```
> out.ia <- c(out.i, out.i - out.a, labels=c("f","i"))
> operm.ia <- cbind(operm.i, operm.i - operm.a,
+                        labels=c("f","i"))
```

First, let us look at the loci for which LOD_f is greater than the 20% genome-wide threshold.

```
> summary(out.ia, perms=operm.ia, alpha=0.2, pvalues=TRUE)

          chr pos lod.f   pval lod.i  pval
c5.loc22    5  22  6.92 0.0000 0.365 0.832
cX.loc58    X  58  3.33 0.0933 0.000 1.000
```

Again, there is strong evidence for a QTL on chromosome 5, but no evidence for a QTL × sex interaction ($\text{LOD}_i \approx 0.4$).

Now, let us look strictly at LOD_i.

```
> summary(out.ia, perms=operm.ia, alpha=0.2, pvalues=TRUE,
+          lodcolumn=2)
```

```
            chr  pos lod.f  pval lod.i    pval
c4.loc65      4 65.0  3.29 0.389  2.80 0.00743
18_72382360  18 48.2  3.07 0.523  1.92 0.06149
```

If we consider the interaction LOD score in isolation, there is good evidence for QTL × sex interactions on chromosomes 4 and 18, but the overall LOD scores, LOD_f, which indicate evidence that the loci have effect in at least one sex, are not large. We are inclined to restrict attention to only those loci for which LOD_f or LOD_a is large, and so view the evidence for QTL × sex interaction on chromosomes 4 and 18 as chance variation, but this may be overly conservative.

It is interesting to compare these results to those obtained by separate analysis of the two sexes. In the separate analyses, we will continue to use the cross as an additive covariate, but sex should not be included, as it will be constant in each sex. We constructed a matrix for the cross factor, `crossX`, in the previous section. We can use `subset` to pull out the relevant individuals from the cross; we also need to subset the covariate matrix.

```
> out.m <- scanone(subset(gutlength, ind = sex==1),
+                  addcovar=crossX[sex==1,])
> out.f <- scanone(subset(gutlength, ind = sex==0),
+                  addcovar=crossX[sex==0,])
```

We may plot the results as follows; see Fig. 7.9.

```
> plot(out.m, out.f, col=c("blue", "red"), ylab="LOD score",
+      alternate.chrid=TRUE)
```

Note particularly the results for chromosome 18, which shows a peak in females but not males. While this is suggestive of a female-specific QTL, we cannot conclude, on the basis of these results alone, that the effect of the locus is different in the two sexes. An assessment of the evidence for a QTL × sex interaction requires the detailed analysis of the joint data, described above.

The sum of the LOD scores for males and females will be similar to LOD_f obtained from the joint analysis, though it is not exactly the same, due to the difference in the treatment of the residual variance. A plot of the differences may be obtained as follows and appears in Fig. 7.10. (The functions `+.scanone` and `-.scanone` are used to add and subtract the LOD scores.)

```
> plot(out.m + out.f - out.i, ylim=c(-0.5,0.5),
+      ylab="LOD(males) + LOD(females) - LODf",
+      alternate.chrid=TRUE)
> abline(h=0, lty=2)
```

To establish the statistical significance of the sex-specific results, we perform permutation tests within each of the males and females. We again need to treat the X chromosome separately and use a stratified permutation test, with individuals stratified by the amount of genotyping that was performed.

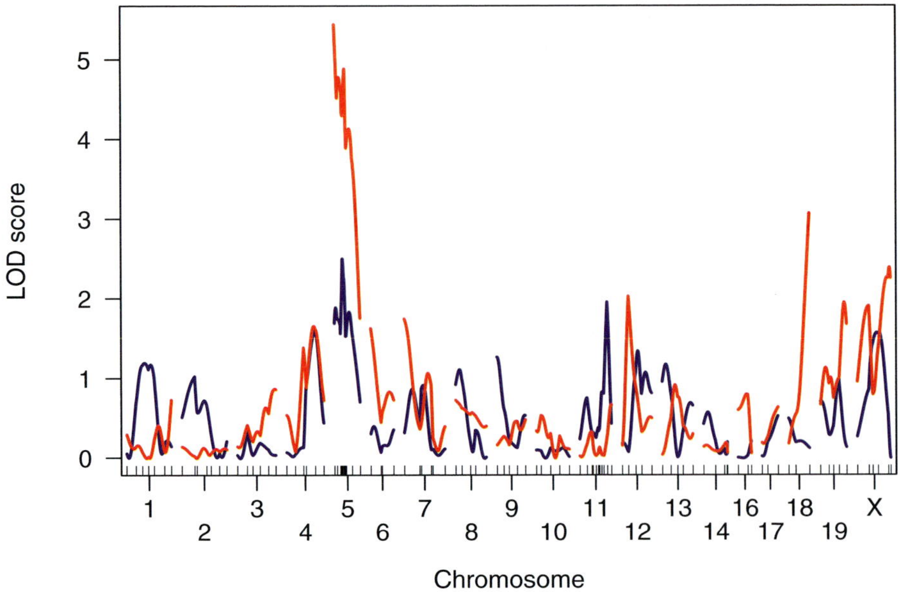

Figure 7.9. Plot of LOD scores for males (in blue) and females (in red) for the `gutlength` data.

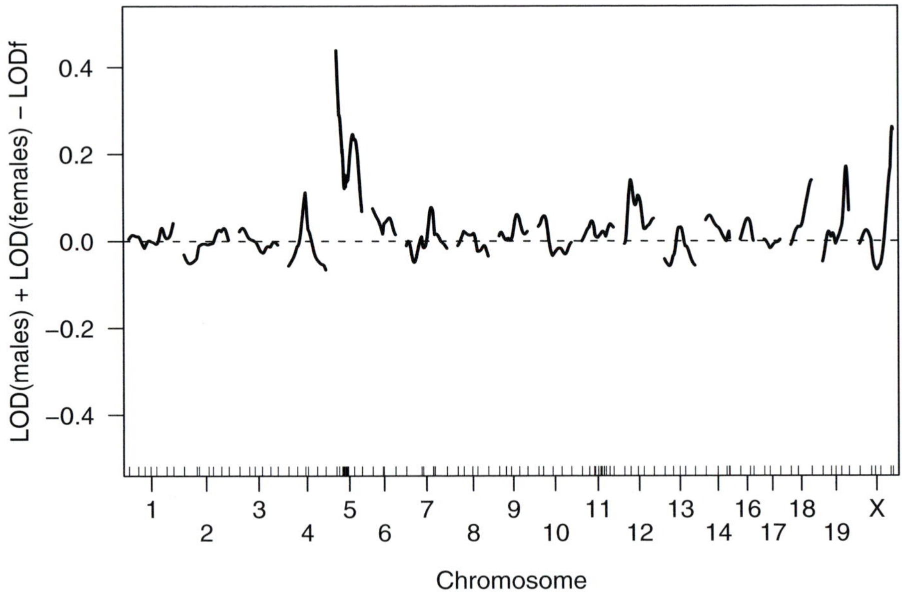

Figure 7.10. Plot of the difference between the sum of the LOD scores from the separate analyses of males and females and LOD_f, from their joint analysis, for the `gutlength` data.

```
> operm.m <- scanone(subset(gutlength, ind = sex==1),
+                    addcovar=crossX[sex==1,], n.perm=1000,
+                    perm.strata=strat[sex==1], perm.Xsp=TRUE)
> operm.f <- scanone(subset(gutlength, ind = sex==0),
+                    addcovar=crossX[sex==0,], n.perm=1000,
+                    perm.strata=strat[sex==0], perm.Xsp=TRUE)
```

Again, to simplify the later summaries, we combine the male and female results.

```
> out.sexsp <- c(out.m, out.f, labels=c("male","female"))
> operm.sexsp <- cbind(operm.m, operm.f,
+                      labels=c("male","female"))
```

The 5% LOD thresholds are the following. Note that the X-chromosome-specific threshold in the males is much lower than the others, as the linkage test concerns just one degree of freedom, rather than two.

```
> summary(operm.sexsp, 0.05)

Autosome LOD thresholds (1000 permutations)
   lod.male lod.female
5%     3.41       3.49

X chromosome LOD thresholds (16118 permutations)
   lod.male lod.female
5%     2.62       3.27
```

The sex-specific results indicate strong evidence for the chromosome 5 locus, and weak evidence for a locus on chromosome 18 in females.

```
> summary(out.sexsp, perms=operm.sexsp, alpha=0.2,
+         pvalues=TRUE, format="allpeaks")

   chr  pos lod.male  pval  pos lod.female  pval
5    5 18.7    2.508 0.302  0.0       5.45 0.000
18  18  0.0    0.499 1.000 48.2       3.08 0.120
```

Note that the chromosome 5 locus is significant in females but not in males; one might conclude that that its effect is different in the two sexes, but our previous results on the QTL × sex interaction indicated no evidence for a sex-difference in the QTL effect. The chromosome 18 locus is now more interesting; it may have effect in females, and our previous results indicated a potential QTL × sex interaction.

Let us complete this study of the `gutlength` data with plots of the estimated effects of the putative QTL as a function of sex, using the function `effectplot`. We first use `sim.geno` to impute the missing data. The averages and SEs in the plots are based on these multiple imputations. Note the use of constructions like `"5@22"` to indicate a "pseudomarker" position (on the grid on which interval mapping was performed) on chromosome 5 at 22 cM. Also, while `effectplot` is generally used to plot the phenotype averages as a function of genotype at putative QTL, we can also split individuals by a covariate. Here `mname1` is the name of the covariate, and since it matches one of the phenotypes in the `gutlength` cross, the `"sex"` phenotype is used.

```
> gutlength <- sim.geno(gutlength, n.draws=128, step=1,
+                        error.prob=0.001)
> par(mfrow=c(2,2))
> effectplot(gutlength, mname1="sex", ylim=c(15.1, 17.2),
+            mname2="4@65", main="Chromosome 4",
+            add.legend=FALSE)
> effectplot(gutlength, mname1="sex", ylim=c(15.1, 17.2),
+            mname2="5@22", main="Chromosome 5",
+            add.legend=FALSE)
> effectplot(gutlength, mname1="sex", ylim=c(15.1, 17.2),
+            mname2="18@48.2", main="Chromosome 18",
+            add.legend=FALSE)
> effectplot(gutlength, mname1="sex", ylim=c(15.1, 17.2),
+            mname2="X@58", main="X chromosome",
+            add.legend=FALSE)
```

The results are in Fig. 7.11. The chromosome 5 locus shows the strongest effect, and its pattern of effect is very similar in the two sexes. The chromosome 18 locus shows little effect in males but some effect in females. The chromosome 4 locus shows some effect in both males and females, but the pattern is different in the two sexes.

7.3 Covariates with non-normal phenotypes

Above, we assumed that the residual phenotypic variation followed a normal distribution. In order to include covariates in the analysis of phenotypes for which the normality assumption for the residual variation is inadequate, one must use an extension of linear regression. There is no obvious extension of the rank-based, nonparametric method to allow covariates. Robust versions of linear regression are available, but we are not aware of any application of such methods to QTL mapping, though one may find that the extended Haley–Knott method (see Sec. 4.2.3) is sufficiently robust.

For binary phenotypes, one may use an extension of logistic regression. Let $\pi_i = \Pr(y_i = 1 | g_i, x_i)$, where y_i is the phenotype (taking values 0 and 1), x_i is an additive covariate, and g_i is the QTL genotype. While one might consider the linear model

$$\pi_i = \mu + \beta_x x_i + \beta_g g_i,$$

this model is unsatisfactory, because the left-hand side takes values between 0 and 1 which the right-hand side need not be between 0 and 1. The use of the

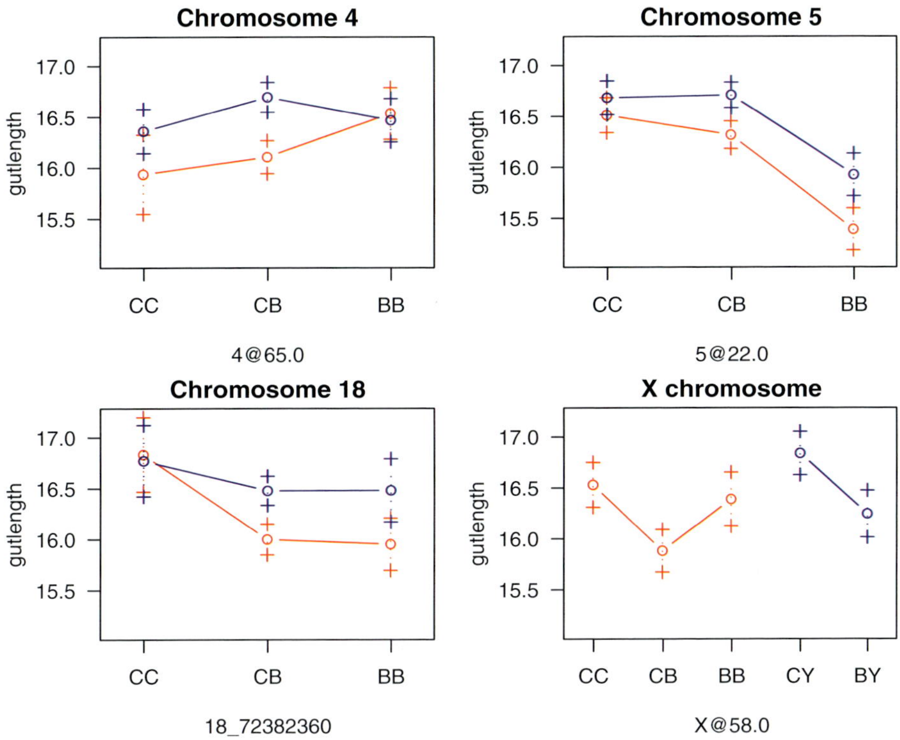

Figure 7.11. Plot of the estimated phenotype averages ± 1 SE as a function of sex (with males in blue and females in red) and genotype, at the positions nearest the peak LOD score on four selected chromosomes, for the `gutlength` data. C and B correspond to the C3HeBFeJ and C57BL/6J alleles, respectively.

logit link function, $\ln[\pi/(1-\pi)]$, fixes this problem. We then have the model

$$\ln[\pi_i/(1-\pi_i)] = \mu + \beta_x x_i + \beta_g g_i$$

Other link functions that transform the probability, π_i, to a scale without bounds may be used. A common example is the *probit* link, $\Phi^{-1}(\pi)$, where Φ is the cumulative distribution function of the standard normal distribution.

The fit of this model is again made more complicated due to the fact that the QTL genotypes, g_i, are generally not observed. However, an EM algorithm to obtain the MLEs of the parameters is relatively straightforward and has been implemented in R/qtl. The use of interactive covariates, and tests of QTL × covariate interactions, are conceptually the same as for the normal model (Sec. 7.2).

The two-part model, described in Sec. 5.3, appropriate for the case of a spike in the phenotype distribution (e.g., mass of gallstones with some individuals exhibiting no gallstones), may also be extended to include covariates, though it is simplest, and little information is likely to be lost, to perform the separate analyses of the binary phenotype (e.g., presence or absence of gallstones) and the conditional quantitative phenotype (e.g., mass of gallstones in those individuals exhibiting gallstones), with each analysis including the relevant covariates.

Example

To illustrate the use of covariates in the analysis of a binary trait, we consider data on neurofibromatosis type 1 (Reilly *et al.*, 2006), included in the R/qtlbook package as the `nf1` data set. The goal was to identify modifiers of the *NPcis* mutation. There are a total of 254 individuals from the backcrosses (C57BL/6J $\times$ A/J) $\times$ C57BL/6J and C57BL/6J $\times$ (C57BL/6J $\times$ A/J), with individuals receiving the *NPcis* mutation from either their mother or father. The `affected` phenotype indicates whether the mice were affected (`1`) or unaffected (`0`) with neurofibromatosis type 1. Mice were genotyped at 106 genetic markers covering the autosomes.

We first need to load the data. It is contained in the `qtlbook` package, and so if that package had not already been loaded, we would first need to type `library(qtlbook)`. The `nf1` data contains one marker with completely missing genotype data; we remove this marker using `drop.nullmarkers`.

```
> data(nf1)
> nf1 <- drop.nullmarkers(nf1)
```

Note the proportion of affected individuals, and that this differs according to the parent-of-origin of the *NPcis* mutation. The function `tapply` is used to get the proportion affected within the two strata defined by the `from.mom` "phenotype."

```
> mean(pull.pheno(nf1,"affected"))

[1] 0.5197

> tapply(pull.pheno(nf1,"affected"),
+          pull.pheno(nf1,"from.mom"), mean)

     0      1
0.6181 0.3909
```

Application of a χ^2 test with the function `chisq.test` demonstrates that this difference is real. (One might also perform Fisher's exact test, using the function `fisher.test`.)

```
> chisq.test(pull.pheno(nf1,"affected"),
+              pull.pheno(nf1,"from.mom"))
```

```
        Pearson's Chi-squared test with Yates' continuity
        correction

data:  pull.pheno(nf1, "affected") and pull.pheno(nf1, "from.mom")
X-squared = 12.00, df = 1, p-value = 0.000533
```

We perform genome scans for the binary phenotype, using parent-of-origin of the *NPcis* mutation first as an additive covariate and then as an interactive covariate; note that R/qtl uses the logit link function. We first run `calc.genoprob` to get the conditional QTL genotype probabilities.

```
> nf1 <- calc.genoprob(nf1, step=1, error.prob=0.001)
> from.mom <- pull.pheno(nf1,"from.mom")
> out.a <- scanone(nf1, model="binary", addcovar=from.mom)
> out.i <- scanone(nf1, model="binary", addcovar=from.mom,
+                  intcovar=from.mom)
```

We further perform permutation tests. As discussed in Sec. 7.2, we need to use `set.seed` to ensure that they are matched.

```
> set.seed(1310709)
> operm.a <- scanone(nf1, model="binary", addcovar=from.mom,
+                    n.perm=1000)
> set.seed(1310709)
> operm.i <- scanone(nf1, model="binary", addcovar=from.mom,
+                    intcovar=from.mom, n.perm=1000)
```

We again combine the results, including the interaction LOD scores, $\text{LOD}_i = \text{LOD}_f - \text{LOD}_a$.

```
> out.all <- c(out.i, out.a, out.i-out.a, labels=c("f","a","i"))
> operm.all <- cbind(operm.i, operm.a, operm.i - operm.a,
+                    labels=c("f","a","i"))
```

We may plot the three LOD scores (LOD_f, LOD_a and LOD_i) as follows; see Fig. 7.12.

```
> plot(out.all, lod=1:3, ylab="LOD score")
```

Only chromosomes 15 and 19 show large values of LOD_f. The chromosome 19 locus shows a clear QTL $\times$ covariate interaction, suggesting that the effect of the locus is modified by the parent-of-origin of the *NPcis* mutation.

```
> summary(out.all, perms=operm.all, alpha=0.2, pvalues=TRUE)

          chr pos lod.f  pval lod.a  pval lod.i  pval
D15Mit111  15  13  2.92 0.110  2.22 0.121 0.703 0.356
D19Mit59   19   0  3.02 0.088  1.40 0.627 1.622 0.043
```

The interaction LOD score for the chromosome 15 locus is not large, but it might be best to test for QTL $\times$ covariate interactions pointwise, rather

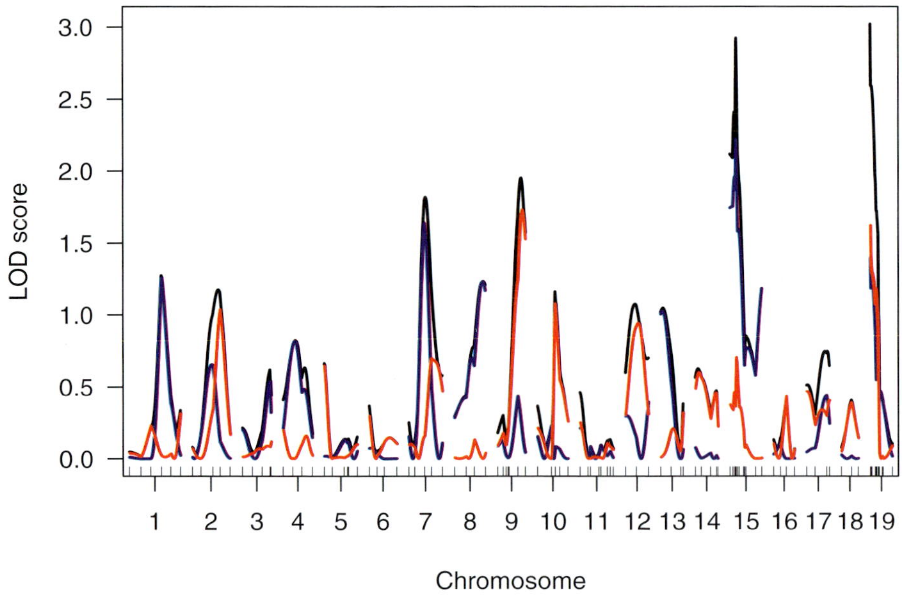

Figure 7.12. Plot of LOD_f (in black), LOD_a (in blue), and LOD_i (in red), for the `nf1` data, with parent-of-origin of the *NPcis* mutation considered as a covariate.

than adjust for the genome scan. That is, we might compare the $\text{LOD}_i = 0.7$ result to its pointwise null distribution, rather than to the distribution of the genome-wide maximum LOD_i under the global null hypothesis. At a specific point, $\text{LOD}_i \times 2\ln(10)$ follows approximately a χ^2 distribution with 1 degree of freedom, under the null hypothesis of no QTL $\times$ covariate interaction, and so the pointwise p-value is the following.

```
> pchisq(0.703 * 2 * log(10), 1, lower=FALSE)
[1] 0.07197
```

It is again of interest to split on the covariate: to perform a genome scan separately in the individuals who received the *NPcis* mutation from their mother and in those who received it from their father.

```
> out.frommom <- scanone(subset(nf1, ind=(from.mom==1)),
+                         model="binary")
> out.fromdad <- scanone(subset(nf1, ind=(from.mom==0)),
+                         model="binary")
```

We again perform permutation tests separately within the two groups.

```
> operm.frommom <- scanone(subset(nf1, ind=(from.mom==1)),
+                         model="binary", n.perm=1000)
> operm.fromdad <- scanone(subset(nf1, ind=(from.mom==0)),
+                         model="binary", n.perm=1000)
```

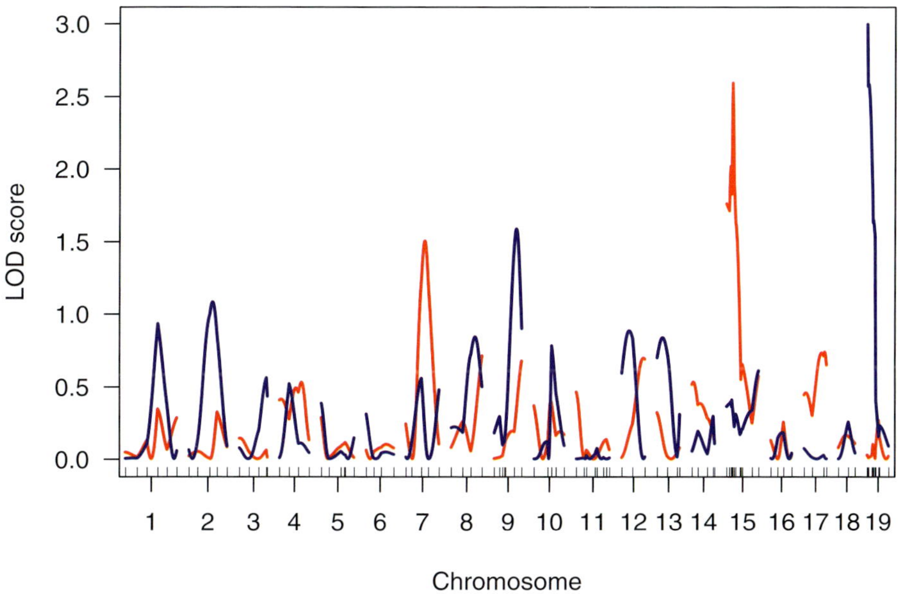

Figure 7.13. LOD scores for the analysis of the `nf1` data, split by parent-of-origin of the *NPcis* mutation, with results for individuals receiving the mutation from their mother and father in red and blue, respectively.

And again we combine the results.

```
> out.bypoo <- c(out.frommom, out.fromdad,
+                labels=c("mom","dad"))
> operm.bypoo <- cbind(operm.frommom, operm.fromdad,
+                      labels=c("mom","dad"))
```

We may plot the results as follows; see Fig. 7.13.

```
> plot(out.bypoo, lod=1:2, col=c("red", "blue"),
+      ylab="LOD score")
```

The chromosome 19 locus has a significant effect in the group receiving the *NPcis* mutation from their father but not in the others; the chromosome 15 locus shows the opposite effect: a large LOD score for the group receiving the mutation from their mother but not for the others.

```
> summary(out.bypoo, perms=operm.bypoo, alpha=0.2, pvalues=TRUE,
+         format="allpeaks")

   chr pos lod.mom  pval pos lod.dad pval
15  15  13   2.590 0.056  66   0.609 1.00
19  19  24   0.265 1.000   0   2.995 0.02
```

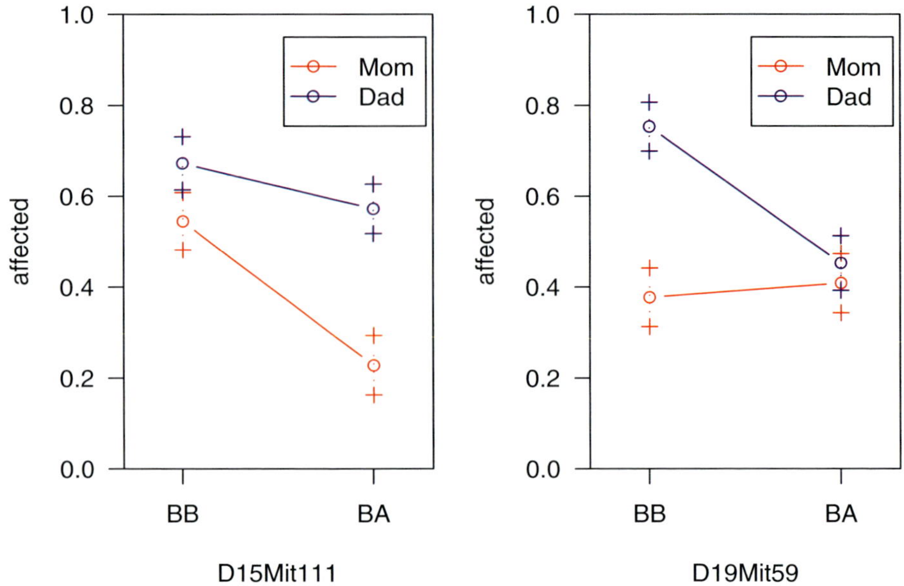

Figure 7.14. Proportion of affecteds as a function of genotype and the parent-of-origin of the *NPcis* mutation, for the `nf1` data.

Finally, let us look at the effects of the inferred QTL. We use `effectplot`; it assumes a continuous outcome, but still gives reasonable results with our binary phenotype. We use `sim.geno` to impute any missing genotype data, and constructions like `"15@13"` to indicate a "pseudomarker" position (on the grid on which interval mapping was performed) on chromosome 15 at 13 cM. Also, while `effectplot` is generally used to plot the phenotype averages as a function of genotype at putative QTL, we can also split individuals by a covariate. Here `mname1` is the name of the covariate, `mark1` is the actual covariate data, and `geno1` gives labels to the levels of the covariate. We use `1-frommom` so that red and blue are attached to "mom" and "dad," respectively.

```
> nf1 <- sim.geno(nf1, n.draws=128, step=1, error.prob=0.001)
> par(mfrow=c(1,2))
> effectplot(nf1, mname1="NPcis", mark1=1-from.mom,
+            geno1=c("Mom", "Dad"), mname2="15@13",
+            ylim=c(0,1))
> effectplot(nf1, mname1="NPcis", mark1=1-from.mom,
+            geno1=c("Mom", "Dad"), mname2="19@0", ylim=c(0,1))
```

The results, in Fig. 7.14, show that the chromosome 19 locus (right panel) has a large effect in the individuals receiving the *NPcis* mutation from their father, with the heterozygotes having a lower chance of being affected, but little effect in the individuals receiving the mutation from their mother. The chromosome 15 locus (left panel of Fig. 7.14) has greater effect in the individuals receiving the mutation from their mother. The chromosome 15 locus appears to have little effect in the individuals receiving the *NPcis* mutation from their father.

7.4 Composite interval mapping

We have so far only discussed single-QTL models: We imagine the presence of a single QTL, and consider each position in the genome, one at a time, as the location of that QTL. Such analysis works well for the identification of loci with clear marginal effect.

In the next two chapters, we will discuss the fit and exploration of multiple-QTL models. The advantages of the simultaneous consideration of multiple QTL are to (a) reduce residual variation and so better detect loci of more modest effect, (b) separate linked QTL, and (c) identify interactions among QTL.

As an initial exploratory step, one may consider a marker near a putative QTL as a covariate in the search for further QTL. The use of markers as covariates fits well into the present chapter, on the use of covariates in QTL mapping, and so we discuss it here.

The chief value of the use of a marker as a covariate is to reduce residual variation and so clarify evidence for further QTL. The marker serves as a proxy for nearby QTL; its inclusion in the model will remove much of the effects of such QTL from what otherwise would appear as residual variation. If there is a large-effect QTL near the marker, the use of the marker as a covariate should increase our power to detect QTL on other chromosomes. One may also include markers as interactive covariates, to identify loci that exhibit an interaction with a locus near the marker.

An extreme case of the use of markers as covariates is the composite interval mapping (CIM) strategy. While the term "composite interval mapping" has been applied to a number of related methods, it is perhaps most commonly applied to a particular strategy implemented in the QTL Cartographer software, which we will describe here and illustrate later.

One first selects a set of markers to serve as covariates. For example, one may use *forward selection* at the markers to identify a set of predetermined size—say seven markers. In forward selection, one considers each marker, one at a time, and chooses the marker, call it $m_{(1)}$, that best predicts the phenotype (that is, gives the smallest residual sum of squares). One then considers all models with $m_{(1)}$ plus one other marker, and finds a second marker, call it $m_{(2)}$, that, when considered with marker $m_{(1)}$, gives the greatest decrease in the residual sum of squares. The process is continued, creating a sequence of nested models of increasing size, to the predetermined number of markers.

Once the set of markers has been chosen, one performs interval mapping (that is, a single-QTL genome scan), with these markers as covariates: One calculates a LOD score comparing the model with the putative QTL in the presence of the covariates to the model with just the covariates. There is one wrinkle: if any of the marker covariates are within some fixed, predetermined distance, d, of the position under test, one compares the model with the QTL and any selected markers that are more than d away from test position to the model with only those selected markers that are more than d away from the test position.

Say $\mathcal{S}$ is the chosen set of marker covariates, and z is the putative QTL position. Then one considers as marker covariates $\mathcal{S}' = \mathcal{S} \setminus (z - d, z + d)$, and then compares the model $\mathcal{S}' \cup \{z\}$ to the model $\mathcal{S}'$. We have abused set notation a bit here, but we hope our meaning is understood.

While the use of markers near putative QTL as covariates in the search for additional loci is a clearly useful exploratory strategy, we recommend against the general use of composite interval mapping. CIM attempts to turn the multidimensional search for QTL into a single-dimensional search by first identifying a subset of covariates. The choice of covariates is critical: if too many or too few markers are chosen, there will be a loss of power to detect QTL. Furthermore, the subsequent scan fails to account for the uncertainty in the choice of relevant marker covariates and can give an overly optimistic view of the precision of localization of QTL.

The ideas underlying composite interval mapping have been influential in the development of more modern approaches for multiple QTL mapping. We prefer to discard composite interval mapping in favor of its more refined descendants, which we will describe in Chap. 9.

Example

To illustrate the use of marker covariates in QTL mapping, we return to the `hyper` data. Let us reload the data, rerun `calc.genoprob`, and rerun the initial genome scan.

```
> data(hyper)
> hyper <- calc.genoprob(hyper, step=1, error.prob=0.001)
> out <- scanone(hyper)
```

We had seen strong evidence for a QTL on chromosome 4. Let us first perform a genome scan, with a marker near the inferred QTL included as an additive covariate. The peak LOD score occurred at 29.5 cM on chromosome 4. We identify the nearest typed marker, pull out its genotype data, and ensure that there is no missing data.

```
> mar <- find.marker(hyper, 4, 29.5)
> g <- pull.geno(hyper)[,mar]
> sum(is.na(g))

[1] 229
```

We see that there is a lot of missing data at that marker. We could use a nearby, fully typed marker, or we could impute the missing data at the marker, and use the imputed genotypes as if they were observed. Let us do the latter. We can use the function `fill.geno` to fill in the genotypes with a single imputation.

```
> g <- pull.geno(fill.geno(hyper))[,mar]
> sum(is.na(g))

[1] 0
```

Because this is a backcross, we can use the `g` vector directly as the covariate. Had this been an intercross, we would need to first create a two-column numeric matrix encoding the genotype data. This may be done in a variety of ways; the following is convenient: one column indicating one of the homozygotes and another indicating the heterozygotes. Again, *we should not do this here*, as we are dealing with a backcross rather than an intercross; we display the following code just as an illustration.

```
> g <- cbind(as.numeric(g==1), as.numeric(g==2))
```

We now perform a genome scan with this marker as an additive covariate.

```
> out.ag <- scanone(hyper, addcovar=g)
```

We may plot the results, along with those from the genome scan without the marker covariate, as follows; see Fig. 7.15.

```
> plot(out, out.ag, col=c("blue", "red"), ylab="LOD score")
```

The evidence for a QTL on chromosome 1 is greatly increased after accounting for the chromosome 4 locus, and, while the LOD curve continues to exhibit two peaks, the distal peak is now strongly favored. Of course, the LOD scores on chromosome 4 shrink to near 0. The shape of the LOD curve for chromosome 6 shows an interesting change, with a second peak near the telomere. There are no other important differences.

One might wish to run a permutation test for the analysis that includes the chromosome 4 marker as a covariate. In such a permutation test, it would be best to omit chromosome 4 from the analysis. The selective genotyping in these data again requires that we use a stratified permutation. While little is likely to be gained from this effort, as we already had strong evidence for a locus on chromosome 1, we will carry out such a permutation test, for completeness.

```
> strat <- (ntyped(hyper) > 100)
> operm.ag <- scanone(hyper, addcovar=g, chr=-4,
+                     perm.strata=strat, n.perm=1000)
```

The 20% and 5% LOD thresholds are as follows. These are not much changed from those obtained for the analysis without any covariates (Sec. 4.3, page 107).

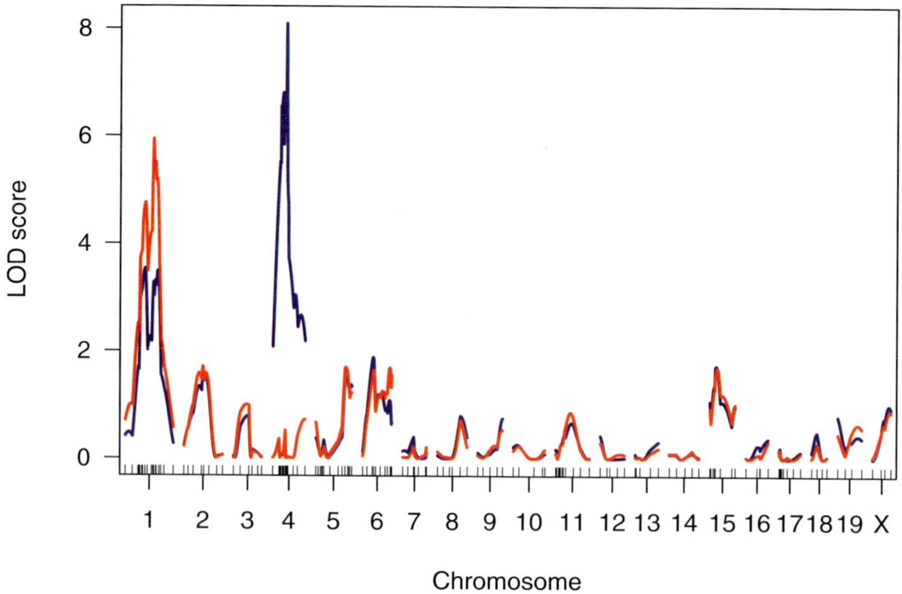

Figure 7.15. LOD scores for the analysis of the `hyper` data, with no covariates (in blue) and with imputed genotypes at D4Mit164 included as an additive covariate (in red).

```
> summary(operm.ag, alpha=c(0.2, 0.05))

LOD thresholds (1000 permutations)
     lod
20% 2.09
5%  2.62
```

As expected, we see strong evidence for a QTL on chromosome 1, and nothing else.

```
> summary(out.ag, perms=operm.ag, alpha=0.2, pvalues=TRUE)

        chr  pos  lod pval
D1Mit94   1 67.8 5.94    0
```

We next consider the marker as an interactive covariate. This allows us to detect loci that show an interaction with the chromosome 4 locus.

```
> out.ig <- scanone(hyper, addcovar=g, intcovar=g)
```

We plot the differences in the LOD scores from the analysis with the marker as an interactive covariate and that with the marker as solely additive; see Fig. 7.16. We see nothing interesting, and so we will not pursue this further.

```
> plot(out.ig - out.ag, ylab="interaction LOD score")
```

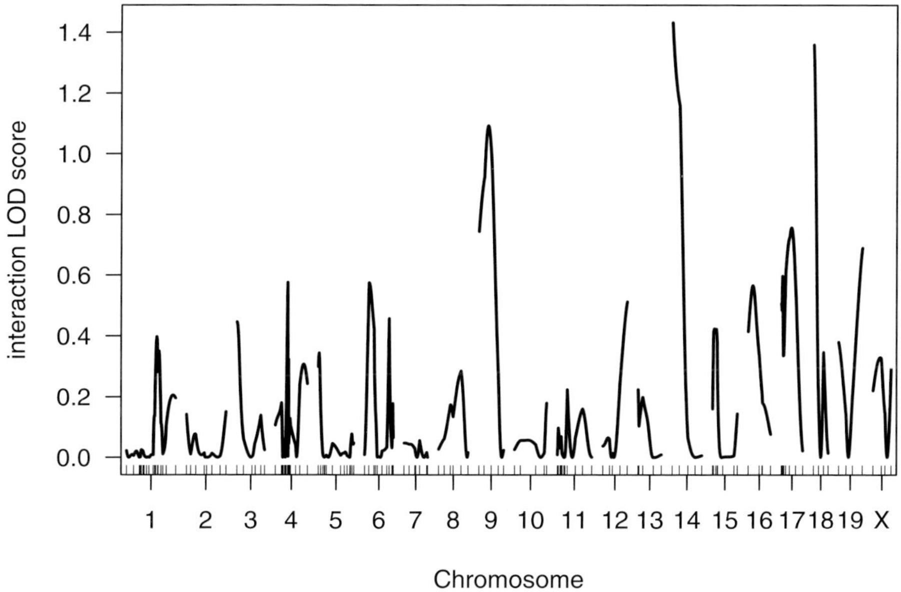

Figure 7.16. Interaction LOD scores, indicating evidence for an interaction between a QTL and the marker D4Mit164, for the `hyper` data.

Finally, let us investigate the use of composite interval mapping (CIM) with these data. The function `cim` is a relatively crude version of the CIM strategy discussed above: forward selection to a fixed number of markers (specified via the argument `n.marcovar`), followed by interval mapping, omitting any marker covariates within some fixed distance of the position under test (specified via the argument `window`, which is twice this distance, meaning the total length of the window to surround the test position).

Of course, forward selection at the markers requires complete marker genotype data, but a selective genotyping strategy was used with the `hyper` data, and so there is a large amount of missing genotype data on many chromosomes. The `cim` function uses a single imputation to fill in any missing genotype data prior to the forward selection procedure.

We will use three marker covariates, and window sizes of 20 and 40 cM, as well as infinity (meaning the entire length of the chromosome).

```
> out.cim.20 <- cim(hyper, n.marcovar=3, window=20)
> out.cim.40 <- cim(hyper, n.marcovar=3, window=40)
> out.cim.inf <- cim(hyper, n.marcovar=3, window=Inf)
```

We plot the results (for selected chromosomes), along with the LOD scores obtained by standard interval mapping, as follows. Note that the function `add.cim.covar` is used to add dots indicating the locations of the selected marker covariates.

```
> chr <- c(1, 2, 4, 6, 15)
> par(mfrow=c(3,1))
> plot(out, out.cim.20, chr=chr, ylab="LOD score",
+      col=c("blue", "red"), main="window = 20 cM")
> add.cim.covar(out.cim.20, chr=chr, col="green")
> plot(out, out.cim.40, chr=chr, ylab="LOD score",
+      col=c("blue", "red"), main="window = 40 cM")
> add.cim.covar(out.cim.40, chr=chr, col="green")
> plot(out, out.cim.inf, chr=chr, ylab="LOD score",
+      col=c("blue", "red"), main="window = Inf")
> add.cim.covar(out.cim.inf, chr=chr, col="green")
```

The results are in Fig. 7.17. Note that the imputation of marker genotype data was performed separately in the three cases, but that otherwise the selection of marker covariates was identical. Randomness in the imputations resulted in some randomness in the choice of marker covariates (the position of the marker on chromosome 1, and whether a locus on chromosome 6 or chromosome 2 was chosen). The CIM results indicate some enhanced evidence for the locations of QTL, but the windowing can give an artifactual improvement in the apparent precision of QTL localization.

7.5 Summary

If a covariate (such as sex or an environmental factor) is associated with the phenotype of interest, its consideration in QTL analyses may reduce residual variation and so give increased power to detect QTL. One should be cautious about the use of secondary phenotypes as covariates, as they are not necessarily independent of genotype. More interesting, however, is the consideration of interactive covariates, in order to investigate potential QTL $\times$ covariate interactions.

While the use of a genetic marker near a large-effect QTL as a covariate, in order to reduce residual variation and so clarify evidence for further QTL, is undoubtedly a good thing, we recommend against the general use of fully-automated composite interval mapping strategies. While composite interval mapping seeks to convert the search for multiple QTL into a single-dimensional scan, we prefer to tackle the multidimensional search for multiple QTL directly. (See Chap. 9.)

7.6 Further reading

Ahmadiyeh *et al.* (2003) may be the first paper to discuss the assessment of QTL $\times$ covariate interactions. See also Solberg *et al.* (2004). The use of covariates in QTL mapping requires a good understanding of multiple linear regression; for that, we recommend Draper and Smith (1998).

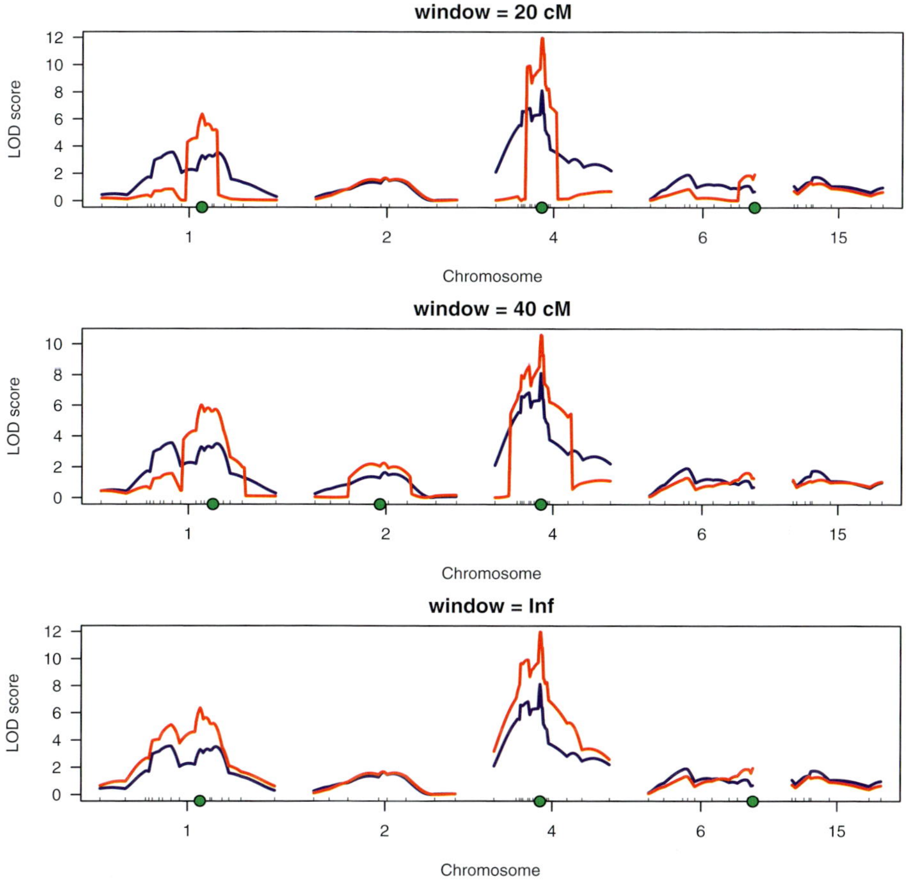

Figure 7.17. Results of composite interval mapping (CIM) for the `hyper` data, with forward selection to three markers and three different choices of window sizes. In each panel, the results from standard interval mapping are in blue, and those from CIM are in red. Selected marker covariates are indicated by green dots.

Composite interval mapping was independently developed by Zeng (1993, 1994); Jansen and Stam (1994); Jansen (1993a). There are important differences in the details of these authors' approaches, but the central idea is the same. We have focused on a particular approach to composite interval mapping that was implemented in QTL Cartographer (Basten *et al.*, 2002).

8

Two-dimensional, two-QTL scans

Most complex traits are understood to be the result of the action of multiple genetic loci. Some loci may be linked, and the effect of some loci may depend on the genotype at others (epistasis). In this chapter, we undertake the first step in the direction of teasing apart the multiple linked or interacting QTL underlying a quantitative trait, by contemplating two-dimensional, two-QTL genome scans.

Thus far we have focused on one-dimensional genome scans, in which one fits all possible single-QTL models. Despite the fact that such genome scans appear to ignore the reality of complex traits (with multiple loci, possibly linked or epistatic), they have worked quite well. Part of this success is attributable to the fact that even if multiple loci are contributing to a trait, they are often on separate chromosomes and have marginal effects large enough to be detectable. Because of the independent assortment of chromosomes (Mendel's second law, that genotypes at loci on different chromosomes are independent), genome scans treating the chromosomes separately give approximately the same results as would be obtained if one had conditioned on additive QTL with small effect on other chromosomes. In this sense, one-dimensional genome scans transcend their one-dimensional nature.

Nevertheless, there are some limitations to adopting a one-dimensional approach to the essentially multidimensional problem of detecting multiple QTL. The joint consideration of multiple QTL has three advantages. First, by taking account of QTL of large effect, one may reduce the residual variation and so better identify QTL of modest effect. Second, the separation of linked QTL is best achieved by comparing the fit of a two-QTL model to the best single-QTL model. (With two distinct peaks in the LOD curve for a given chromosome, one may be tempted to infer the presence of two QTL, and there is some literature considering the shape of LOD curves to infer the presence of multiple linked loci, an effort that we term the *phrenology* of LOD curves. It is far better to simply assess the improvement in fit that comes from a two-QTL model.) Finally, epistasis between QTL may only be assessed via models that explicitly consider multiple QTL.

K.W. Broman, Ś. Sen, *A Guide to QTL Mapping with R/qtl*,
Statistics for Biology and Health, DOI 10.1007/978-0-387-92125-9_8,

The obvious next step, beyond a one-dimensional genome scan, is consideration of all possible two-locus QTL models in a two-dimensional genome scan. This allows us to identify potential interactions among QTL and to assess evidence for linked QTL. The interpretation of the results of two-dimensional genome scans is less straighforward than those of one-dimensional scans. That is the subject of this chapter.

We begin by considering two-dimensional genome scans in the context of normally distributed traits. We then discuss the analysis of binary traits. The special treatment of the X chromosome is covered briefly. Finally, we discuss methods to handle covariates.

8.1 The normal model

Imagine the presence of precisely two QTL, and consider each pair of positions in the genome as the putative locations for those QTL. Let (s,t) denote the pair of positions, and consider the following four models; for now we assume that the residual variation is normally distributed with constant variance.

$$\begin{aligned} \mathrm{H}_f\colon\ & y = \mu + \beta_1 q_1 + \beta_2 q_2 + \gamma(q_1 \times q_2) + \epsilon \\ \mathrm{H}_a\colon\ & y = \mu + \beta_1 q_1 + \beta_2 q_2 + \epsilon \\ \mathrm{H}_1\colon\ & y = \mu + \beta_1 q_1 + \epsilon \\ \mathrm{H}_0\colon\ & y = \mu + \epsilon \end{aligned} \tag{8.1}$$

In the full model (H_f), the two QTL are allowed to interact. In the additive model (H_a), they are assumed to act additively (i.e., the effect of each locus is the same, no matter the genotype at the other locus). We also consider all possible single-QTL models (i.e., the results of the single-QTL genome scan), and the null model (H_0), that there are no QTL.

The fit of these models requires that we take account of the missing genotype data at the putative QTL, making use of the genotype data at linked markers. This may again be accomplished in multiple ways; the four methods discussed in Chap. 4 (maximum likelihood via the EM algorithm, Haley–Knott regression, the extended Haley–Knott method, and multiple imputation) may all be readily extended to the two-QTL case. In the case that the two putative QTL are on the same chromosome, the multiple imputation approach requires that genotypes at the two positions be drawn from their joint distribution, but this is easily done (see Sec. D.3). The other methods require the calculation of the joint genotype probabilities, $\Pr(q_1, q_2|\boldsymbol{M})$, where $\boldsymbol{M}$ is the observed multipoint marker genotype data. If one assumes no crossover interference and no genotyping errors, and if there is complete genotype data at an intervening marker, then the QTL genotypes will be conditionally independent, given the marker data, and so

$$\Pr(q_1, q_2|\boldsymbol{M}) = \Pr(q_1|\boldsymbol{M}) \times \Pr(q_2|\boldsymbol{M}). \tag{8.2}$$

If we wish to allow for genotyping errors, or if there is no fully typed marker between the putative QTL, then the product rule in equation (8.2) will not apply. However, with the assumption of no crossover interference, the joint distribution may be calculated efficiently via the hidden Markov model technology (see Sec. D.4).

Turning now to the summary and interpretation of two-dimensional, two-QTL genome scans, let $l_f(s,t)$ denote the $\log_{10}$ likelihood for the full model with QTL at s and t, $l_a(s,t)$ denote the $\log_{10}$ likelihood for the additive model with QTL at s and t, $l_1(s)$ denote the $\log_{10}$ likelihood for the single-QTL model with the QTL at s, and l_0 denote the $\log_{10}$ likelihood under the null (with no QTL).

We may immediately define several LOD scores, comparing the fit of the four models.

$$\begin{aligned}
\text{LOD}_f(s,t) &= l_f(s,t) - l_0 \\
\text{LOD}_a(s,t) &= l_a(s,t) - l_0 \\
\text{LOD}_i(s,t) &= l_f(s,t) - l_a(s,t) \\
\text{LOD}_1(s) &= l_1(s) - l_0
\end{aligned}$$

LOD_f measures the improvement in the fit of the full two-locus model over the null model, and indicates evidence for at least one QTL, with allowance for interaction. Similarly, LOD_a measures the improvement in the fit of the two-locus additive model, and indicates evidence for at least one QTL, assuming no interaction. LOD_i measures the improvement in the fit of the full model over that of the additive model, and so indicates evidence for interaction. LOD_1 is simply the LOD score from the one-dimensional, single-QTL genome scan.

These LOD scores can be difficult to interpret, as LOD_f and LOD_a will be large on any chromosome with clear evidence for a QTL. That is, if the initial genome scan indicated strong evidence for a QTL at position s_0, so that $\text{LOD}_1(s_0)$ is large, then $\text{LOD}_f(s_0,t)$ and $\text{LOD}_a(s_0,t)$ will be large for all t.

Most important, in the consideration of the results of a two-dimensional, two-QTL genome scan, is an assessment of evidence for a second QTL: does the two-QTL model provide sufficiently improved fit over the best single-QTL model? We find it valuable to focus on an individual chromosome or pair of chromosomes. Consider a pair of chromosomes (j,k), including the case $j=k$, and let $c(s)$ denote the chromosome for position s. We now consider the maximum LOD scores over that pair of chromosomes.

$$\begin{aligned}
M_f(j,k) &= \max_{c(s)=j, c(t)=k} \text{LOD}_f(s,t) \\
M_a(j,k) &= \max_{c(s)=j, c(t)=k} \text{LOD}_a(s,t) \\
M_1(j,k) &= \max_{c(s)=j \text{ or } k} \text{LOD}_1(s)
\end{aligned}$$

So $M_f(j,k)$ is the $\log_{10}$ likelihood ratio comparing the full model with QTL on chromosomes j and k to the null model, and $M_a(j,k)$ is the analogous thing for the additive model. Note that the pair of positions at which the full model is maximized may be different from the pair of positions at which the additive model is maximized. $M_1(j,k)$ is the $\log_{10}$ likelihood ratio comparing the model with a single QTL on either chromosomes j or k to the null model.

We derive three further LOD scores from the above.

$$\begin{aligned} M_i(j,k) &= M_f(j,k) - M_a(j,k) \\ M_{fv1}(j,k) &= M_f(j,k) - M_1(j,k) \\ M_{av1}(j,k) &= M_a(j,k) - M_1(j,k) \end{aligned}$$

$M_i(j,k)$ is the $\log_{10}$ likelihood ratio comparing the full model with QTL on chromosomes j and k to the additive model with QTL on chromosomes j and k, and so indicates evidence for an interaction between QTL on chromosomes j and k, assuming that there is precisely one QTL on each chromosome (or, for $j = k$, that there are two QTL on the chromosome).

$M_{fv1}(j,k)$ is the $\log_{10}$ likelihood ratio comparing the full model with QTL on chromosomes j and k to the single-QTL model, with a single QTL on either chromosome j or k. Thus, it indicates evidence for a second QTL, allowing for the possibility of epistasis.

$M_{av1}(j,k)$ is the $\log_{10}$ likelihood ratio comparing the additive model with QTL on chromosomes j and k to the single-QTL model, with a single QTL on either chromosome j or k. Thus, it indicates evidence for a second QTL, assuming no epistasis.

We focus particularly on M_{fv1} and M_{av1}, concerning evidence for a second QTL. M_{fv1} allows for interactions between the QTL, and so will enable the detection of loci with limited marginal effect. However, in the absence of a strong interaction, M_{av1} would give greater power to detect a second QTL, as the additional degrees of freedom in the M_{fv1} statistic results in a greater threshold for significance.

In order to identify loci of interest from the results of a two-dimensional, two-QTL genome scan, we consider the distributions of the five LOD scores ($M_f(j,k)$, $M_{fv1}(j,k)$, $M_i(j,k)$, $M_a(j,k)$ and $M_{av1}(j,k)$) under the global null hypothesis, that there are no QTL. Through either computer simulation or a permutation test, we may estimate quantiles of the null distributions of these LOD scores, to be used as thresholds. We use the following rule: A pair of chromosomes (j,k) is reported as interesting if either of the following holds:

$$\begin{cases} M_f(j,k) \ge T_f \ \text{ and } \ [M_{fv1}(j,k) \ge T_{fv1} \ \text{or} \ M_i(j,k) \ge T_i] \\ M_a(j,k) \ge T_a \ \text{ and } \ M_{av1}(j,k) \ge T_{av1} \end{cases} \tag{8.3}$$

We are inclined to ignore $M_i(j,k)$ in this rule (i.e., setting $T_i = \infty$) and use a common significance level ($\alpha = 5$ or 10%) for the four remaining thresholds.

The thresholds can be obtained by a permutation test (see below), but this is extremely time-consuming. For a mouse backcross, we suggest the 5%

thresholds $(T_f, T_{fv1}, T_i, T_a, T_{av1}) = (6.0, 4.7, 4.4, 4.7, 2.6)$. For a mouse intercross, we suggest $(T_f, T_{fv1}, T_i, T_a, T_{av1}) = (9.1, 7.1, 6.3, 6.3, 3.3)$. These are the estimated 95th percentiles of the null distributions of the corresponding LOD scores, obtained by 10,000 simulations of crosses with 250 individuals, markers at a 10 cM spacing, and analysis by Haley–Knott regression.

While we are recommending that the thresholds be derived under the global null hypothesis of no QTL, our interest is really in evidence for a second QTL, over and above that for a single QTL. Thus, we really should be considering the distributions of these statistics in the presence of a single QTL: how large will M_{fv1} and M_{av1} be, if there exists precisely one QTL? However, this is not a simple matter, and the answer would likely depend, to some extent, on the location and effect of the hypothesized QTL. And so we make the assumption that the distributions of M_{fv1} and M_{av1} are approximately the same, whether there exists a single QTL or no QTL.

Our separate treatment of individual pairs of chromosomes may be confusing initially. Our goal, with this approach, is to identify multiple pairs of linked or interacting QTL across the genome, in the same way that one seeks, in a traditional one-dimensional genome scan, to identify QTL on multiple chromosomes, rather than just the single locus giving the largest LOD score. Our test statistics M_{fv1} and M_{av1} are equivalent to the usual sort of likelihood ratio statistics for comparing a model with two QTL to a model with a single QTL.

Example

As an illustration, we return to the `hyper` data (see Sec. 2.3). Let us load the data and run `calc.genoprob` to calculate the conditional QTL genotype probabilities, given the available marker data. We use a more coarse grid, `step=2.5`, for the sake of computational speed.

```
> library(qtl)
> data(hyper)
> hyper <- calc.genoprob(hyper, step=2.5, err=0.001)
```

The two-dimensional, two-QTL scan is accomplished with the function `scantwo`, which operates much like `scanone`, though computation time is greatly increased. By default, the analysis is performed by maximum likelihood via the EM algorithm (`method="em"`). We use `verbose=FALSE` to suppress tracing information.

```
> out2 <- scantwo(hyper, verbose=FALSE)
```

The output has class `"scantwo"`, and so may be plotted with `plot.scantwo` and summarized with `summary.scantwo`. Each of these functions includes considerable flexibility. First, let us plot the two-dimensional scan results for selected chromosomes.

```
> plot(out2, chr=c(1,4,6,7,15))
```

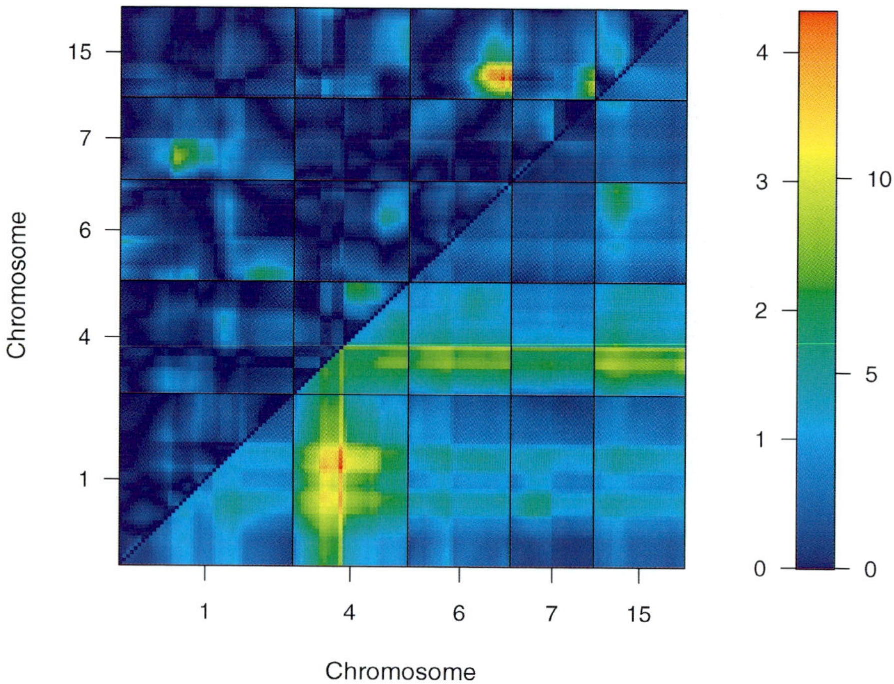

Figure 8.1. LOD scores, for selected chromosomes, from a two-dimensional, two-QTL genome scan with the `hyper` data. LOD_i is displayed in the upper left triangle; LOD_f is displayed in the lower right triangle. In the color scale on the right, numbers to the left and right correspond to LOD_i and LOD_f, respectively.

The result is shown in Fig. 8.1. By default, LOD_i is plotted in the upper left triangle, and LOD_f is plotted in the lower right triangle. In the color scale on the right, the numbers on the left and right correspond to LOD_i and LOD_f, respectively. Note that LOD_f (lower right triangle) is large for chromosome 4 considered with any other chromosome. These "tails" in LOD_f are due to the strong evidence for a QTL on chromosome 4, and the fact that LOD_f compares the full two-QTL model to the null model, and so is large in the presence of evidence for at least one QTL. Further note the evidence for an interaction between loci on chromosomes 6 and 15. (LOD_i, in the upper left triangle, is large.)

In place of LOD_f, we prefer to focus on LOD_{fv1}, which indicates evidence for a second QTL, allowing for epistasis. Above, we had defined $M_{fv1}(j,k) = M_f(j,k) - M_1(j,k)$, for a pair of chromosomes (j,k). We now consider $\text{LOD}_{fv1}(s,t) = \text{LOD}_f(s,t) - M_1(j,k)$, though replacing negative values with 0. $\text{LOD}_{av1}(s,t)$ may be defined similarly.

We may plot LOD_{fv1} in the lower right triangle, in place of LOD_f, with the argument `lower="cond-int"`, as follows. (One may also use `lower="fv1"`.)

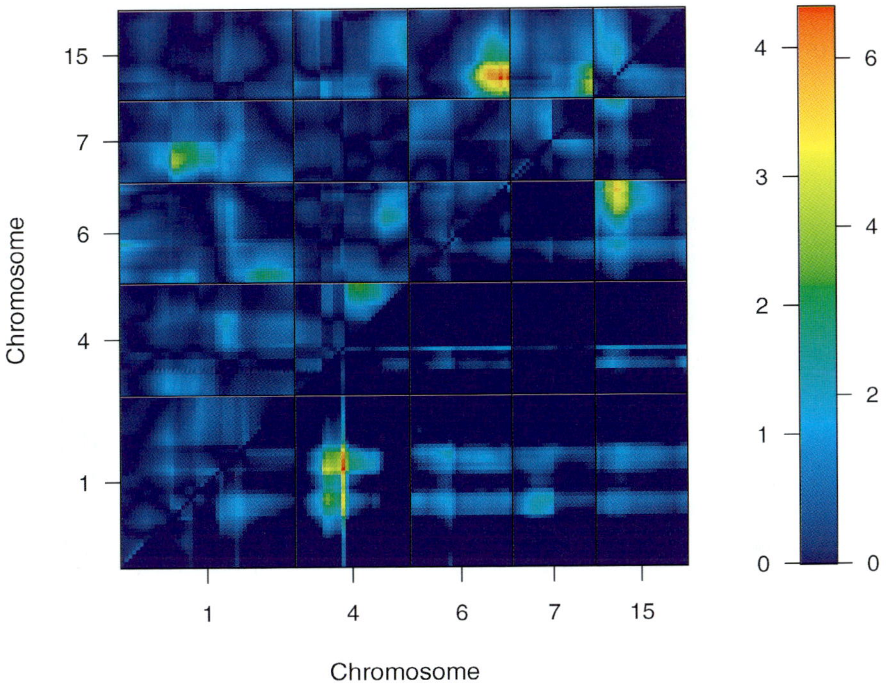

Figure 8.2. LOD scores, for selected chromosomes, from a two-dimensional, two-QTL genome scan with the `hyper` data. LOD_i is displayed in the upper left triangle; LOD_{fv1} is displayed in the lower right triangle. In the color scale on the right, numbers to the left and right correspond to LOD_i and LOD_{fv1}, respectively.

```
> plot(out2, chr=c(1,4,6,7,15), lower="cond-int")
```

In the plot of LOD_{fv1}, in Fig. 8.2, the lower right triangle is cleaned up considerably. We now see strong evidence for a pair of QTL on chromosomes 1 and 4, and on chromosomes 6 and 15. (That is, for these pairs, the fit of the two-QTL model, with one QTL on each chromosome, considerably improves on the fit of the best single-QTL model, with a single QTL on one or the other chromosome.) The tails that came from the chromosome 4 locus have been eliminated, making the figure more easily interpreted.

One may also plot LOD_a and/or LOD_{av1}, and what is plotted in the upper triangle may be modified in the same way as we modified what was plotted in the lower right triangle. For example, we may plot LOD_a and LOD_{av1} as follows. To get LOD_{av1} in the lower right triangle, we may either use `lower="cond-add"` or `lower="av1"`.

```
> plot(out2, chr=c(1,4,6,7,15), upper="add", lower="cond-add")
```

The result is displayed in Fig. 8.3. Note that LOD_a, like LOD_f, has the same "tails" problem (with large LOD scores appearing on chromosome 4

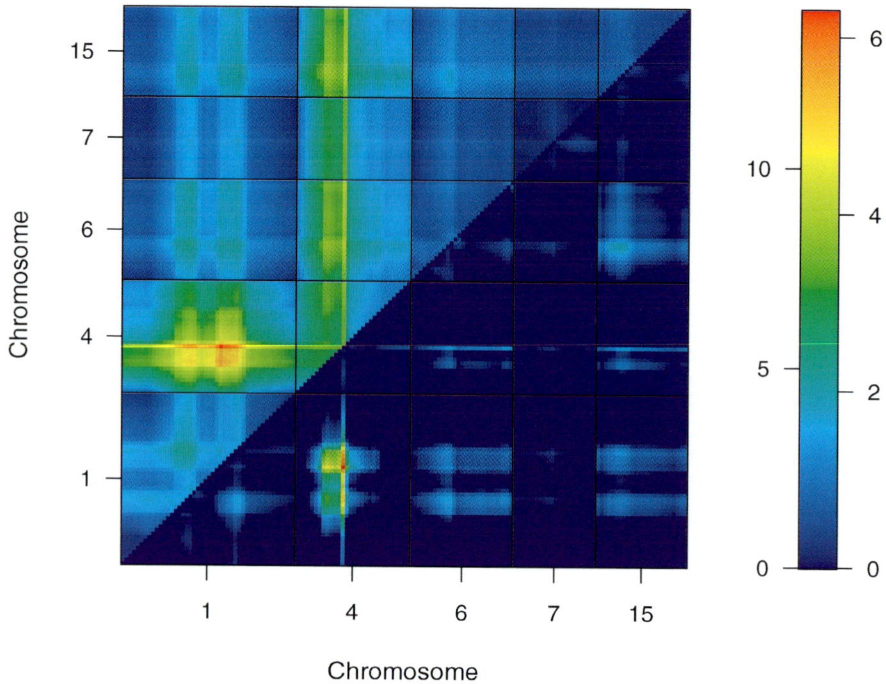

Figure 8.3. LOD scores, for selected chromosomes, from a two-dimensional, two-QTL genome scan with the `hyper` data. LOD_a is displayed in the upper left triangle; LOD_{av1} is displayed in the lower right triangle. In the color scale on the right, numbers to the left and right correspond to LOD_a and LOD_{av1}, respectively.

together with any other chromosome). Also note that, for chromosomes 6 and 15, the evidence for a second QTL has disappeared. These loci are seen as important only when they are allowed to interact.

Recall the two peaks in the LOD curve for chromosome 1 with the `hyper` data. (For example, see Fig. 4.6 on page 88.) We are thus particularly interested in evidence for a second QTL on chromosome 1. We may plot LOD_{av1} and LOD_{fv1} for chromosome 1 as follows.

```
> plot(out2, chr=1, lower="cond-int", upper="cond-add")
```

The results, in Fig. 8.4, indicate little evidence for a second QTL on chromosome 1. Inclusion of a second locus increases the $\log_{10}$ likelihood by ~ 1.5.

The plots of the results of the two-dimensional, two-QTL scan are not so simple to interpret as those of a one-dimensional, single-QTL scan. Thus, we place greater reliance on the numeric summaries from `summary.scantwo`. Use of `summary(out2)` would produce a table giving a single row for each pair of chromosomes. With 20 chromosomes, that is a table with 210 rows. That is an unwieldy amount of information to sift through, and so it is best to provide a

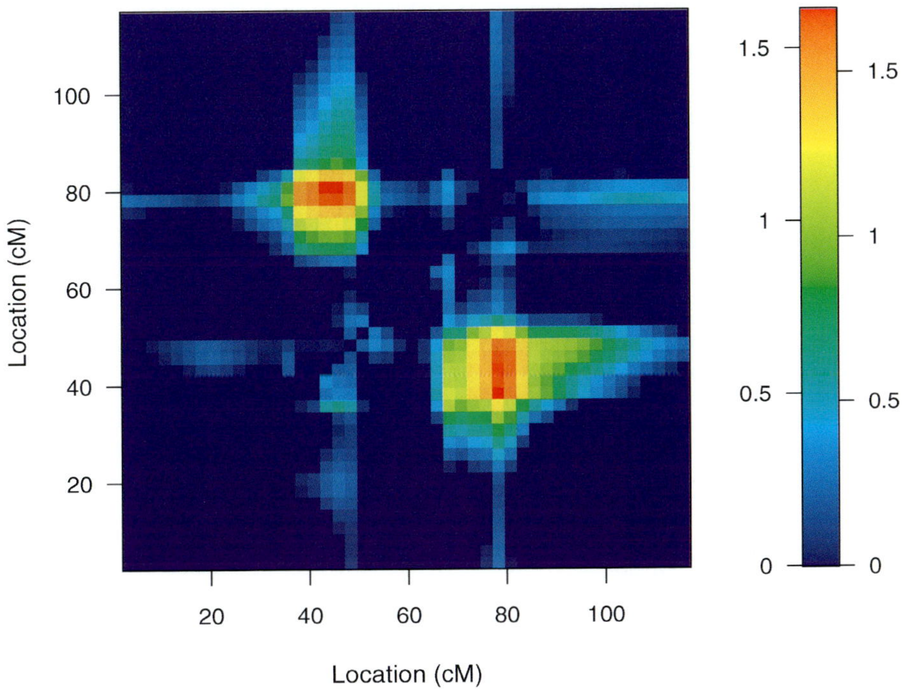

Figure 8.4. LOD scores, for chromosome 1, from a two-dimensional, two-QTL genome scan with the `hyper` data. LOD_{av1} is displayed in the upper left triangle; LOD_{fv1} is displayed in the lower right triangle. In the color scale on the right, numbers to the left and right correspond to LOD_{av1} and LOD_{fv1}, respectively.

set of thresholds, $(T_f, T_{fv1}, T_i, T_a, T_{av1})$; then, only those chromosomes satisfying the rule in equation (8.3) (page 216) will be displayed.

For example, we may use the thresholds $(T_f, T_{fv1}, T_i, T_a, T_{av1}) = (6.0, 4.7, 4.4, 4.7, 2.6)$, suggested by simulations.

```
> summary(out2, thresholds=c(6.0, 4.7, 4.4, 4.7, 2.6))
```

```
        pos1f pos2f lod.full lod.fv1 lod.int     pos1a pos2a
c1:c4    68.3    30    14.24     6.6   0.305      68.3  30.0
c6:c15   60.0    18     7.27     5.4   3.458      25.0  20.5
        lod.add lod.av1
c1:c4     13.94    6.30
c6:c15     3.81    1.95
```

There is clear evidence for a pair of QTL on chromosomes 1 and 4, and this is true whether or not an interaction is allowed ($M_{fv1} = 6.6$ and $M_{av1} = 6.3$). We see little evidence for an interaction between these loci ($M_i = 0.3$). Note that (`pos1f`, `pos2f`) indicate the estimated positions of the QTL (on chromosomes 1 and 4, respectively) under the full model, while (`pos1a`, `pos2a`) are the estimated positions under the additive model. These are allowed to be different, though for the chromosomes (1,4) pair, the likelihoods for the full and additive models happened to be maximized at the same pair of positions.

The summary also indicates strong evidence for the pair of QTL on chromosomes 6 and 15, though only if an interaction is allowed ($M_{fv1} = 5.4$ and $M_{av1} = 1.9$), and here the full model was maximized at a different pair of positions than the additive model. In the summary, the evidence for interaction, $M_i = 3.5$ is the difference $M_f - M_a$, or equivalently $M_{fv1} - M_{av1}$, with the full and additive models allowed to be maximized at different positions.

As mentioned above, we are inclined to ignore the value of M_i, by setting $T_i = \infty$. This is accomplished as follows, though here the same summary is obtained.

```
> summary(out2, thresholds=c(6.0, 4.7, Inf, 4.7, 2.6))

        pos1f pos2f lod.full lod.fv1 lod.int     pos1a pos2a
c1:c4    68.3    30    14.24     6.6   0.305      68.3  30.0
c6:c15   60.0    18     7.27     5.4   3.458      25.0  20.5
        lod.add lod.av1
c1:c4     13.94    6.30
c6:c15     3.81    1.95
```

The thresholds above were obtained by simulation of a backcross with a genome modelled after that of the mouse, markers at a 10 cM spacing, and with Haley–Knott regression used for the analysis. While these may serve as a reasonably general guide, it would be better to use a permutation test with the observed data, so that the results would take account of the phenotype distribution, marker density, and pattern of missing genotype data. A permutation test may be accomplished with `scantwo` by specifying the argument `n.perm`, though the required computation time may be daunting.

Because a selective genotyping strategy was used for the `hyper` data, it is best to use a stratified permutation test, splitting the individuals into two strata according to the amount of genotype data available, and permuting the phenotypes relative to the genotypes separately within the strata. This may be accomplished by creating a vector indicating the two strata, to be specified via the `perm.strata` argument in `scantwo`. Thus, we perform the permutation test for the two-dimensional, two-QTL genome scan as follows.

```
> strat <- (nmissing(hyper) > 50)
> operm2 <- scantwo(hyper, n.perm=1000, perm.strat=strat)
```

The permutation results have class `"scantwoperm"`. We may use `summary.scantwoperm` to get estimated thresholds.

```
> summary(operm2)

bp (1000 permutations)
```

```
     full  fv1  int  add  av1  one
5%   6.13 4.86 4.34 4.64 3.04 2.82
10%  5.63 4.48 3.95 4.27 2.73 2.48
```

The 5% thresholds are similar to those presented previously, though T_{av1} is a bit larger.

Because the computation time for the permutations can be on the order of 100 hours, one may want to split the calculations across multiple computers. In doing so, one should be careful about the seed for the random number generator. This is saved as the object `.Random.seed` in the R workspace. If multiple sets of permutations are run in parallel from the same R workspace, one may unintentionally use the same seed, and so obtain identical results, for each set of permutations. Thus, one should call `set.seed` separately for each set.

The permutations above could be run in two batches using the following pieces of code.

```
> set.seed(85842518)
> operm2a <- scantwo(hyper, n.perm=500, perm.strat=strat)
> save(operm2a, file="perm2a.RData")

> set.seed(85842519)
> operm2b <- scantwo(hyper, n.perm=500, perm.strat=strat)
> save(operm2b, file="perm2b.RData")
```

The two bits can then be combined with the `c.scantwoperm` function.

```
> load("perm2a.RData")
> load("perm2b.RData")
> operm2 <- c(operm2a, operm2b)
```

The same technique may also be used for a permutation test in a one-dimensional genome scan, and there is a function `c.scanoneperm` for combining multiple batches of permutation replicates.

We may include the permutation results in the call to `summary.scantwo`, so that the thresholds are calculated automatically, and so that p-values may be calculated. In place of thresholds, we must provide significance levels. We may ignore the M_i values by giving a significance level of 0 (which corresponds to $T_i = \infty$). Using 20% significance levels, we obtain the following.

```
> summary(out2, perms=operm2, alphas=c(0.2, 0.2, 0, 0.2, 0.2),
+           pvalues=TRUE)

        pos1f pos2f lod.full  pval lod.fv1  pval lod.int pval
c1:c4    68.3  30.0    14.24 0.000    6.60 0.003   0.305 1.00
c3:c3    37.2  44.7     4.53 0.493    3.75 0.348   0.215 1.00
c6:c15   60.0  18.0     7.27 0.013    5.40 0.023   3.458 0.25
```

```
        pos1a pos2a lod.add  pval lod.av1  pval
c1:c4    68.3  30.0   13.94 0.000    6.30 0.000
c3:c3    37.2  44.7    4.32 0.092    3.53 0.011
c6:c15   25.0  20.5    3.81 0.205    1.95 0.444
```

Note that, if we had wanted to consider a threshold on the interaction LOD score, M_i, we would have instead typed `alphas=rep(0.2,5)`, or, as short hand, simply `alphas=0.2`.

In addition to the loci on chromosomes 1, 4, 6 and 15, we see evidence for a pair of linked, additive QTL on chromosome 3. But these are very tightly linked (separated by just 7.5 cM), and so may be an artifact.

Two-dimensional, two-QTL scans often show artifacts along the diagonal (that is, for two tightly linked QTL), particularly in cases in which two putative loci are not separated by a typed marker. One thus may find value in the utility function `clean.scantwo`, which replaces LOD scores, for pairs of positions that are not separated by at least one marker, with 0, and so eliminates the bulk of such artifacts. This will be done automatically within `scantwo` if one uses the argument `clean.output=TRUE`, and this is particularly recommended for permutations.

We may plot the two-dimensional scan results for chromosome 3, zeroing out the LOD scores for pairs of positions that are not separated by a marker, as follows.

```
> plot(clean(out2), chr=3, lower="cond-add")
```

In the results, in Fig. 8.5, large blue squares along the diagonal correspond to pairs of positions that are not separated by a marker. Note that the peak in the LOD surface remains. This may also be seen in the summary of the "cleaned" results:

```
> summary(clean(out2), perms=operm2,
+           alphas=c(0.2, 0.2, 0, 0.2, 0.2))

       pos1f pos2f lod.full lod.fv1 lod.int     pos1a pos2a
c1:c4   68.3  30.0    14.24    6.60   0.305      68.3  30.0
c3:c3   37.2  44.7     4.53    3.75   0.215      37.2  44.7
c6:c15  60.0  18.0     7.27    5.40   3.458      25.0  20.5
       lod.add lod.av1
c1:c4    13.94    6.30
c3:c3     4.32    3.53
c6:c15    3.81    1.95
```

Let us study the effects of the putative linked loci on chr 3 via the functions `plot.pxg` and `effectplot`. The `effectplot` function requires imputed genotype data, and so we first run `sim.geno`, though only for chromosome 3.

```
> hyperc3 <- sim.geno(subset(hyper, chr=3), step=2.5,
+                      error.prob=0.001, n.draws=256)
```

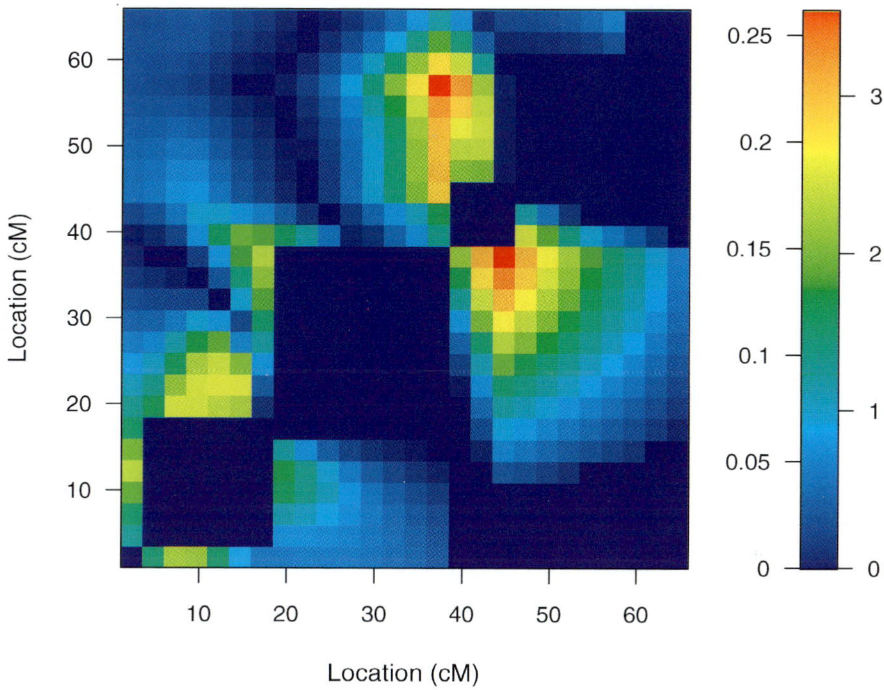

Figure 8.5. LOD scores, for chromosome 3, from a two-dimensional, two-QTL genome scan with the `hyper` data, with values for pairs of positions that are not separated by a marker replaced by 0. LOD_i is displayed in the upper left triangle; LOD_{av1} is displayed in the lower right triangle. In the color scale on the right, numbers to the left and right correspond to LOD_i and LOD_{av1}, respectively.

We now use `find.marker` to find the names of the markers nearest the two inferred QTL, for use in `plot.pxg`. For the function `effectplot`, we may consider "pseudomarkers" (that is, the positions on the grid between markers), which we refer may refer to directly by their chromosome and cM position. For example, we use `"3@37.2"` to refer to the pseudomarker closest to 37.2 cM on chromosome 3.

```
> mar <- find.marker(hyperc3, "3", c(37.2, 44.7))
```

Now we make the plots, which appear in Fig. 8.6. Note the use of `ylim` to adjust the y-axis limits in `effectplot`, so that the limits in the two plots correspond.

```
> par(mfrow=c(1,2))
> plot.pxg(hyperc3, marker=mar)
> effectplot(hyperc3, mname1="3@37.2", mname2="3@44.7",
+            ylim=range(pull.pheno(hyperc3,1)))
```

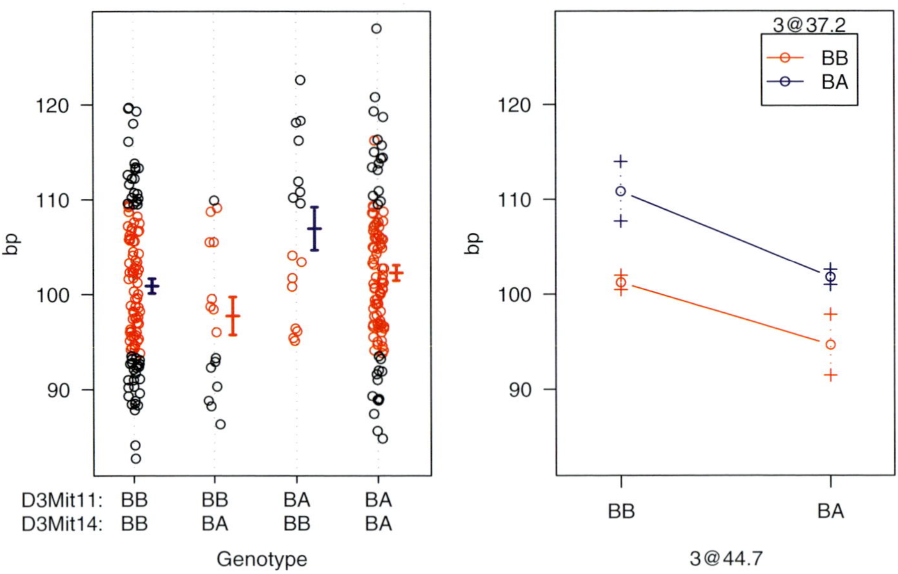

Figure 8.6. The effect of two putative linked QTL on chromosome 3 on the blood pressure phenotype in the `hyper` data. Left: a dot plot of the phenotype as a function of marker genotypes, with black dots corresponding to observed genotypes and red dots corresponding to missing (and so imputed) genotypes. Right: estimated phenotype averages for each of the four two-locus genotype groups, at the inferred locations of the two putative QTL.

Figure 8.6 may require a bit of study. In the left panel, with the output from `plot.pxg`, the red dots correspond to individuals missing genotypes at one or both markers (their two-locus genotypes are imputed from the available data), while the black dots correspond to individuals with genotype data at both markers. Note that the two recombinant classes have quite different phenotype averages: individuals that are BB at D3Mit11 and BA at D3Mit14 have low blood pressure, while those that are BA at D3Mit11 and BB at D3Mit14 have high blood pressure. The same feature is seen in the output from `effectplot`: the two QTL appear to have effects of opposite sign, but of approximately the same magnitude. Two linked QTL that have effects of opposite sign are said to be linked in *repulsion*. With QTL linked in repulsion, the region will exhibit little marginal effect, but if both loci are considered, the two may stand out clearly. With QTL linked in *coupling* (having effects of the same sign), on the other hand, a marginal effect will be apparent, but it will be difficult to separate the two loci.

We should, of course, also study the effects of the loci on chromosomes 1, 4, 6, and 15. We will again use `sim.geno` to impute the missing genotype data, and `effectplot` to plot the estimated phenotype averages as a function of QTL genotypes, considering just two QTL at a time.

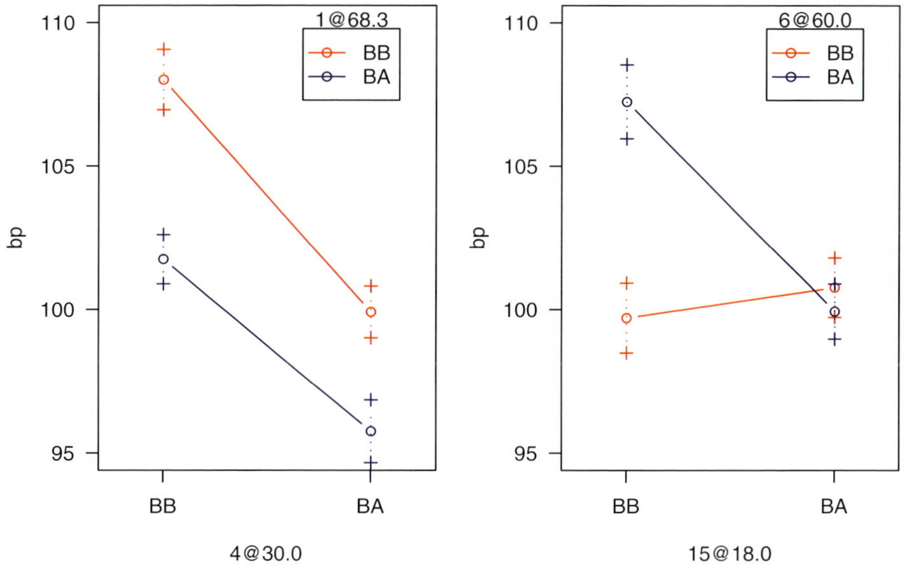

Figure 8.7. Estimated average blood pressure as a function of genotype at loci on chromosomes 1 and 4 (left panel) or chromosomes 6 and 15 (right panel), for the `hyper` data.

```
> hypersub <- sim.geno(subset(hyper, chr=c(1,4,6,15)), step=2.5,
+                      error.prob=0.001, n.draws=256)
> par(mfrow=c(1,2))
> effectplot(hypersub, mname1="1@68.3", mname2="4@30",
+            ylim=c(95, 110))
> effectplot(hypersub, mname1="6@60", mname2="15@18",
+            ylim=c(95, 110))
```

As seen in Fig. 8.7, the QTL on chromosomes 1 and 4 are seen to act approximately additively: the effect of the chromosome 4 locus is the same for each of the two genotypes at the chromosome 1 locus, and vice versa. Also, both loci have effects of the same sign: BB individuals have higher blood pressure than BA individuals.

The chromosome 6 and 15 loci, on the other hand, show clear epistasis: the chromosome 15 locus has effect only in the presence of the BA genotype at chromosome 6. Similarly, the chromosome 6 locus has effect only in the presence of the BB genotype at chromosome 15. In fact, only individuals that are BB at chromosome 15 and BA at chromosome 6 show high blood pressure; the other three genotype groups have similar, low blood pressure levels.

8.2 Binary traits

While the discussion in the previous section assumed a continuously varying phenotype, with residual variation following a normal distribution, the same techniques may be applied in the case of a binary phenotype.

In the full model (with two interacting QTL), we allow a separate penetrance for each two-locus genotype. That is, $\Pr(y = 1|q_1, q_2)$ is estimated separately for each possible pair of QTL genotypes (q_1, q_2). If complete genotype data were available at the two putative QTL, the penetrance would be estimated by the proportion of individuals with that particular two-locus genotype that were affected.

In the additive QTL model, on the other hand, one must introduce a link function, as for the use of covariates with a binary trait (Sec. 7.3). In R/qtl, the *logit* link function, $\ln[\pi/(1-\pi)]$, is used. And so, taking $\pi = \Pr(y = 1|q_1, q_2)$, the additive model is then

$$\ln[\pi/(1-\pi)] = \mu + \beta_1 q_1 + \beta_2 q_2 \tag{8.4}$$

The meaning of "additive" (and so the meaning of "interaction" or "epistasis") depends on the particular link function, just as, for a continuously varying phenotype, the meaning depends on the choice of transformation. With the logit link, we assume, in the additive model, that the effect of a change in the genotype at the first QTL is to modify the log odds for being affected by a constant multiplicative factor, independent of the genotype at the second QTL. (The *odds* corresponding to a probability π is the ratio $\pi/(1-\pi)$. And so with a probability $\pi = 2/3$, the odds are 2:1.)

Aside from this need for a link function, the two-dimensional, two-QTL genome scan for a binary phenotype is directly analogous to that for a continuously varying phenotype with the normality assumption. We calculate the same sorts of LOD scores, and we use the same techniques to interpret the results. The link function is used to accommodate the fact that the penetrance, $\Pr(y = 1|q_1, q_2)$, takes values between 0 and 1, while the right-hand side of equation (8.4) need not be between 0 and 1. The link function transforms the penetrance so that both sides of equation (8.4) are unbounded.

Example

We return to the `nf1` data of Reilly *et al.* (2006), concerning neurofibromatosis type 1 (see Sec. 7.3). This is a backcross, with all individuals carrying the *NPcis* mutation, which may have come from their mother or from their father.

We must load the R/qtlbook package, and then the data, and then we drop one marker with completely missing genotype data.

```
> library(qtlbook)
> data(nf1)
> nf1 <- drop.nullmarkers(nf1)
```

In the single-QTL scan results (see Fig. 7.13 on page 203), we saw a big difference between individuals receiving the *NPcis* mutation from their mother (which showed linkage to chromosome 15) and those receiving the mutation from their father (which showed linkage to chromosome 19). Thus, we will perform the two-dimensional scan separately in these groups.

We first calculate the conditional QTL genotype probabilities (we use a coarse grid, for the sake of computational speed), and pull out the indicator for the origin of the *NPcis* mutation from the phenotype data.

```
> nf1 <- calc.genoprob(nf1, step=2.5, error.prob=0.001)
> frommom <- pull.pheno(nf1,"frommom")
```

Now we run the two dimensional, two QTL genome scans. Note the use of `model="binary"` to indicate analysis for a binary trait, which must take values 0 and 1. Also note that R/qtl uses maximum likelihood via the EM algorithm; while Haley–Knott regression or multiple imputation could also be used to deal with missing genotype data for a binary trait, these approaches have not yet been implemented in R/qtl.

```
> out.frommom <- scantwo(subset(nf1, ind=(frommom==1)),
+                         model="binary")
> out.fromdad <- scantwo(subset(nf1, ind=(frommom==0)),
+                         model="binary")
```

Before proceeding further, let us perform permutation tests, so that we may make sense of the results. The computations take a very long time, and so one may wish to split the calculations across multiple computers (see Sec. 8.1, page 223).

```
> operm.frommom <- scantwo(subset(nf1, ind=frommom==1),
+                          model="binary", n.perm=1000)
> operm.fromdad <- scantwo(subset(nf1, ind=frommom==0),
+                          model="binary", n.perm=1000)
```

Significance thresholds may be obtained with the `summary.scantwoperm` function. Note that these are a bit smaller than those obtained in Sec. 8.1 (for the `hyper` data, with a normal model).

```
> summary(operm.frommom)

affected (1000 permutations)
    full  fv1  int  add  av1
5%  5.68 4.46 4.13 4.54 2.28
10% 5.39 4.11 3.82 4.09 2.06

> summary(operm.fromdad)

affected (1000 permutations)
    full  fv1  int  add  av1
5%  5.58 4.40 4.11 4.48 2.25
10% 5.26 4.09 3.74 4.08 2.02
```

Let us study the results for the subset receiving the *NPcis* mutation from their mother. We will use a significance level of $\alpha = 0.2$ as a cutoff, ignoring the interaction LOD score, M_i.

```
> summary(out.frommom, perms=operm.frommom, pvalues=TRUE,
+           alphas=c(0.2, 0.2, 0, 0.2, 0.2))

        pos1f pos2f lod.full  pval lod.fv1  pval lod.int  pval
c7:c17     45     3     5.74 0.046    4.25 0.076    3.62 0.151
              pos1a pos2a lod.add  pval lod.av1 pval
c7:c17           40    43    2.11 0.915   0.622    1
```

We see some evidence for interacting QTL on chromosomes 7 and 17. Note that linkage was previously seen to chromosome 15; it does not show up in this summary, as no one locus, when added to the chromosome 15 locus, sufficiently improves the fit. The chromosome 7 and 17 loci appear interesting only when considered together.

We may plot the two-dimensional scan results for chromosomes 7, 15 and 17 as follows. The result is in Fig. 8.8. LOD_i appears in the upper left triangle, and LOD_{fv1} appears in the lower right.

```
> plot(out.frommom, chr=c(7,15,17), lower="cond-int")
```

Let us now turn to the results for the subset receiving the `NPcis` mutation from their father. We again use a significance level of $\alpha = 0.2$ as a cutoff.

```
> summary(out.fromdad, perms=operm.fromdad, pvalues=TRUE,
+           alphas=c(0.2, 0.2, 0, 0.2, 0.2))

        pos1f pos2f lod.full  pval lod.fv1  pval lod.int pval
c9:c19     58     0     5.52 0.055    2.53 0.933   0.504    1
              pos1a pos2a lod.add  pval lod.av1  pval
c9:c19         55.5     0    5.02 0.018    2.03 0.095
```

We see evidence for a QTL on chromosome 9, in addition to the QTL previously identified on chromosome 19. The loci show no evidence of interaction. We plot LOD_i and LOD_{av1} for this pair of chromosomes as follows. See Fig. 8.9. The evidence for an interaction is negligible.

```
> plot(out.fromdad, chr=c(9, 19), lower="cond-add")
```

Let us complete our investigations with plots of the effects of the putative QTL pairs. We will use `effectplot`, though it is not ideal for a binary outcome. We first use `sim.geno` to perform multiple imputations of the genotypes at the putative QTL.

```
> nf1.fm <- sim.geno(subset(nf1, chr=c(7,17), ind=(frommom==1)),
+                      step=2.5, error.prob=0.001, n.draws=256)
> nf1.fd <- sim.geno(subset(nf1, chr=c(9,19), ind=(frommom==0)),
+                      step=2.5, error.prob=0.001, n.draws=256)
```

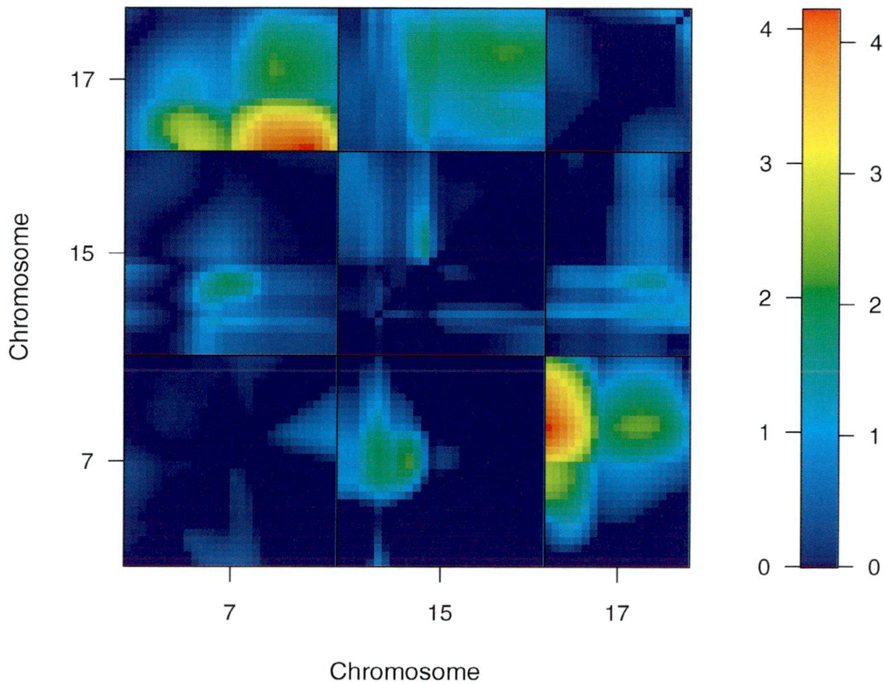

Figure 8.8. LOD scores, for selected chromosomes, from a two-dimensional, two-QTL genome scan with the `nf1` data, for those individuals receiving the *NPcis* mutation from their mother. LOD_i is displayed in the upper left triangle; LOD_{fv1} is displayed in the lower right triangle. In the color scale on the right, numbers to the left and right correspond to LOD_i and LOD_{fv1}, respectively.

We now create the plots, which appear in Fig. 8.10. We refer to pseudomarkers (that is, the positions on the grid between markers), directly by their chromosome and cM position.

```
> par(mfrow=c(1,2))
> effectplot(nf1.fm, mname1="7@45", mname2="17@3",
+            ylim=c(0,1))
> effectplot(nf1.fd, mname1="9@55.5", mname2="19@0",
+            ylim=c(0,1))
```

In the left panel of Fig. 8.10, we see that, for the individuals receiving the *NPcis* mutation from their mother, those who are BB at both the chromosome 7 and 17 loci are largely unaffected, while in the other three groups, about half are affected. The right panel shows that, for the individuals receiving the mutation from their father, the putative QTL on chromosomes 9 and 19 display approximately additive effects. (However, recall that with the logit link, additivity is in terms of log odds and so would not give parallel lines in this plot.)

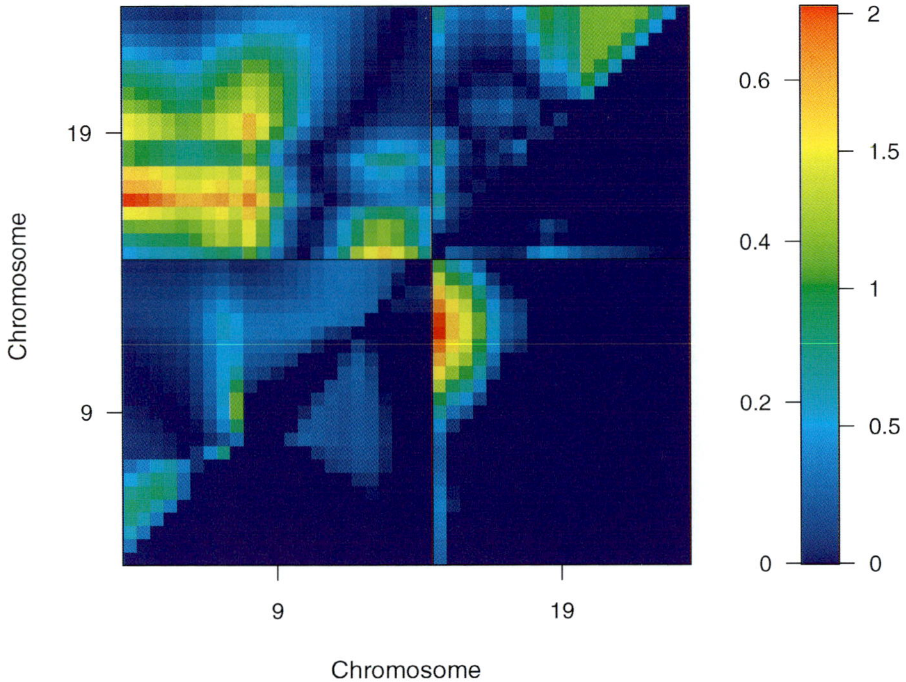

Figure 8.9. LOD scores, for selected chromosomes, from a two-dimensional, two-QTL genome scan with the `nf1` data, for those individuals receiving the *NPcis* mutation from their father. LOD_i is displayed in the upper left triangle; LOD_{av1} is displayed in the lower right triangle. In the color scale on the right, numbers to the left and right correspond to LOD_i and LOD_{av1}, respectively.

8.3 The X chromosome

If you thought the discussion of the X chromosome in single-QTL analysis (Sec. 4.4) was painful, you may wish to skip this section. In a two-dimensional, two-QTL genome scan, the two-dimensional scan within the X chromosome needs to be treated specially, and also the scans for the X chromosome against each autosome require special treatment. There are several technical difficulties.

First, as in the case of the single-QTL scan, additional covariates may be needed under the null hypothesis (see Table 4.3 on page 112), in order to avoid spurious linkage to the X chromosome as a result of sex- or cross-direction-differences in the phenotype. These will be needed for the case of two QTL on the X chromosome, and for the case of one QTL on the X chromosome and one on an autosome. Moreover, in the latter situation, we require the results of the single-QTL scan on the autosomes with these covariates included, for the comparison of the two-QTL models to a single-QTL model.

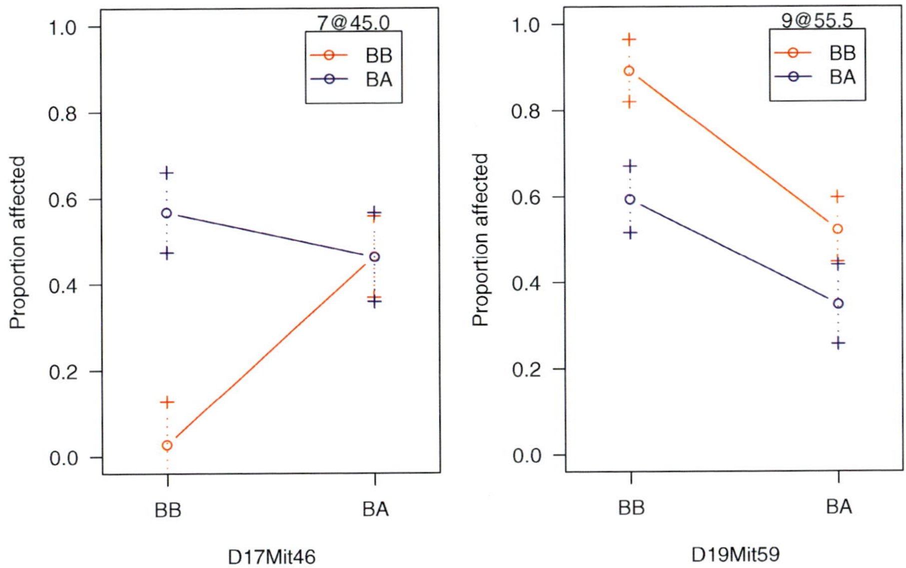

Figure 8.10. Estimated proportions of affected individuals as a function of genotype at two putative QTL for individuals receiving the *NPcis* mutation from their mother (left panel) or from their father (right panel), for the `nf1` data.

Table 8.1. Observed two-locus genotypes on the X chromosome, for a backcross with both sexes.

	QTL 1			
QTL 2	AA	AB	AY	BY
AA	*	*		
AB	*	*		
AY			*	*
BY			*	*

Second, in the case of two QTL on the X chromosome, we must recognize that not all two-locus genotypes will be observed. For example, consider the case of a backcross with both sexes. Females will have genotype AA or AB at a locus on the X chromosome, and males will be hemizygous A or B. In a single-QTL scan on the X, we consider these four groups, AA:AB:AY:BY. But in the consideration of two QTL on the X chromosome, only eight of the sixteen two-locus genotypes can actually be observed. (See Table 8.1.) While this is not a complex issue, the extension of algorithms for use with the X chromosome does require some care.

Finally, the degrees of freedom associated with our various LOD score statistics can vary greatly among the three possible cases: both putative QTL are on autosomes, one QTL is on an autosome and one is on the X chromosome, or both QTL are on the X chromosome. For example, consider our

Table 8.2. The number of estimated parameters for each model and the number of degrees of freedom for each test statistic, for a two-QTL scan in the case of an intercross with both sexes and both directions, according to the chromosome types for the two putative QTL.

	No. parameters				No. df				
chr type	f	a	1	0	f	fv1	i	a	av1
A:A	9	5	3	1	8	6	4	4	2
A:X	18	8	6	3	15	12	10	8	2
X:X	12	9	6	3	9	3	3	6	3

most complex situation, of an intercross with both sexes and both cross directions. The number of parameters for each of the four possible models and the number of degrees of freedom for each of the five possible test statistics, are displayed in Table 8.2. In the comparison of the full model (with two interacting QTL) to the best single-QTL model, with the M_{fv1} statistic, there are six degrees of freedom if both putative QTL reside on autosomes, three degrees of freedom if both putative QTL reside on the X chromosome, and twelve degrees of freedom if one QTL is on an autosome and one is on the X chromosome.

Thus, in assessing the statistical significance of the results of a two-dimensional, two-QTL scan, one must consider these three cases (illustrated in Fig. 8.11) separately, just as the autosomes and the X chromosome were considered separately in the single-QTL analysis (Sec. 4.4.2). We must admit that the details on how this should be done have not yet been worked out. For single-QTL analysis, we considered the autosomal and X chromosome genetic lengths. In the two-dimensional, two-QTL scan, one might consider the *areas* of the relevant regions.

Example

Let us briefly return to the `gutlength` data, described in Sec. 7.1. These data concern an intercross with both sexes and both directions.

We first load the data (which are contained in the R/qtlbook package), and run `calc.genoprob`. We will use a 5 cM grid, for the sake of speed, and because we will be taking only a cursory look at the results.

```
> data(gutlength)
> gutlength <- calc.genoprob(gutlength, step=5,
+                            error.prob=0.001)
```

We now run `scantwo`, using the defaults (maximum likelihood via the EM algorithm for the normal model).

```
> out.gl <- scantwo(gutlength)
```

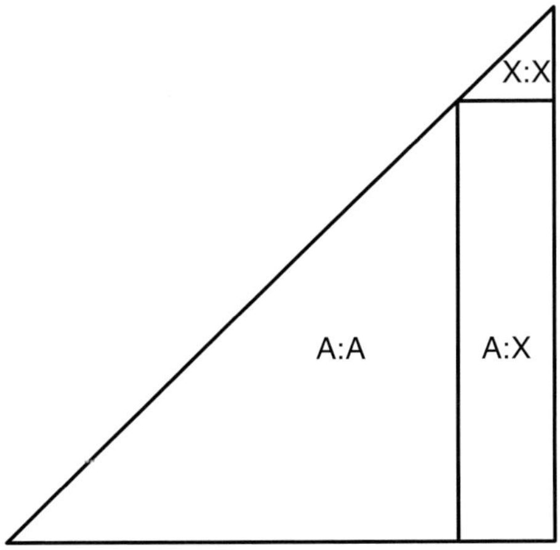

Figure 8.11. Regions in a two-dimensional, two-QTL scan, with the X chromosome included, that require separate treatment.

A plot of LOD_{fv1} and LOD_i is obtained as follows. (See Fig. 8.12.) We use `alternate.chrid=TRUE` so that the chromosome IDs may be more easily distinguished.

```
> plot(out.gl, lower="cond-int", alternate.chrid=TRUE)
```

Note the high LOD scores for the X chromosome when considered with any other chromosome: the segments for the X chromosome, on the top and right edges of the plot, really stand out. This is largely due to the fact that the degrees of freedom for these LOD score statistics are much larger for the A:X case (see Table 8.2). These portions of the two-dimensional scan require separate significance thresholds.

If we apply the simulation-derived significance thresholds for an intercross, cited in Sec. 8.1, we obtain the following summary.

```
> summary(out.gl, thresholds=c(9.1, 7.1, 6.3, 6.3, 3.3))

      pos1f pos2f lod.full lod.fv1 lod.int      pos1a pos2a
c5:cX    20    55    11.62    5.19   1.671         20    60
cX:cX    70    75     7.82    4.51   0.839         70    75
      lod.add lod.av1
c5:cX    9.95    3.52
cX:cX    6.98    3.67
```

We see evidence for QTL on chromosomes 5 and X (seen previously in the single-QTL scan; see Fig. 7.5 on page 188), and possibly two linked QTL on the X chromosome. But the significance thresholds we have used are likely inappropriate, and so these results should be viewed with a great deal of skepticism.

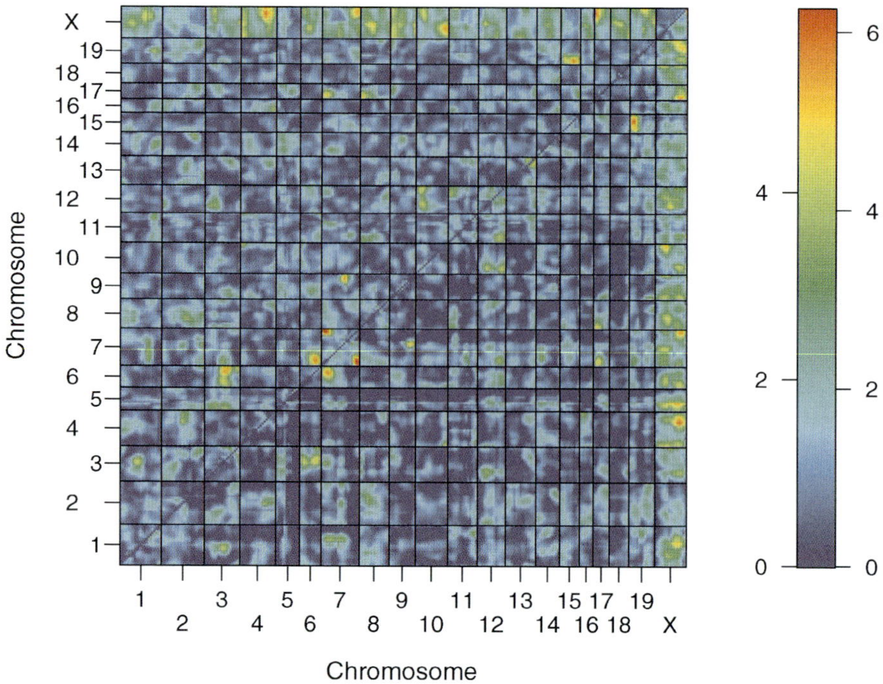

Figure 8.12. LOD scores from a two-dimensional, two-QTL genome scan with the `gutlength` data. LOD_i is displayed in the upper left triangle; LOD_{fv1} is displayed in the lower right triangle. In the color scale on the right, numbers to the left and right correspond to LOD_i and LOD_{fv1}, respectively.

8.4 Covariates

As discussed in Chap. 7, it may be useful to take account of covariates, such as sex, in QTL analyses. If a covariate is associated with the phenotype, its consideration will reduce residual variation and so can enhance our ability to detect QTL. Additive covariates may be easily incorporated into a two-dimensional, two-QTL genome scan. With inclusion of an additive covariate, x, the four models in equation (8.1) (page 214) become the following.

$$
\begin{aligned}
\text{H}_f&\colon\ y = \mu + \beta_x x + \beta_1 q_1 + \beta_2 q_2 + \gamma(q_1 \times q_2) + \epsilon \\
\text{H}_a&\colon\ y = \mu + \beta_x x + \beta_1 q_1 + \beta_2 q_2 + \epsilon \\
\text{H}_1&\colon\ y = \mu + \beta_x x + \beta_1 q_1 + \epsilon \\
\text{H}_0&\colon\ y = \mu + \beta_x x + \epsilon
\end{aligned}
\tag{8.5}
$$

LOD scores may be calculated as before (see Sec. 8.1), and the interpretation of the two-dimensional scan results are essentially unchanged. (As discussed in Sec. 7.1, one should be cautious about the use of secondary phenotypes as covariates, as they will not necessarily be independent of QTL genotype.)

On the other hand, interactive covariates in a two-dimensional scan can be cumbersome, and the results are not easy to interpret. In R/qtl, interactive covariates in `scantwo`, indicated via the `intcovar` argument, are allowed to interact with both QTL, as well as with the QTL × QTL interaction. And so the four models become the following.

$$\begin{aligned}
\text{H}_f\colon\ & y = \mu + \beta_x x + \beta_1 q_1 + \beta_2 q_2 + \gamma_{12}(q_1 \times q_2) + \\
& \qquad \gamma_{x1}(x \times q_1) + \gamma_{x2}(x \times q_2) + \gamma_{x12}(x \times q_1 \times q_2) + \epsilon \\
\text{H}_a\colon\ & y = \mu + \beta_x x + \beta_1 q_1 + \beta_2 q_2 + \gamma_{x1}(x \times q_1) + \gamma_{x2}(x \times q_2) + \epsilon \\
\text{H}_1\colon\ & y = \mu + \beta_x x + \beta_1 q_1 + \gamma_{x1}(x \times q_1) + \epsilon \\
\text{H}_0\colon\ & y = \mu + \beta_x x + \epsilon
\end{aligned} \tag{8.6}$$

The five LOD scores may be calculated as before, but the interpretation is rather different, as the LOD scores incorporate evidence for QTL × covariate interactions. Consider, for example, a binary covariate, such as sex, coded as $x = 0$ (female) or 1 (male). The interaction LOD score, LOD_i, being large indicates evidence for a QTL × QTL interaction in at least one of the two sexes.

We seldom make use of interactive covariates in a two-dimensional, two-QTL genome scan. While they may be useful, we prefer to postpone further investigation of QTL × covariate interactions to the more general exploration of multiple-QTL models, to be discussed in the next chapter.

Example

We return to the `gutlength` data, considered in the previous section and described in Sec. 7.1. First, we construct the covariates that we had used in Chap. 7.

```
> cross <- as.numeric(pull.pheno(gutlength,"cross"))
> frommom <- as.numeric(cross < 3)
> forw <- as.numeric(cross == 1 | cross == 3)
> sex <- as.numeric(pull.pheno(gutlength,"sex") == "M")
> crossX <- cbind(frommom, forw, frommom*forw)
> x <- cbind(sex, crossX)
```

Now, we perform the two-dimensional, two-QTL genome scan, with these additive covariates. It may take quite a bit of time.

```
> out.gl.a <- scantwo(gutlength, addcovar=x)
```

Let us plot the differences in the LOD scores obtained with and without the use of the additive covariates. We use `allow.neg=TRUE` so that the range of values includes negative numbers. Note that the differences are calculated via the function `-.scantwo`.

```
> plot(out.gl.a - out.gl, allow.neg=TRUE, alternate.chrid=TRUE)
```

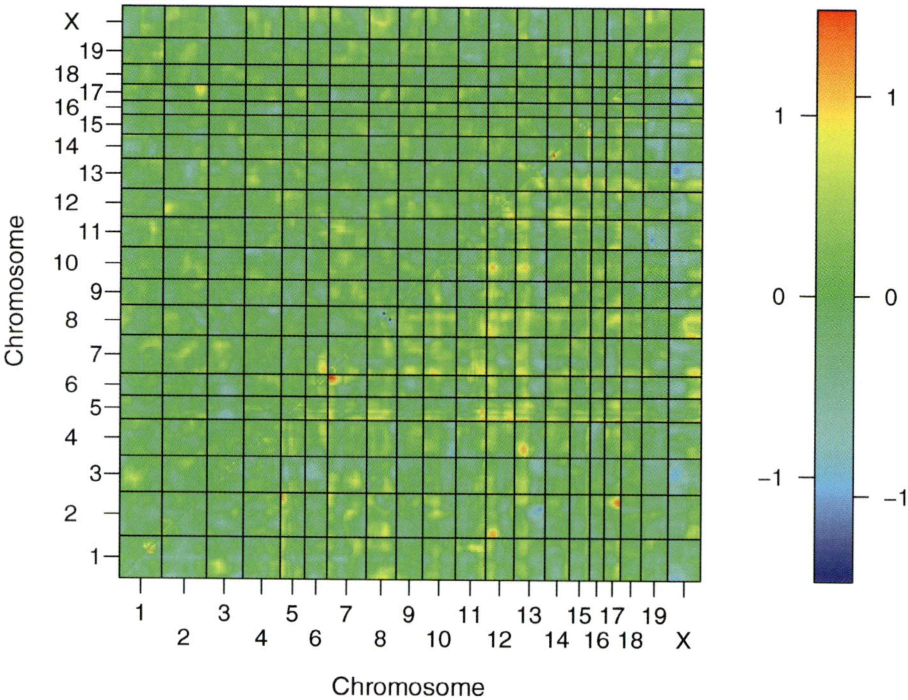

Figure 8.13. Differences in LOD scores calculated with and without the use of covariates, from a two-dimensional, two-QTL genome scan with the `gutlength` data. Differences in LOD_i are displayed in the upper left triangle; and differences in LOD_f are displayed in the lower right triangle. In the color scale on the right, numbers to the left and right correspond to LOD_i and LOD_f, respectively.

Note that the LOD scores have gone up and down by as much as one unit (Fig. 8.13). However, the location of the maximum LOD score is the same, and has not changed much with the inclusion of the covariates. (We use the function `max.scantwo`, for pulling out the peak with maximum LOD score from a two-dimensional, two-QTL genome scan.)

```
> max(out.gl)

      pos1f pos2f lod.full lod.fv1 lod.int      pos1a pos2a
c5:cX    20    55     11.6    5.19    1.67         20    60
      lod.add lod.av1
c5:cX    9.95    3.52
```

```
> max(out.gl.a)

      pos1f pos2f lod.full lod.fv1 lod.int      pos1a pos2a
c5:cX    20    55     11.6    5.06    1.44         20    60
      lod.add lod.av1
c5:cX    10.2    3.61
```

8.5 Summary

Most QTL will be seen in the results of the traditional, single-QTL genome scan. However, two-dimensional, two-QTL genome scans, while computationally intensive, provide the opportunity to identify additional QTL, particularly those involved in epistatic interactions. In addition, the two-dimensional scan can give more clear evidence regarding linked QTL. This is particularly true in the case that two QTL are linked in *repulsion* (having effects of opposite sign): such QTL will show little marginal effect, but may stand out clearly when the two loci are considered jointly.

The interpretation of the results of a two-dimensional scan is not simple. We have described one approach to summarize the results, with an emphasis on measuring evidence for the presence of a second QTL. It is important to emphasize that the results of a two-dimensional scan need to be considered in combination with the initial single-QTL analysis. Moreover, the findings should be considered preliminary, to be used as a starting point for the fit and exploration of multiple-QTL models.

The X chromosome presents additional difficulties in the context of a two-dimensional genome scan. The varying degrees of freedom attached to different LOD scores (in the case of two QTL on autosomes, two QTL on the X chromosome, or one QTL on an autosome and one on the X chromosome) imply that separate significance thresholds will be required, but how this should be done remains a matter for research.

It is often important to consider covariates within QTL analyses. The inclusion of additive covariates in a two-dimensional, two-QTL genome scan is simple. Interactive covariates, however, are difficult to work with in this context, and so we have seldom made use of them in two-dimensional scans.

8.6 Further reading

Haley and Knott (1992) were the first to propose an exhaustive search over all two-QTL models. Sugiyama *et al.* (2001) applied the approach to the `hyper` data, which we have used extensively as an example. Sen and Churchill (2001) helped to solidify the use of two-dimensional, two-QTL scans as a general technique. Ljungberg *et al.* (2004) described an algorithm for identifying the optimal pair of loci in a two-dimensional scan without performing an exhaustive search.

9

Fit and exploration of multiple-QTL models

The majority of efforts for QTL mapping have used a hypothesis testing approach. For example, in single-QTL analyses (Chap. 4), one considers each genomic position, one at a time, and asks the question, "Is there a QTL here?" A primary focus is on the adjustment for the number of tests (i.e., for the scan across the genome), to control the rate of false positive declarations of linkage.

A two-dimensional, two-QTL genome scan (Chap. 8) largely gets around the principal weaknesses of single-QTL analysis (of separating linked QTL and identifying interacting QTL), but again this is a hypothesis testing approach. One considers a pair of putative QTL and asks, "Are there QTL here and here?"

These approaches work surprisingly well, largely due to the independent assortment of chromosomes, but they are formally correct only if a phenotype is affected by no more than two QTL. However, we expect complex traits to be affected by multiple loci. The single- and two-QTL analysis methods indicate individual pieces of the complex genetic architecture that underlies the phenotype. To properly weigh the evidence for the QTL, one should ultimately bring these pieces together in a single coherent model.

The primary goal of QTL mapping, to identify the set of loci that contribute to the phenotypic variation, thus does not fit well into the sequence of yes-or-no questions that forms the hypothesis testing framework. Rather, QTL mapping is best viewed as a *model selection* or *variable selection* problem: what set of loci (and QTL × QTL interactions) are best supported by the data?

In this chapter, we describe the key aspects of the model selection approach to multiple QTL mapping. We focus primarily on classical approaches to the problem, though we briefly describe approaches using Bayesian statistical inference, primarily to emphasize the advantages and disadvantages of the Bayesian approach. We further describe the R/qtl functions for fitting and exploring multiple QTL models.

K.W. Broman, Ś. Sen, *A Guide to QTL Mapping with R/qtl*,
Statistics for Biology and Health, DOI 10.1007/978-0-387-92125-9_9,

We will focus on the case of a continuously varying quantitative trait with normally distributed residual variation. The ideas may be extended for use with alternate types of phenotypes, including binary traits, censored survival times, and phenotype distributions exhibiting a spike, but we will not pursue such extensions here.

9.1 Model selection

As discussed in Chap. 1, the QTL mapping problem can be split into two parts: the missing data problem and the model selection problem. The missing data problem arises from the fact that, while QTL may reside anywhere in the genome, we observe individuals' genotypes only at discrete landmarks, the genetic markers. We thus use the observed marker genotype data to infer the genotypes at all intervening locations. This problem, while it remains an annoyance, has been well-solved in a number of ways (including maximum likelihood via the EM algorithm, multiple imputation and Haley–Knott regression); it concerns the fit of a QTL model in the case that genotypes at the putative QTL are not observed.

The more important aspect of QTL mapping is the model selection problem. Imagine one could observe complete genotype data on each individual. For example, in a backcross, imagine that we knew, at every polymorphism at which the parental strains differ, which individuals were homozygous and which were heterozygous. We are still confronted with the difficult problem of identifying the subset of loci that truly influence the phenotype, and of how they combine together to produce the phenotype. By *model*, we mean a defined set of genetic loci (and, potentially, QTL $\times$ QTL interactions), and in *model selection*, we seek to identify such a set of QTL that are well supported by the observed data.

As with hypothesis testing, in selecting a set of loci that are viewed to influence the phenotype, one can make two types of errors: we may miss important loci (false negatives), and we may include extraneous loci (false positives). Unlike hypothesis testing, here we can make both errors at the same time.

QTL mapping projects are initiated for a variety of different purposes. In evolutionary studies, one may be particularly interested in the distribution of QTL effects and of the relative contribution of epistatic effects. In agriculture, one may be primarily interested in obtaining information that will guide future selection experiments. In biomedical experiments, the ultimate goal is to identify the gene or genes (and perhaps even the individual mutations) that contribute to phenotypic differences, in order to gain a better understanding of the mechanism of disease and to identify potential targets for therapeutic drugs.

We focus primarily on QTL mapping for biomedical research. In this context, we are particularly concerned with avoiding false positives, so that downstream experiments to fine-map and ultimately identify the causal gene or genes will not be conducted in vain. We thus view the goal of model selection for QTL mapping to be to control the rate of inclusion of extraneous loci, while identifying as many true QTL as possible.

Knowledge of epistatic interactions can be important, as they may influence the downstream fine-mapping experiments. For example, experiments to dissect a QTL often begin with the construction of a congenic strain, in which a genomic segment from one strain is introgressed into another strain, creating an inbred strain that is homozygous A everywhere except for one genomic segment, where it is homozygous B. Information on epistatic interactions may indicate that one type of congenic (in which a segment from the B strain is isolated within the A background) may have no effect, while the other (in which the A segment is isolated within the B background) may have a large effect.

However, we are generally not so concerned with precisely identifying the interactions between QTL, but rather we seek to identify the major players and want to ensure that the presence of interactions does not prevent us from identifying important loci. False inference of the presence of an interaction that does not exist (or of the absence of an interaction that does exist) is not so bad as missing an important locus or including an extraneous one.

While there is a large literature on the problem of model selection in linear regression, there are some important differences in the QTL mapping problem. First, model selection in regression has often focused on the minimization of prediction error: one expects that all covariates have some effect on the outcome, but by eliminating some covariates from the prediction model, one allows some bias but eliminates a great deal of variance, and so ultimately obtains improved predictions. In QTL mapping, however, we are not interested in prediction, but rather in identifying the important elements of the model. Second, in QTL mapping we are confronted with a continuum of potential covariates (putative QTL locations along the chromosomes), though with a great deal of missing covariate information. Related to this point, we do not expect to identify the exact site of a QTL, but want to pick loci that are not far from the true QTL. Finally, the correlation among the covariates has a special structure in QTL mapping: from chromosome to chromosome, the genotypes are independent. Moreover, along a chromosome, they display a very simple correlation structure. In the case of no crossover interference, genotypes along a chromosome form a Markov chain, so that, given the genotypes at a particular locus, the genotypes at sites to the left of the locus are conditionally independent of the genotypes at the sites to the right of the locus. While this conditional independence does not hold in the presence of crossover interference, the correlation structure remains relatively simple, and in either case it is effectively known. With this simple correlation structure among the potential covariates, procedures that have been found to perform poorly in other contexts may actually work quite well for the QTL mapping problem.

In defining a model selection procedure for QTL mapping, one must make four distinct choices. First, one must choose a *class of models*, such as strictly additive QTL models, models that allow pairwise interactions between QTL, or models that allow interactions of any order. Second, one must define a method for *model fit*; that is, for a particular QTL model, with QTL in fixed positions and a defined set of epistatic interactions, how will the missing genotype data be overcome to define the fit of the model? Third, we need a method for *model search*. The space of models will be far too large for the models to be considered exhaustively; we need a procedure for exploring the model space to identify good ones, recognizing that we will only be able to consider small slices through the model space. Finally, and most importantly, we need a method for *model comparison*. In forming a model comparison procedure, it can be useful to imagine that we could fit all possible models; which model would we then choose? Larger models will provide an improved fit, and so we must balance quality of fit with model complexity. How much of an improvement in fit should be required in order to incorporate an additional QTL into the model?

We will discuss these four aspects of a model selection procedure in the following subsections. We will then bring these separate streams of thought back together, to emphasize the tradeoffs accompanied by any set of choices.

9.1.1 Class of models

The first choice, in forming a model selection procedure, is of the class of models. The simplest class is that of strictly additive QTL models.

$$y = \mu + \sum_j \beta_j q_j + \epsilon$$

With such an additive model, the effect of a QTL is constant, independent of the genotypes at other loci.

One may also wish to include models with pairwise interactions.

$$y = \mu + \sum_j \beta_j q_j + \sum_{j,k} \beta_{jk} q_j q_k + \epsilon$$

In doing so, we generally enforce a hierarchy, under which the inclusion of a QTL $\times$ QTL interaction requires the inclusion of main effects for each QTL. One advantage of such a hierarchical structure is that the coding of QTL effects becomes immaterial. The model space is also restricted. The enforcement of such a hierarchy is similar to the choice, in single-QTL analysis in an intercross, to always include both degrees of freedom at any QTL, and so not explore strictly dominant or recessive models, or models in which the two alleles at a locus are strictly additive. The hierarchical structure can make it difficult to identify loci with important interactions but limited marginal

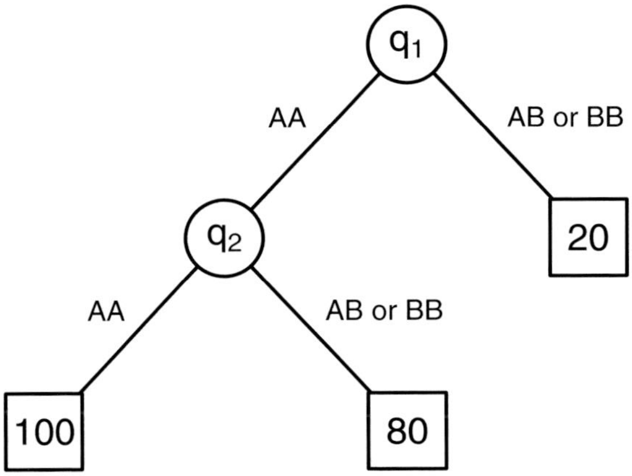

Figure 9.1. An example of a regression tree. This is a type of decision tree that describes a QTL model. Individuals with genotype AB or BB at QTL 1 have average phenotype 20, individuals with genotype AA at both QTL 1 and QTL 2 have average phenotype 100, and individuals with genotype AA at QTL 1 and either AB or BB at QTL 2 have average phenotype 80.

effects since, in an intercross, one brings in all eight degrees of freedom for a pair of interacting loci.

One can, of course, expand the model space to include higher-order interactions, including three-way interactions (in which the strength of interaction between two loci depends on the genotype at a third locus), four-way interactions, and so on. Evidence for higher-order interactions has been observed, particularly for the case of a QTL $\times$ QTL $\times$ covariate interaction, but this expansion of the model space can come at a severe price, in the form of a greater false positive rate or a decrease in power (if the false positive rate is controlled). In an intercross, it is assuredly difficult to distinguish a true three-way interaction from three QTL with only pairwise interactions, particularly because of the 1:2:1 segregation of alleles at a locus, which makes many three-locus genotypes quite rare and so potentially unobserved in a cross of ordinary size. Thus, while we admit that an investigation of three-way interactions in a backcross may be fruitful, we believe that it is likely best to focus, in an intercross, on only pairwise interactions.

Related to the class of models is what might be called the "form of expression" of a model. Here we have in mind regression trees, such as that in Fig. 9.1. This is a form of decision tree, in which terminal nodes define the average phenotype for individuals with a particular multilocus QTL genotype.

The class of regression trees is equivalent to the class of ordinary linear regression models that allow all orders of interactions, but the form in which such models are expressed is radically different. The simple regression tree in Fig. 9.1 would be cumbersome to write with a linear model, but then an additive QTL model would make a more complex regression tree.

Regression trees have not been used much in genetics, but the natural way in which complex interactions can be expressed through regression trees does make them intriguing. However, the enormous model space is forbidding.

One final point: it is sometimes necessary to restrict the model space in order to avoid artifacts or to speed up computations. For example, we often assume that any interval between markers contains at most one QTL. One may further require that any two QTL are separated by at least two markers. This is done because a model with two tightly linked QTL in repulsion (that is, having effects of opposite signs) can occasionally provide a spuriously good fit. This is particularly the case when there are a few individuals with outlying phenotype values. It is related to difficulties along the diagonal in a two-dimensional, two-QTL genome scan.

Another way to restrict model space is to assume that putative QTL must reside at the marker locations. In the case of high-density genotype data, this can be a reasonable approximation, and can allow great speed-up in the computations, particularly if the marker genotype data are relatively complete. Related to this is the density of the grid (i.e., the step size) in standard interval mapping: a more coarse grid gives less precise results but the calculations are much faster.

9.1.2 Model fit

A central part of any QTL mapping procedure is a method for fitting QTL models. In the case of complete genotype data at the putative QTL, one would use linear regression. (Recall that we are focusing, in this chapter, on continuously varying quantitative traits with normally distributed residual variation.) The fit of a model could be described by the residual sum of squares, $\text{RSS}(\gamma) = \sum_i (y_i - \hat{y}_i)^2$, where γ denotes a model, y_i is the phenotype for individual i and $\hat{y}_i$ is the corresponding fitted value under the model. Equivalently, one could use the LOD score, $\text{LOD}(\gamma) = (n/2)\log_{10}[\text{RSS}_0/\text{RSS}(\gamma)]$, where $\text{RSS}_0 = \sum_i (y_i - \bar{y})^2$ is the residual sum of squares for the null model. Again, the good models are those for which $\text{RSS}(\gamma)$ is small and so $\text{LOD}(\gamma)$ is large.

In general, one will have no genotype data at the putative QTL, and will need to use data at linked markers to infer the QTL genotypes. This is the missing data problem, which we have now rolled into the model selection problem.

The various methods for fitting a single-QTL model (described in Chap. 4) are all readily extended to the case of a multiple-QTL model, though with different degrees of difficulty. The relative advantages and disadvantages of the different methods largely remain, though the multiple imputation approach is finally given an opportunity to shine in the context of multiple-QTL models.

The simplest method is Haley–Knott regression. We replace the indicators for QTL genotypes with their expected values, given the available marker data, and so perform linear regression of the phenotype on the multiple QTL genotype *probabilities*. The great advantage of this approach is computational speed: we perform a single regression for each QTL model of interest. The disadvantage of the approach is that it again fails to make complete use of the available genotype data, and it can give spuriously large evidence for linkage in the presence of selectively genotyped data. But in the case of relatively dense markers and relatively complete marker genotype data, Haley–Knott regression is clearly the preferred method.

The multiple imputation approach extends to the case of multiple-QTL models without modification, and while with single- and two-QTL models the computation time for the imputations and for the fixed set of linear regressions to be performed at each putative QTL or QTL pair weakened the value of the approach, the imputations are performed just once, and so in the exploration of multiple-QTL models, the multiple imputation approach is quite valuable.

The extended Haley–Knott method can be applied for the case of multiple-QTL models, but it has not yet been implemented in software. While we expect that its speed and robustness properties will carry over to the case of multiple-QTL models, this remains to be tested.

The extension of standard interval mapping (maximum likelihood via an EM algorithm) to the case of multiple QTL models is called multiple interval mapping (MIM). The theory is straightforward, though there are some practical difficulties. We seek to fit a model of the form

$$y = X\beta + \epsilon$$

where X is a matrix of QTL genotype indicators (and possibly indicators for QTL $\times$ QTL interactions), which are not observed. The EM algorithm to derive maximum likelihood estimates under this model involves first calculating the expected value of the X and $X'X$ matrices, given the observed phenotype and marker genotype data, and given current estimates for the parameters β and σ:

$$\begin{aligned} Z^{(s)} &= \mathrm{E}(X|y, M, \hat{\beta}^{(s-1)}, \hat{\sigma}^{(s-1)}) \\ W^{(s)} &= \mathrm{E}(X'X|y, M, \hat{\beta}^{(s-1)}, \hat{\sigma}^{(s-1)}) \end{aligned}$$

With $Z^{(s)}$ and $W^{(s)}$ in hand, the M-step of the EM algorithm is relatively easy. Updated estimates of the model parameters, β, are obtained by solving the normal equations, with X and $X'X$ replaced by their expected values:

$$W^{(s)}\hat{\beta}^{(s)} = \left[Z^{(s)}\right]' y$$

The updated estimate of the residual SD is obtained as follows.

$$\hat{\sigma}^{(s)} = \sqrt{\left(y'y - y'Z^{(s)}\hat{\beta}^{(s)}\right)/n}$$

It is the E-step (calculating Z and W) that is difficult. Part of the problem, particularly for calculating W, is one of bookkeeping: handling arbitrary models with an arbitrary number of QTL and an arbitrary set of epistatic interactions. But with that aside, there remains the difficulty, in the case of p QTL, of summing over the 2^p possible multilocus QTL genotypes in a backcross (or 3^p in an intercross). This must be done for each individual and at each iteration of the EM algorithm, and so for models with a large number of QTL, the computational demands can be great. One technique that has been used is to trim multilocus QTL genotypes that have low prior probability, say < 0.001. By prior probability, we mean the probability of a multilocus genotype pattern conditional on the available marker genotype data (but not conditional on the observed phenotypes and the current estimates of the model parameters). This trimming can greatly reduce the number of possible multilocus QTL genotypes; the result is an approximation to the true likelihood, but it can give a great gain in computational speed.

With any of the four methods for fitting a QTL model (which we will denote γ), we will characterize model fit by the LOD score: the $\log_{10}$ likelihood ratio comparing the model γ to the null model (denoted $\emptyset$).

9.1.3 Model search

The model space is far too large for all models to be considered exhaustively. If we restrict ourselves to the marker positions and to additive QTL models, with 100 markers there are $2^{100} \approx 10^{30}$ models. If we assume that there are no more than 25 QTL but allow pairwise interactions between QTL, there are about 10^{113} possible models.

The enormous size of the model space thus requires a search algorithm, so that we may identify the good models even though we may visit only a minuscule proportion of the possible models. Let us focus initially on additive QTL models.

The simplest algorithm is forward selection. We begin by considering all possible single-QTL models, and we pick the best of these (i.e., that giving the largest LOD score), say $\gamma_1 = \{\lambda_1\}$. We then consider all two-QTL models that include our first QTL, and pick the locus that gives the greatest increase in LOD, to obtain $\gamma_2 = \{\lambda_1, \lambda_2\}$. Next, we consider all three-QTL models that include the first two QTL identified, to obtain $\gamma_3 = \{\lambda_1, \lambda_2, \lambda_3\}$. We continue in this way, building a nested sequence of models of increasing size. In the case of 100 putative QTL locations and additive QTL models, we would visit a set of 5051 models (out of the total of $\sim 10^{30}$ models).

Forward selection has a rather poor reputation in the statistical literature, as once a term enters the model, it is never removed. And in many cases one will find that, even with an extremely large data set, a single covariate that does not belong in the model may mimic a pair of covariates that do belong in a way that the single false covariate will always enter the model before either of the pair of covariates that truly belong. However, with the simple correlation structure among genotypes at putative QTL locations, one can show that in the case of additive QTL models, this problem will not occur, at least for very large data sets.

The reverse of forward selection is backward elimination. One begins with a large model (perhaps obtained by forward selection), and drops the covariates from the model one at a time, at each step dropping the covariate that results in the smallest decrease in the LOD score. Thus we construct a nested sequence of models of decreasing size.

One may similarly construct stepwise algorithms that include some forward and some backward steps. There are also numerous randomized algorithms (including Markov chain Monte Carlo, simulated annealing, and genetic algorithms) that take a random walk through model space, and generally do not require a strict improvement in the LOD score at each step.

In the case that pairwise interactions between loci are to be considered, a forward selection algorithm would also consider the addition of an interaction between the loci in the current model, and of a new locus that interacts with one of the loci in the current model. One may also wish to include two-step jumps, in which at each step of forward selection, a two-dimensional scan is performed, and so two novel QTL (whether additive or interacting) may be brought into the model together. This would be particularly useful for identifying a pair of interacting loci that show no marginal effects, or for a pair of loci linked in repulsion (having effects of opposite signs); in both of these cases, the two loci may not appear interesting individually, and so would likely be missed by a single-step forward selection algorithm.

It can be useful, in a search algorithm for multiple QTL mapping, to consider a refinement of the QTL locations at each step of a stepwise algorithm, as the likely position for a QTL may change as additional QTL are brought into the model. A simple but effective algorithm is formed by refining the location of each QTL in the model one at a time, keeping the locations of all other QTL fixed. The QTL are not moved between chromosomes, and the order of the QTL within a chromosome is not altered, and so in refining the position of a given QTL, one scans across its chromosome, or along the interval defined by the positions of the flanking QTL, if there are multiple QTL on the chromosome. We would generally consider the QTL in a random order, and would iterate the refinement process until no further change in QTL position is observed.

In general, we would prefer the most exhaustive search possible, though more extensive searches are accompanied by greater computation time and may give no improvement, in that the optimal model might be found with a simpler and less computationally intensive search.

Our preferred approach is to use forward selection to a model of moderate size (say 10 or 20 QTL), followed by backward elimination all the way to the null model. The chosen model would be that which optimized the model comparison criterion (see the next section), among all models visited. A simple and effective search algorithm, allowing for the possibility of pairwise interactions (implemented in the `stepwiseqtl` function in R/qtl; see Sec. 9.3.7), is the following.

1. Start by performing a single-QTL genome scan, and choose the position giving the largest LOD score.
2. With a fixed QTL model in hand:
   a. Scan for an additional additive QTL.
   b. For each QTL in the current model, scan for an additional interacting QTL.
   c. If there are ≥ 2 QTL in the current model, consider adding one of the possible pairwise interactions.
   d. Optionally perform a two-dimensional, two-QTL scan, seeking to add a pair of novel QTL, either additive or interacting.
   e. Step to the model that gives the largest value for the model comparison criterion, among those considered at the current step.
3. Refine the locations of the QTL in the current model.
4. Repeat steps 2 and 3 up to a model with some predetermined number of loci.
5. Perform backward elimination, all the way back to the null model. At each step, consider dropping one of the current main effects or interactions; move to the model that maximizes the model comparison criterion, among those considered at this step. Follow this with a refinement of the locations of the QTL.
6. Finally, choose the model having the largest model comparison criterion, among all models visited.

In this forward/backward algorithm, it is likely best to build up to an overly large model and then prune it back. Note that there is no "stopping rule;" the chosen model is that which optimizes the model comparison criterion, among all models visited. The search can be time consuming, particularly if a two-dimensional scan is performed at each forward step. Such two-dimensional scans may be useful for identifying QTL linked in repulsion (having effects of opposite sign) or interacting QTL with limited marginal effects, but our limited experience suggests that they are not necessary; important linked or interacting QTL pairs can be picked up in the forward selection to a large model, and will be retained in the backward elimination phase.

9.1.4 Model comparison

By far the most important aspect of the QTL mapping problem is the criterion for choosing among possible QTL models. For models of the same size, we would choose that with the largest LOD score (i.e., that with the maximum likelihood). However, the LOD score can always be increased by adding an additional QTL to a model. Thus, we must seek some balance between the fit of a model (represented by the LOD score), and the complexity of the model (the number of QTL and interactions). Our goal is to define a criterion that will appropriately balance the false positive and false negative rates. In particular, we seek to control the false positive rate (that is, the rate of inclusion of extraneous loci) at some chosen level and then identify as many true QTL as possible.

We will not attempt to discuss this issue in detail, but rather will focus on a single approach that we consider most appropriate for the QTL mapping problem in biomedical research.

Much of the literature on model selection criteria has focused on minimizing prediction error; while these approaches may also work well for identifying the important players (which is our goal), in general they tend to produce overly large models, as the inclusion of a few extraneous covariates will not weaken predictions so much as missing a few important covariates. Also, much work has focused on the asymptotic behavior of the criteria (that is, the case of an infinitely large sample), but in the large sample case, some important features of the data (such as the number of potential covariates) are no longer seen to matter, and so the asymptotic behavior of a procedure can be a poor guide to its small sample performance.

Let us begin by considering additive QTL models. We prefer to use a penalized LOD score.

$$\text{pLOD}_a(\gamma) = \text{LOD}(\gamma) - T|\gamma| \tag{9.1}$$

where γ denotes a model, $|\gamma|$ is the number of QTL in the model, and T is a penalty on the size of the model.

We seek a penalty T that will control the rate of inclusion of extraneous loci at some chosen rate (e.g., 5%). We would like the rate to be controlled no matter the true model, but consider, in particular, the case of the null model, $\emptyset$, and that we perform a single-QTL genome scan. The penalized LOD for the null model is $\text{pLOD}_a(\emptyset) = 0$, since the LOD score is defined relative to the null model and $|\emptyset| = 0$. The penalized LOD for a single-QTL model is $\text{pLOD}_a(\{\lambda\}) = \text{LOD}(\lambda) - T$, where $\{\lambda\}$ is the model with a single QTL at position λ and $\text{LOD}(\lambda)$ is the LOD score from a single-QTL scan at position λ.

We would choose the model $\{\lambda\}$ over the null model when $\text{LOD}(\lambda) > T$. This suggests setting T to be the 95th percentile of the genome-wide maximum LOD score under the global null hypothesis (of no QTL anywhere), which may be estimated via a permutation test. This extends the usual procedure for single-QTL analysis to a criterion useful for choosing among additive QTL models of any size. The choice of penalty guarantees the control of the false positive rate at the target level only for the case that the truth is the null model and that the search considers models with no more than one QTL. But computer simulation experiments indicate that the false positive rate is maintained reasonably well for larger models and for more extensive searches.

To extend this approach to the case of pairwise interactions among QTL, we could apply separate penalties on the main effects and the interactions, T_m and T_i:

$$\text{pLOD}(\gamma) = \text{LOD}(\gamma) - T_m|\gamma|_m - T_i|\gamma|_i \tag{9.2}$$

where $|\gamma|_m$ is the number of QTL in the model and $|\gamma|_i$ is the number of interaction terms. We will focus on the case that a hierarchical structure is imposed on the model, with the inclusion of an interaction requiring the inclusion of both of the corresponding main effects.

We take T_m to be the significance threshold from a single-QTL genome scan, as before. With this choice, the extended criterion in equation (9.2) is equivalent to that in equation (9.1) if one restricts the search to additive QTL models.

We are left with the choice of the penalty on interaction terms. We can apply the same sort of logic that led us to the penalty on main effects. Imagine there are two additive QTL, and that we perform a two-dimensional, two-QTL genome scan. We could then define the interaction penalty as follows:

$$T_i^H = \text{95th percentile of } [\max_{\lambda_1,\lambda_2} \text{LOD}_f(\lambda_1,\lambda_2) - \max_{\lambda_1,\lambda_2} \text{LOD}_a(\lambda_1,\lambda_2)] \tag{9.3}$$

where LOD_f and LOD_a are the LOD scores for full and additive two-QTL models, respectively (see Chap. 8). With this choice, we would control the rate of inclusion of an extraneous interaction at our target rate. While ideally one would determine the distribution of $\max \text{LOD}_f - \max \text{LOD}_a$ in the presence of two additive QTL, it is not likely to be too different from the distribution under the null hypothesis of no QTL, and so we may use a permutation test in a two-dimensional genome scan to estimate T_i. We call this choice the heavy penalty, T_i^H, as we will introduce a light penalty next.

If the logic in the previous paragraph is reasonable, why not extend it? Imagine that there is a single QTL, and that we perform a two-dimensional, two-QTL scan. To control the rate of inclusion of a false interacting QTL, we define a light interaction penalty, T_i^L, through the equation

$$T_m + T_i^L = \text{95th percentile of } [\max_{\lambda_1,\lambda_2} \text{LOD}_f(\lambda_1,\lambda_2) - \max_{\lambda} \text{LOD}_1(\lambda)] \tag{9.4}$$

where LOD_f is the LOD score for the full model, with two interacting QTL, and LOD_1 is the LOD score for a single-QTL model. With the interaction penalty defined in this way, we ensure that, in the case of a single QTL,

Table 9.1. Estimated penalties, derived by computer simulation, for a genome modeled after the mouse and with markers at a 10 cM spacing.

	Cross	
Penalty	Backcross	Intercross
T_m	2.69	3.52
T_i^H	2.62	4.28
T_i^L	1.19	2.69

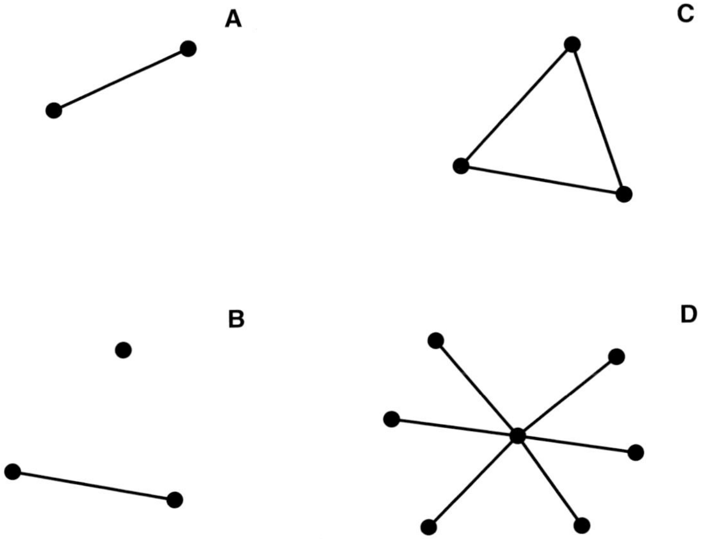

Figure 9.2. Graphical depictions of QTL models, with nodes corresponding to QTL and edges indicating interactions. **A**: Two interacting QTL. **B**: Three QTL, of which two interact. **C**: Three QTL with all pairs interacting. **D**: A seven-QTL model, with one QTL exhibiting pairwise interactions with each of the other QTL.

the rate of inclusion of a second, interacting QTL is controlled at the chosen rate. In the presence of two additive QTL, the interaction term will be falsely included at a much higher rate, but we are less concerned about such errors and seek principally to control the rate of false positive QTL.

In thinking about these penalties, and the difference between the heavy and light interaction penalties, it is good to have some particular numbers in mind. Estimated penalties, derived via computer simulation of backcrosses and intercrosses with genomes modeled after that of the mouse and with markers at a 10 cM spacing, are shown in Table 9.1. The light interaction penalty is *much* smaller than the heavy interaction penalty.

It is useful, in the consideration of QTL models with possible pairwise interactions (and with our imposed hierarchical structure that requires the inclusion of both main effects along with any interaction term), to depict such models as graphs, with nodes (i.e., dots) corresponding to QTL and edges (i.e., line segments between QTL) indicating interactions. Consider Fig. 9.2. Model A has two interacting QTL. Model B has three QTL, of which two interact. Model C has three QTL with all possible pairwise interactions. In model D, there are seven QTL, with one QTL interacting with each of the other six.

In the analysis of QTL data with the exclusive use of the light interaction penalty, T_i^L, we often identified models like that in Fig. 9.2D, with a single QTL interacting with many others. This seems implausible. Computer simulation experiments confirmed this sentiment: in the case of an additive QTL model with many QTL, the exclusive use of the light interaction penalty will often result in the identification of an extraneous locus, interacting with many or all of the true loci. The problem is that, with the light interaction penalty, we imagine that the truth is a "dot" (one QTL), and we control the rate of inclusion of an extraneous "pin" (an additional, interacting QTL). But in the presence of multiple QTL, an extraneous locus can purchase its entry into the chosen model via numerous lightly weighted interactions: we apply too small a penalty to multipronged pins.

An *ad hoc* modification of our penalized LOD score can eliminate this undesired behavior. Consider the graph corresponding to a particular model. (For example, imagine that the four parts of Fig. 9.2 formed a single model, with a total of 15 QTL and 11 pairwise interactions.) For each connected component of the graph (that is, for each cluster of interacting QTL), we apply a single light interaction penalty and give all other interactions the heavy penalty. (For the model formed from all parts of Fig. 9.2, there would be 15 main effect penalties, T_m, four light interaction penalties, T_i^L, and seven heavy interaction penalties, T_i^H.) This approach serves as a compromise between using only heavy interaction penalties (which would result in low power to detect interacting loci) and only light interaction penalties (which can lead to a high rate of extraneous loci).

The use of this model comparison criterion is guaranteed to control the rate of inclusion of extraneous loci only in the presence of one or two QTL, and with the model search not extending beyond two QTL. Moreover, we should expect that the inclusion of extraneous loci will increase with the size of the model, as in the presence of many QTL, there are many more ways to incorporate an additional extraneous interacting locus (i.e., to attach an extraneous "pin" to the model). But such behavior can probably not be eliminated and is not unreasonable; having one extraneous locus among nine identified QTL is not so bad as one extraneous locus among three identified QTL.

9.1.5 Further discussion

A model selection procedure for QTL mapping has four parts: a choice of the class of models, a method for model fit, a method for model search, and a criterion for comparing models. We prefer to keep these pieces distinct. In particular, we are opposed to the use of "stopping rules," which combine model search and model comparison; rather, a model comparison criterion should be chosen, imagining one could visit all possible models in the class under consideration, and the aim of the model search algorithm should be to optimize this criterion.

Some strategies seek to restrict the search over models in order to increase power. The argument is that, if a smaller number of models are visited, the adjustment for the range of models visited will be less severe and so a more permissive model comparison criterion may be used. We prefer instead to restrict the class of models (e.g., focusing solely on additive QTL models). If the truth is approximately additive, greater power to detect QTL (and a reduced false positive rate) can then be achieved by not allowing the possibility of interactions. However, if there exist large interactions and important loci with limited marginal effect, a search over additive models will have low power to detect such QTL.

The model comparison criterion is by far the most important aspect of a model selection procedure. Still, the model search procedure remains important. One will ideally identify optimal models with the shortest computation time. In the case of a single phenotype, a more extensive search may be feasible. But if a model selection approach is to be applied to many phenotypes, a reduced search may be required in order that the computations may be performed in a reasonable amount of time.

The penalized LOD criterion defined in Sec. 9.1.4 fails to consider covariates and QTL $\times$ covariate interactions. It is a simple matter to include a fixed set of additive covariates, but if one wishes to choose among a larger set of covariates or if one seeks to identify potential QTL $\times$ covariate interactions, the criterion would need to be modified, with special penalties for such terms.

In addition, the X chromosome generally requires special treatment. As described in Sec. 4.4, the potentially different number of parameters for a QTL on the X chromosome requires an X-chromosome-specific threshold, and so X-linked QTL may require a separate penalty. Similarly, epistatic interactions between QTL on the X chromosome, and between a QTL on the X chromosome and one on an autosome, also require special penalties (see Sec. 8.3).

Finally, linked QTL may deserve special treatment, as one may wish to be more permissive in identifying multiple linked QTL. The existence of multiple linked QTL can be extremely important in defining the course of fine-mapping experiments. Our penalized LOD criterion is quite strict in requiring strong evidence for an additional QTL, in order to control the rate of inclusion of extraneous loci. A more liberal approach for linked QTL could be valuable.

9.2 Bayesian QTL mapping

Our description of model selection in QTL mapping has followed a relatively traditional approach (though with some important differences). However, there has been much interest in, and important developments of, Bayesian methods for QTL mapping, for the most part relying on Markov chain Monte Carlo (MCMC). While we view a complete description of the Bayesian methods and their application as beyond the scope of this book, we would be lax to not include at least a brief description of these methods. In this section, we seek to highlight some of the advantages and disadvantages of the Bayesian methods for multiple QTL mapping.

Let y denote the phenotype, M the marker genotypes, q the unknown QTL genotypes, γ a QTL model (possibly with interactions), λ the locations of the QTL, and μ all other model parameters (including QTL effects and the residual SD). The classical and Bayesian methods rely on the same likelihood function

$$\begin{aligned} L(\lambda, \mu, \gamma | y, M) &= \Pr(y | M, \lambda, \mu, \gamma) \\ &= \sum_q \Pr(y, q | M, \lambda, \mu, \gamma) \\ &= \sum_q \Pr(q | M, \lambda, \gamma) \Pr(y | q, \mu, \gamma) \end{aligned}$$

Note that, given the QTL genotypes q, the likelihood separates into two parts. $\Pr(q|M, \lambda, \gamma)$ concerns the relationship between the marker and QTL genotypes; $\Pr(y|q, \mu, \gamma)$ is the phenotype model.

In the classical approach, one maximizes over λ and μ (QTL positions and effects) to obtain the likelihood for a QTL model.

$$L(\gamma | y, M) = \max_{\lambda,\mu} L(\lambda, \mu, \gamma | y, M)$$

One chooses among QTL models, γ, by considering this likelihood, penalized for model complexity.

In the Bayesian approach, one specifies a prior distribution on QTL models and on QTL positions and effects, $\Pr(\lambda, \mu, \gamma)$. The prior is intended to capture one's initial uncertainty in the state of nature. Inference is then conducted through the posterior distribution, given the data.

$$\Pr(\gamma, \lambda, \mu | y, M) \propto L(\lambda, \mu, \gamma | y, M) Pr(\lambda, \mu, \gamma)$$

In particular, one may consider the marginal posterior on QTL models, averaging (i.e., integrating) out QTL positions and effects.

There are thus two key distinctions between the classical and Bayesian approaches to QTL mapping. First, in the classical approach one maximizes over QTL positions and effects, while in the Bayesian approach one averages over these unknown parameters, using a suitable prior. Second, in the classical approach one considers the maximized likelihood for a model, penalized by model complexity, while in the Bayesian approach, one specifies a prior distribution on QTL models and then considers the posterior distribution given the data.

The key technical issue in the Bayesian approach concerns the calculation of the posterior distribution. The distribution is too complex to be described directly, and so we instead sample from the distribution and use the distribution across samples as an approximation to the posterior. Independent random samples are not feasible, and so we form a Markov chain whose stationary distribution is the target posterior distribution. (This is called Markov chain Monte Carlo, MCMC.) Let $\theta = (\lambda, \mu, \gamma)$. We form a Markov chain $\theta_1, \theta_2, \theta_3, \ldots$ (that is, a sequence of random draws such that the distribution of θ_i depends only on θ_{i-1} and not on the entire history), whose limiting distribution is the target posterior, $\lim_{i \to \infty} \Pr(\theta_i) = \Pr(\theta | y, M)$.

There are a variety of techniques for constructing such a Markov chain. While constructing such a chain is relatively easy, great care is required to identify a chain with appropriate mixing properties (to reduce serial dependence and ensure rapid convergence to the stationary distribution). Such details are often confusing to the novice, and so we will omit them. The essence of the approach is that one obtains a sequence of draws, $\theta_1, \ldots, \theta_k$, which we may view as a dependent sample from the posterior distribution, $\Pr(\theta|y, M)$. We may then derive a number of valuable summaries, including the posterior distribution of the number of QTL, the posterior probability that a particular genomic location is involved in the phenotype, and the posterior probability for a particular QTL model.

In the classical approach, we consider a quite strict definition for a QTL model, with the QTL in defined positions. In the Bayesian approach, with the locations of QTL varying across MCMC samples, it is best to soften one's view of a QTL model, and speak instead of a QTL pattern, such as "two QTL on chromosome 1, one on chromosome 4, and one on each of chromosomes 6 and 15, with the QTL on chromosomes 6 and 15 interacting." One may approximate the posterior probability of such a pattern by the proportion of MCMC samples for which the QTL model fits that pattern. One might then choose the pattern with the largest estimated posterior probability.

The Bayesian approach to QTL mapping has a number of advantages. It unifies all aspects of the problem (model fit, search and comparison), it provides a more natural expression of uncertainty in the results (particularly regarding the chance that a particular locus contributes to the phenotype), it more fully captures uncertainty (e.g., the estimated QTL effects take account of uncertainty in QTL positions), and extensions to include QTL $\times$ covariate interactions or alternate phenotype models are relatively straightforward.

The key difficulty concerns the specification of the prior distribution (particularly regarding the number of QTL and the number of interactions). In a sense, the choice of such a prior is equivalent to specification of the target false positive rate (e.g., increasing the expected number of QTL or interactions in the prior will lead to higher false positive rates), but the exact relationship is not easy to anticipate. In addition, a particular QTL model may be seen in only a small proportion of the MCMC samples, and not in the best possible light (as the QTL effects are also sampled). This is not a problem, if one focuses on the posterior probability for specific features of the underlying genetic architecture (such as the chance that a particular locus is involved), but it can make it difficult to define more complex features of the QTL model and so to compare the results of the Bayesian analysis to the results of the classical approach to the problem.

We prefer to say that the classical and Bayesian approaches to QTL mapping are complementary, with different features and different goals, but it may be that we are just being nice to our Bayesian colleagues or failing to admit the defeat of the classical approach.

In summary, the Bayesian approach to QTL mapping can provide a more satisfying set of results, with a more natural expression of the uncertainty in the inferential statements, but the specification of prior distributions is more difficult than the specification of false positive rates (as required for the classical approach). Moreover, the construction of appropriate MCMC algorithms requires great care, and use of MCMC in practice may require considerable training.

9.3 Multiple QTL mapping in R/qtl

R/qtl contains a variety of functions for the fit and exploration of multiple-QTL models, using the classical approach described in Sec. 9.1. [For Bayesian QTL mapping, consider the R/qtlbim package (Yandell *et al.*, 2007).] We will first give a brief overview of the available functions. In the following subsections, we provide a detailed illustration of their use.

Only multiple imputation and Haley–Knott regression are currently implemented for the fit of a multiple-QTL model. The use of multiple interval mapping (MIM) and extended Haley–Knott regression, once implemented, will follow that of Haley–Knott regression, with any instances of `method="hk"` replaced by `method="em"` (for MIM) or `method="ehk"` (for extended Haley–Knott).

The two most basic functions are `makeqtl` and `fitqtl`. The function `makeqtl` is used to create a "QTL object" (of class `"qtl"`); this specifies the locations of a set of putative QTL to be considered. The function `fitqtl` is used to fit a defined QTL model, with QTL in fixed positions and with a defined set of covariates and potentially QTL × QTL and QTL × covariate interactions. The form of the QTL model is specified through a formula, such as `y~Q1+Q2+Q3*Q4`. The `fitqtl` function is particularly important, as it can be used to obtain estimates of QTL effects. One may also perform a "drop-one-QTL-at-a-time" analysis, to assess the support for individual loci and interactions.

The function `refineqtl` is used to refine the locations of QTL in the context of a multiple QTL model. It uses an iterative algorithm with the aim of obtaining the maximum likelihood estimates of the QTL positions. If the function is called with `keeplodprofile=TRUE`, one may then use the function `plotLodProfile` to plot LOD profiles for each QTL, again in the context of the multiple-QTL model, as is commonly used in multiple interval mapping.

With the function `addqtl`, one may scan for a single QTL to be added to a multiple-QTL model; with `addpair`, one may scan for an additional pair of QTL to be added. The output of these functions is of the forms produced by `scanone` and `scantwo`, and so one may use the corresponding plot and summary functions to inspect the results. The functions `addqtl` and `addpair` make use of a more basic and highly flexible function, `scanqtl`, for performing general, multidimensional scans in the context of a multiple-QTL model. The output of `scanqtl` can be quite complicated to interpret, and, for most users, `addqtl` and `addpair` are sufficient, and so we will not discuss the use of `scanqtl` in this book.

The function `addint` may be used to test all possible pairwise interactions among the QTL in a multiple-QTL model.

There are several functions for manipulating a QTL object (created by `makeqtl`). The function `addtoqtl` is used to add additional QTL to an object, `dropfromqtl` is used to remove QTL from an object, `replaceqtl` is used to move QTL to new positions, and `reorderqtl` is used to change the order of loci within a QTL object.

Finally, the function `stepwiseqtl` provides a fully automated model selection algorithm, using the search algorithm described in Sec. 9.1.3 to optimize the penalized LOD score criterion described in Sec. 9.1.4.

9.3.1 `makeqtl` and `fitqtl`

We again return to the `hyper` data of Sugiyama *et al.* (2001) (see Sec. 2.3). We will use multiple imputation, as Haley–Knott regression performs poorly in the case of selectively genotyping, which was used for these data.

First we need to load the package and the data.

```
> library(qtl)
> data(hyper)
```

We must first run `sim.geno` to perform the imputations. We'll use 128 imputations; this is insufficient for the current data, which has extensive missing genotype information, but suffices to illustrate the methods. In practice, it is a good idea to repeat the analysis with independent imputations. If the results are much changed, increase the number of imputations. We will perform calculations on a 2 cM grid; a finer grid would provide more precise results but at the expense of greater computation time.

```
> hyper <- sim.geno(hyper, step=2, n.draws=128, err=0.001)
```

The results of `scanone` and `scantwo`, which we won't revisit, indicated QTL on chr 1, 4, 6 and 15, with an interaction between the QTL on chr 6 and 15, and the possibility of a second QTL on chr 1. (See Sec. 4.2.1 and 8.1.) We will begin by fitting this four-QTL model. (We take the QTL locations from the `scantwo` results on page 221.) The function `makeqtl` is used to create a QTL object; it pulls out the imputed genotypes at the selected locations.

```
> qtl <- makeqtl(hyper, chr=c(1, 4, 6, 15),
+                pos=c(68.3, 30, 60, 18))
```

Note that if you type the name of the QTL object, you get a brief summary. The QTL locations are not exactly as requested, as we are using a different step size.

```
> qtl
```

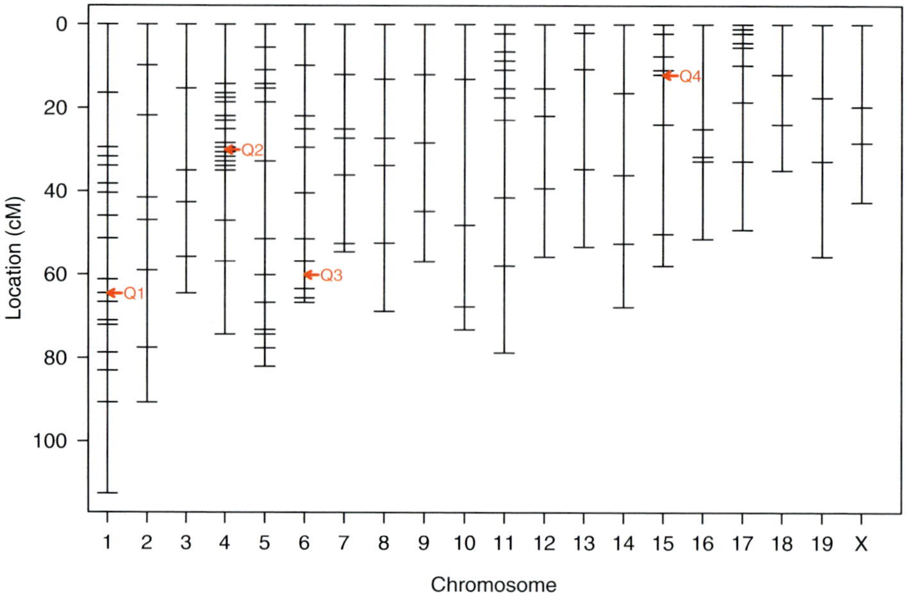

Figure 9.3. Locations of the QTL object `qtl` on the genetic map for the `hyper` data.

```
        name chr  pos n.gen
Q1   1@67.8    1 67.8     2
Q2   4@30.0    4 30.0     2
Q3   6@60.0    6 60.0     2
Q4  15@17.5   15 17.5     2
```

Also, there is a plot function for displaying the locations of the QTL on the genetic map. See Fig. 9.3.

```
> plot(qtl)
```

We may now use `fitqtl` to fit the model (with QTL in fixed positions). We use a model formula to indicate the model; in particular, we use `Q3*Q4` to indicate that QTL 3 and 4 should interact. (More on this shortly; see page 263.) The function `summary.fitqtl` is used to get a summary of the results.

```
> out.fq <- fitqtl(hyper, qtl=qtl, formula=y~Q1+Q2+Q3*Q4)
> summary(out.fq)

Full model result
----------------------------------
Model formula: y ~ Q1 + Q2 + Q3 + Q4 + Q3:Q4
```

```
        df    SS       MS   LOD  %var Pvalue(Chi2) Pvalue(F)
Model    5  5870 1174.01 21.92 33.22            0         0
Error 244 11799    48.36
Total 249 17669

Drop one QTL at a time ANOVA table:
-----------------------------------
               df Type III SS    LOD   %var F value
1@67.8          1    1319.608  5.755  7.469  27.289
4@30.0          1    2940.393 12.079 16.642  60.807
6@60.0          2    1615.617  6.967  9.144  16.705
15@17.5         2    1464.060  6.350  8.286  15.138
6@60.0:15@17.5  1    1174.205  5.150  6.646  24.282
                     Pvalue(Chi2)         Pvalue(F)
1@67.8         0.0000002629244384 0.000000375579930
4@30.0         0.0000000000000876 0.000000000000183
6@60.0         0.0000001079672044 0.000000158669280
15@17.5        0.0000004468012174 0.000000634616856
6@60.0:15@17.5 0.0000011153038687 0.000001536448359
```

The initial table indicates the overall fit of the model; the LOD score of 21.9 is relative to the null model (with no QTL). The sums of squares (`SS`) and mean square (`MS`) are as from an analysis of variance. The percent variance explained (`%var`) is the estimated proportion of the phenotype variance explained by all terms in the model.

In the second table, each locus is dropped from the model, one at a time, and a comparison is made between the full model and the model with the term omitted. If a QTL is dropped, any interactions it is involved in are also dropped, and so the loci on chr 6 and 15 are associated with 2 degrees of freedom, as the 6×15 interaction is dropped when either of these QTL is dropped.

The results indicate strong evidence for all of these loci as well as for the interaction. Let us briefly describe the columns in the table. Most important are the LOD scores, which are the $\log_{10}$ likelihood ratios comparing the full model (with all terms) to the reduced models (with one term omitted); the "Type III sums of squares" indicates the increase in the sum of the squared residuals accompanied by omitting the term. The F statistics are the ratio of the mean square (the sum of squares divided by the degrees of freedom) to the error mean square from the first table. The percent variance explained (`%var`) is the estimated proportion of the phenotypic variance explained by that term. Two pointwise p-values are provided. The first, `Pvalue(Chi2)`, is based on the LOD score, with the assumption that $\text{LOD} \times (2 \ln 10)$ follows a χ^2 distribution with the appropriate degrees of freedom. The second, `Pvalue(F)`, is based on the F statistic. Both p-values should be considered with caution, as they are *pointwise* and so do not account for the search over the genome that led us to the current model. If the `summary.fitqtl` function is called with `pvalues=FALSE`, the two columns of p-values are omitted.

The "drop-one" analysis is particularly valuable for studying the support for the individual terms in the model. Another important use of `fitqtl` is to get estimated QTL effects. This is obtained with the use of `get.ests=TRUE`. We may use `dropone=FALSE` to suppress the drop-one analysis.

```
> out.fq2 <- fitqtl(hyper, qtl=qtl, formula=y~Q1+Q2+Q3*Q4,
+                   dropone=FALSE, get.ests=TRUE)
> summary(out.fq2)

Full model result
----------------------------------
Model formula: y ~ Q1 + Q2 + Q3 + Q4 + Q3:Q4

       df    SS      MS   LOD  %var Pvalue(Chi2) Pvalue(F)
Model   5  5870 1174.01 21.92 33.22            0         0
Error 244 11799   48.36
Total 249 17669

Estimated effects:
-----------------
                       est     SE       t
Intercept         101.5642 0.4533 224.042
1@67.8             -4.7428 0.9159  -5.178
4@30.0             -7.0314 0.9191  -7.650
6@60.0              3.8161 0.9545   3.998
15@17.5            -2.3571 0.9197  -2.563
6@60.0:15@17.5     -9.0567 1.8941  -4.781
```

The estimated effects are derived by coding the homozygous and heterozygous genotypes (in a backcross) as -0.5 and $+0.5$, respectively. Thus, the estimated effect for a QTL is the difference between the phenotype averages for the heterozygotes and homozygotes. The `hyper` data come from the backcross $(\text{B} \times \text{A}) \times \text{B}$, with A and B being the A/J and C57Bl/6J mouse strains. The estimated effects for the chr 1, 4 and 15 loci being negative indicates that the A allele results in a decrease in blood pressure (i.e., heterozygotes, AB, have lower blood pressure than homozygous, BB). (And note that the A strain has lower blood pressure than the B strain.) The chr 6 locus is a so-called *transgressive* QTL; the A allele is associated with an increase in blood pressure. (See also Fig. 8.7 on page 227.)

The interaction effect for the loci on chr 6 and 15 is large and negative. It is based on the product of the genotype columns for the two QTL. It can be interpreted as the difference in the effect of the chr 6 locus, according to whether an individual is heterozygous or homozygous at the chr 15 locus (and vice versa).

In the above, we have used multiple imputation for the calculations. To use Haley–Knott regression for such calculations, one must first call `calc.genoprob` (to calculate the conditional QTL genotype probabilities, given the marker data) rather than `sim.geno`. Then, in the call to `makeqtl`, one must use the argument `what="prob"`, to define QTL based on the genotype probabilities rather than the imputations. Finally, in the calls `fitqtl` (and the related functions, described below), one must use the argument `method="hk"`.

Model formulas deserve further explanation. Such formulas are used widely in R (see the R help file for `formula`); R/qtl uses a restricted version. First note that we always write "`y ~`" (to be read as "the phenotype y is modeled as...") at the beginning of a QTL formula. QTL are indicated by their numeric index with the QTL object (`Q1`, `Q2`, etc.). Interactions between QTL may be specified using a colon; for example, `Q3:Q4` indicates that the interaction between QTL 3 and 4 should be included, and `Q1:Q3:Q4` indicates the three-way interaction between QTL 1, 3 and 4. An asterisk is similar, but indicates that all lower order interactions should also be included. For example, the term `Q3*Q4` is equivalent to `Q3+Q4+Q3:Q4`, and the term `Q1*Q3*Q4` indicates all first-order terms, all pairwise interactions, and the three-way interaction. That is, `Q1*Q3*Q4` is equivalent to `Q1+Q3+Q4+Q1:Q3+Q1:Q4+Q3:Q4+Q1:Q3:Q4`.

In all cases, we enforce a hierarchy in the QTL models, so that the inclusion of a pairwise interaction requires the inclusion of both of the corresponding main effects, and the inclusion of a three-way interaction requires the inclusion of the main effects and all pairwise interactions.

Covariates may be included in the QTL model fit by `fitqtl`. As with `scanone` and `scantwo`, the covariates must be strictly numeric. (See Chap. 7 and Sec. 8.4.) And here the set of covariates, which will be indicated through the argument `covar`, must form a matrix (or data frame). Rather than separately indicating additive and interactive covariates, one refers to covariates and QTL $\times$ covariate interactions in the model formula. Refer to the covariates in the formula by their column names.

9.3.2 `refineqtl`

The function `refineqtl` allows us to get improved estimates of the locations of the QTL. The position for each QTL is varied, one at a time, keeping all other QTL locations fixed, and keeping the chromosome assignments and order of QTL fixed. The process is iterated to convergence. We use `verbose=FALSE` to suppress the display of tracing information.

```
> rqtl <- refineqtl(hyper, qtl=qtl, formula=y~Q1+Q2+Q3*Q4,
+                   verbose=FALSE)
```

The output is a modified QTL object, with loci in new positions. We can type the name of the new QTL object to see the new locations.

```
> rqtl
```

```
      name chr  pos n.gen
Q1  1@67.8   1 67.8     2
Q2  4@30.0   4 30.0     2
Q3  6@66.7   6 66.7     2
Q4 15@17.5  15 17.5     2
```

The locus on chr 6 changed position slightly. Let us use `fitqtl` to assess the improvement in fit; we'll skip the drop-one analysis.

```
> out.fq3 <- fitqtl(hyper, qtl=rqtl, formula=y~Q1+Q2+Q3*Q4,
+                   dropone=FALSE)
> summary(out.fq3)

Full model result
----------------------------------
Model formula: y ~ Q1 + Q2 + Q3 + Q4 + Q3:Q4

       df    SS      MS   LOD  %var Pvalue(Chi2) Pvalue(F)
Model   5  5893 1178.64 22.03 33.35            0         0
Error 244 11776   48.26
Total 249 17669
```

The LOD score comparing the full model to the null model has increased by 0.1, from 21.9 to 22.0.

By default, `refineqtl` saves the LOD traces at the last iteration, which can then be plotted with `plotLodProfile`, as follows.

```
> plotLodProfile(rqtl, ylab="Profile LOD score")
```

The LOD profiles in Fig. 9.4 are similar to the usual LOD curves, but instead of comparing a model with a single QTL at a particular position to the null model, we compare, at each position for a given QTL, the model with the QTL of interest at that particular position (and with the positions of all other QTL fixed at their maximum likelihood estimates) to the model with the QTL of interest omitted (and with the positions of all other QTL fixed at their maximum likelihood estimates). For the loci on chr 6 and 15, the 6×15 interaction is omitted when either of the two loci is omitted.

And so in the LOD profile for the locus on chr 1, we compare the four-QTL model (including the 6 × 15 interaction and with the position of the chr 1 QTL varying but with the other three QTL fixed at their estimated locations) to the three-QTL model with the chr 1 locus omitted. In the LOD profile on chr 15, we compare the four-QTL model (with the position of the chr 15 locus varying but with the other three QTL fixed at their estimated locations) to the three-QTL additive model (that is, without the chr 15 locus and without the 6 × 15 interaction). Note that the maximum LOD for each of the LOD profiles should be the value observed in the drop-one analysis from `fitqtl`.

These profile LOD curves are useful for the assessment of both the evidence for the individual QTL and the precision of localization of each QTL, but note

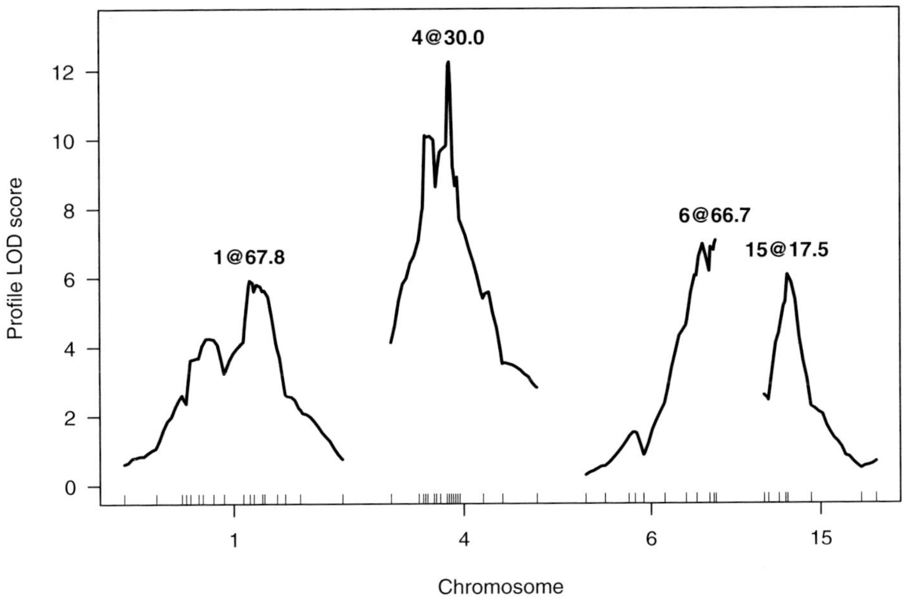

Figure 9.4. LOD profiles for a four-QTL model with the `hyper` data.

that they fail to take account of the uncertainty in the location of the other QTL in the model.

The functions `lodint` and `bayesint` (see Sec. 4.5) can be used to derive approximate confidence intervals for the locations of the QTL, using the LOD profiles calculated by `refineqtl`. However, these should be viewed with some caution, as they are calculated assuming that the locations of the other QTL are known without error (that is, they fail to account for the uncertainty in the locations of the other QTL), and the performance of these approximate intervals in the context of a multiple-QTL model is not well understood. Thus, they are sure to be overly liberal (that is, their coverage probabilities are likely less than 95%).

To calculate a 1.5-LOD support interval and an approximate 95% Bayesian credible interval for the QTL on chr 4, we refer to the QTL, in `lodint` and `bayesint`, by its numeric index (in this case, 2).

```
> lodint(rqtl, qtl.index=2)

          chr  pos   lod
D4Mit288    4 28.4  9.83
c4.loc30    4 30.0 12.24
D4Mit80     4 31.7  9.18

> bayesint(rqtl, qtl.index=2)
```

```
           chr  pos   lod
D4Mit164     4 29.5 12.17
c4.loc30     4 30.0 12.24
D4Mit178     4 30.6 11.55
```

These intervals are *much* more narrow than the intervals calculated in the context of a single-QTL model (see Sec. 4.5). The 1.5-LOD support interval is 3.3 cM long (versus 12.0 cM), and the approximate 95% Bayesian credible interval is. 1.1 cM long (versus 13.1 cM).

9.3.3 addint

The function `addint` is used to test, one at a time, all possible QTL × QTL interactions that are not already included in a model. For our model with loci on chr 1, 4, 6 and 15, and with a 6×15 interaction, we consider each of the five other possible pairwise interactions, and compare the base model (with the four QTL and just the 6×15 interaction) to the model with the additional interaction included.

The syntax of the function is similar to that of `fitqtl`. The output is a table of results similar to that provided by the drop-one analysis of `fitqtl`. As with `fitqtl`, by default two columns of pointwise p-values are provided: one based on the assumption that, under the null hypothesis, LOD × $(2 \ln 10)$ follows a χ^2 distribution with the appropriate degrees of freedom, and a second based on the F statistic. As in `fitqtl`, the p-values should be considered with caution, as they are *pointwise* and so do not account for the search over the genome that led us to the current model. To save space, we will omit the p-values from the tabular results via `pvalues=FALSE`.

```
> addint(hyper, qtl=rqtl, formula=y~Q1+Q2+Q3*Q4, pvalues=FALSE)

Model formula: y ~ Q1 + Q2 + Q3 + Q4 + Q3:Q4

Add one pairwise interaction at a time table:
--------------------------------------------
                 df Type III SS      LOD      %var F value
1@67.8:4@30.0     1   61.380843 0.283709 0.347394   1.273
1@67.8:6@66.7     1    1.402998 0.006468 0.007940   0.029
1@67.8:15@17.5    1   71.144546 0.328975 0.402653   1.477
4@30.0:6@66.7     1   64.916064 0.300094 0.367402   1.347
4@30.0:15@17.5    1   16.225653 0.074853 0.091832   0.335
```

There is little evidence for any of these interactions.

Different results are obtained if we use as the formula `y~Q1+Q2+Q3+Q4` (that is, omitting the 6×15 interaction).

```
> addint(hyper, qtl=rqtl, formula=y~Q1+Q2+Q3+Q4, pvalues=FALSE)
```

```
Model formula: y ~ Q1 + Q2 + Q3 + Q4

Add one pairwise interaction at a time table:
--------------------------------------------
                 df Type III SS       LOD       %var F value
1@67.8:4@30.0    1    60.30574   0.25455    0.34131   1.147
1@67.8:6@66.7    1    11.62534   0.04898    0.06580   0.220
1@67.8:15@17.5   1    86.59440   0.36589    0.49009   1.650
4@30.0:6@66.7    1    39.81677   0.16793    0.22535   0.756
4@30.0:15@17.5   1    58.92350   0.24870    0.33349   1.120
6@66.7:15@17.5   1  1115.46429   4.91316    6.31314  23.113
```

The 6×15 interaction is also tested, and the LOD scores for the other interactions are somewhat different, as they concern comparisons between the four-locus additive model and the model with that one interaction added.

9.3.4 `addqtl`

The `addqtl` function is used to scan for an additional QTL, to be added to the model. By default, the new QTL is strictly additive.

```
> out.aq <- addqtl(hyper, qtl=rqtl, formula=y~Q1+Q2+Q3*Q4)
```

The output of `addqtl` has the same form as that from `scanone`, and so we may use the same summary and plot functions. For example, we can identify the genome-wide maximum LOD score with `max.scanone`.

```
> max(out.aq)

        chr  pos  lod
D5Mit31   5 66.7 1.58
```

We may plot the results with `plot.scanone`; see Fig. 9.5.

```
> plot(out.aq, ylab="LOD score")
```

The LOD scores compare the base model to the model with one additional QTL. There is a suggestion of an additional QTL on chr 5, but the evidence is not strong.

We may also use `addqtl` to scan for an additional QTL, interacting with one of the current loci. This is done by including the additional QTL in the model formula, with the relevant interaction term. For example, let's scan for an additional QTL interacting with the chr 15 locus.

```
> out.aqi <- addqtl(hyper, qtl=rqtl,
+                   formula=y~Q1+Q2+Q3*Q4+Q4*Q5)
```

We plot the results as follows; see Fig. 9.6.

```
> plot(out.aqi, ylab="LOD score")
```

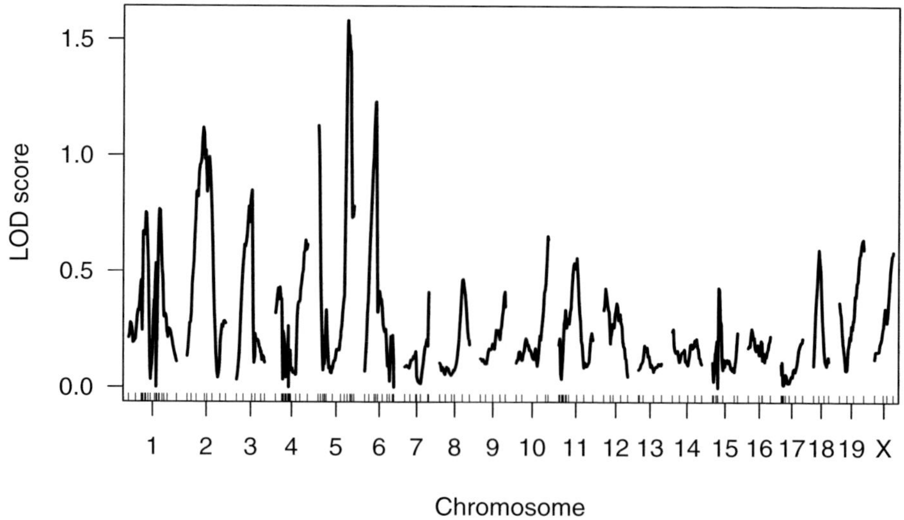

Figure 9.5. LOD curves for adding one QTL to the four-QTL model, with the `hyper` data.

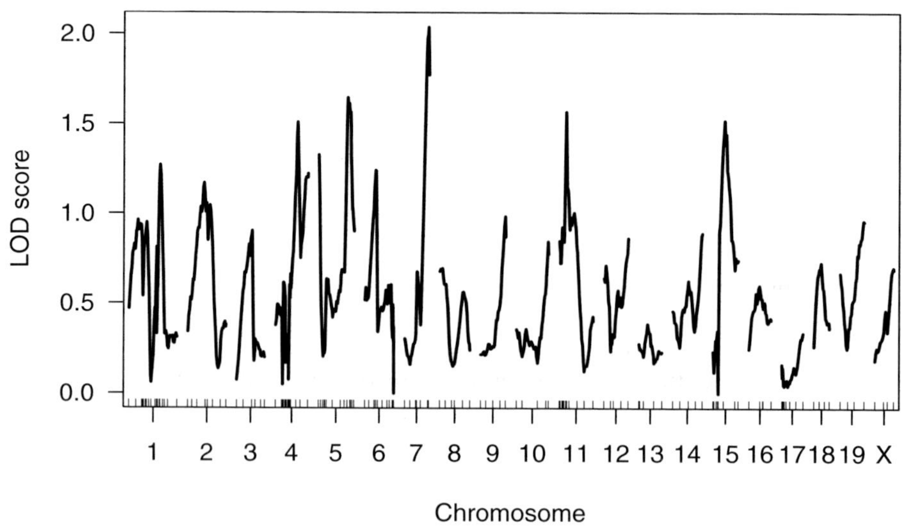

Figure 9.6. LOD curves for adding one QTL, interacting with the chr 15 locus, to the four-QTL model, with the `hyper` data.

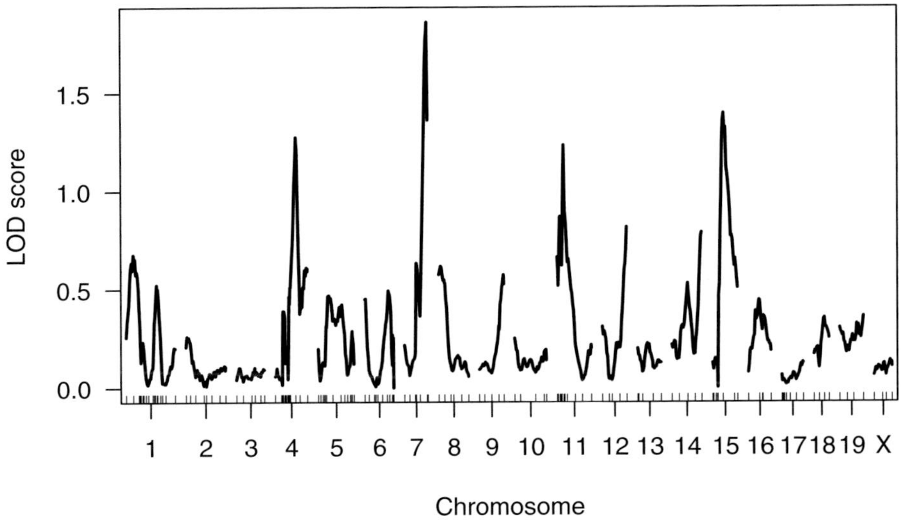

Figure 9.7. Interaction LOD curves in the scan for an additional QTL, interacting with the chr 15 locus, to be added to the four-QTL model, with the `hyper` data.

Also of interest are the LOD scores for the interaction between the chr 15 locus and the new locus being scanned, which are the differences between the LOD scores in `out.aqi` and `out.aq`. See Fig. 9.7.

```
> plot(out.aqi - out.aq, ylab="LOD score")
```

There is nothing particularly exciting in either of these plots (though there is a suggestion of a QTL on chr 7, interacting with the QTL on chr 15).

9.3.5 `addpair`

The function `addpair` is similar to `addqtl`, but it performs a two-dimensional scan to seek a pair of QTL to add to a multiple-QTL model. By default, `addpair` performs a two-dimensional scan analogous to that of `scantwo`: for each pair of positions for the two putative QTL, it fits both an additive model and a model including an interaction between the two QTL.

Recall that in the single-QTL analysis with the `hyper` data, there were two peaks in the LOD curve on chr 1, indicating that there may be two QTL on that chromosome. In the context of our multiple-QTL model, the LOD profile on chr 1 (see Fig. 9.4) still shows two peaks, though the distal peak is more prominent.

We may use `addpair` to investigate the possibility of a second QTL on chr 1. To do so, we omit the chr 1 locus from our formula, and perform a two-dimensional scan just on chr 1.

```
> out.ap <- addpair(hyper, qtl=rqtl, chr=1, formula=y~Q2+Q3*Q4,
+                   verbose=FALSE)
```

The output is of the same form as that produced by `scantwo`, and so we may use the same summary and plot functions.

```
> summary(out.ap)

      pos1f pos2f lod.full lod.fv1 lod.int      pos1a pos2a
c1:c1  43.3  73.3     7.83    1.98   0.458       45.3  77.3
      lod.add lod.av1
c1:c1    7.37    1.52
```

There is little evidence for an interaction, and the LOD score comparing the model with two additive QTL on chr 1 to that with a single QTL on chr 1 is 1.52, indicating relatively weak evidence for a second QTL on chr 1.

Let us also plot the results. We'll focus on the evidence for a second QTL on the chromosome, displaying LOD_{fv1} (evidence for a second QTL, allowing for an interaction) and LOD_{av1} (evidence for a second QTL, assuming the two are additive). See Fig 9.8.

```
> plot(out.ap, lower="cond-int", upper="cond-add")
```

There is a good deal of flexibility in the way that `addpair` may be used. As in `addqtl`, where one can scan for loci that interact with a particular locus in the model, we can use `addpair` to scan for an additional pair, with any prespecified set of interactions.

For example, we may retain the loci on chr 1, 4 and 6, and scan for an additional pair of interacting loci, one of which also interacts with chr 6. This would be useful for assessing evidence for an additional QTL interacting with the chr 15 locus, but allowing the location of the locus on chr 15 to vary. We use the formula `y~Q1+Q2+Q3+Q5*Q6+Q3:Q5`, as we will omit the chr 15 locus (`Q4`), scan for an additional interacting pair (`Q5*Q6`), and allow the first QTL in the additional pair to interact with the chr 6 locus (`Q3`). Note that the positions of the chr 1, 4 and 6 loci are assumed known. A three-dimensional scan could be performed with the `scanqtl` function, but we do not discuss such searches in this book.

To save time, we will focus just on chr 7 and 15.

```
> out.ap2 <- addpair(hyper, qtl=rqtl, chr=c(7,15), verbose=TRUE,
+                    formula=y~Q1+Q2+Q3+Q5*Q6+Q3:Q5)
```

Because we are using a special formula here, with one of the new QTL interacting with the chr 6 locus, the results are similar to, but not quite the same as, those from `scantwo`. Rather than fitting an additive and an interactive model at each pair of positions, we fit just the single model specified in the formula. And note that as the formula is not symmetric in `Q5` and `Q6`, we must do a full 2-dimensional scan, rather than just scan the triangle. (That is, we need `Q5` and `Q6` assigned to chromosomes (7,15) as well as (15,7).)

The summary of the results are somewhat different here. For each pair of chromosomes, a set of three LOD scores are presented. `lod.2v0` compares the

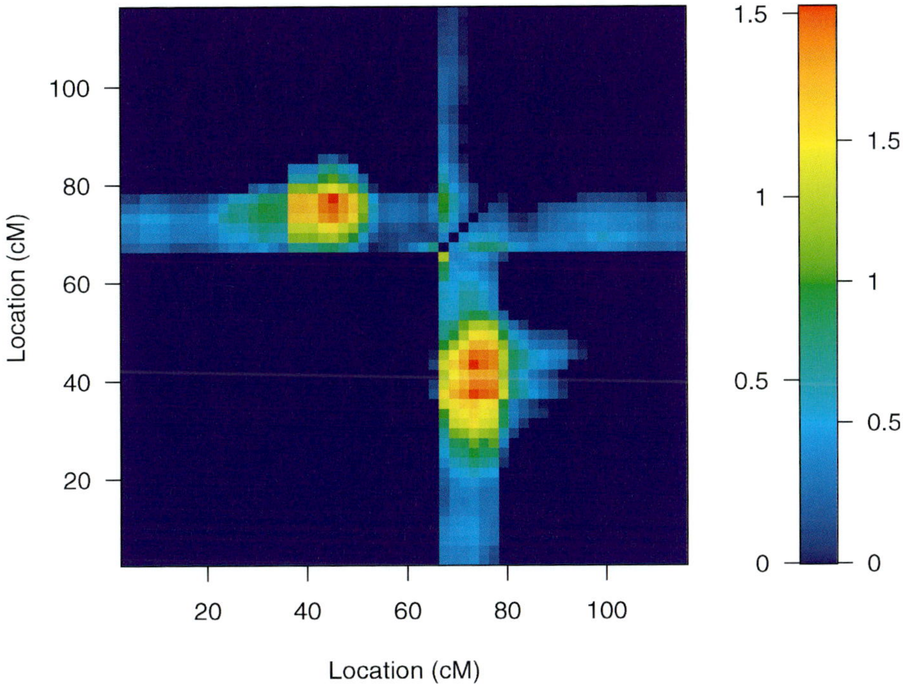

Figure 9.8. Results of a two-dimensional, two-QTL scan on chr 1, in the context of a model with additional QTL on chr 4, 6 and 15, and a 6×15 interaction, with the `hyper` data. LOD_{av1} is in the upper left triangle, and LOD_{fv1} is in the lower right triangle. In the color scale on the right, numbers to the left and right correspond to LOD_{av1} and LOD_{fv1}, respectively.

full model to the model with neither of the two new QTL included, `lod.2v1b` compares the full model to the model with the first of the two new QTL omitted, and `lod.2v1a` compares the full model to the model with the second QTL omitted. When a QTL is omitted, any interactions involving that QTL are also omitted.

```
> summary(out.ap2)

        pos1a pos2a lod.2v0 lod.2v1b lod.2v1a
c7 :c7   29.1  25.1    2.89     2.54     1.55
c7 :c15  51.1  15.5    3.84     2.51     2.51
c15:c7   17.5  53.1    8.08     7.72     2.01
c15:c15  17.5  31.5    7.59     6.26     1.52
```

Note that, because of the lack of symmetry in the formula we used in `addpair`, separate results are provided for the two cases `c7:c15` (in which the chr 7 locus interacts with the chr 6 locus) and `c15:c7` (in which the chr 15 locus interacts with the chr 6 locus). The `c15:c7` row is most interesting, but lod.2v1a is 2.01, indicating little evidence for a chr 7 locus. (Note that lod.2v1a here concerns both the chr 7 locus and the 7×15 interaction.) This is the same as the peak on chr 7 in Fig. 9.6, in which we scanned for an additional locus, interacting with the chr 15 locus. While here we allowed the location of the chr 15 locus to vary, the estimated location was 17.5 cM, as before.

With this sort of addpair output, the thresholds argument should have length just 1 or 2 (which is different from the usual case for summary.scantwo). Rows will be retained if lod.2v0 is greater than thresholds[1] and either of lod.2v1a or lod.2v1b is greater than thresholds[2]. (If a single thresholds is given, we assume that thresholds[2]==0.) Note that, of the other arguments to summary.scantwo, all but allpairs is ignored.

The plot of the output from addpair, in the case of a special formula, is also different from the usual scantwo plot.

```
> plot(out.ap2)
```

The plot, shown in Fig. 9.9, contains LOD scores comparing the full five-QTL model to the three-QTL model (having loci on chr 1, 4 and 6). The x-axis corresponds to the first of the new QTL (Q5), which is the one that interacts with the chr 6 locus. The y-axis corresponds to the second of the new QTL (Q6). Clearly, the QTL interacting with the chr 6 locus wants to be on chr 15.

Note that the lower and upper arguments to plot.scantwo are ignored in this case.

9.3.6 Manipulating qtl objects

Our analysis of the hyper data, above, did not indicate much evidence for any further QTL. If we had seen evidence for additional loci, we would want to add them to the QTL object and repeat our explorations with fitqtl, addint, addqtl, and addpair.

The functions addtoqtl, dropfromqtl and replaceqtl can be used to facilitate such analyses. Rather than recreating a QTL object from scratch with makeqtl, one can use addtoqtl to add an additional locus to a QTL object that was previously created. For example, if we were satisfied with the evidence for an additional QTL on chr 1, it could be added to the QTL object rqtl as follows. We use print to simultaneously assign the result to an object and print it.

```
> print( rqtl2 <- addtoqtl(hyper, rqtl, 1, 43.3) )

      name chr  pos n.gen
Q1  1@67.8   1 67.8     2
Q2  4@30.0   4 30.0     2
Q3  6@66.7   6 66.7     2
```

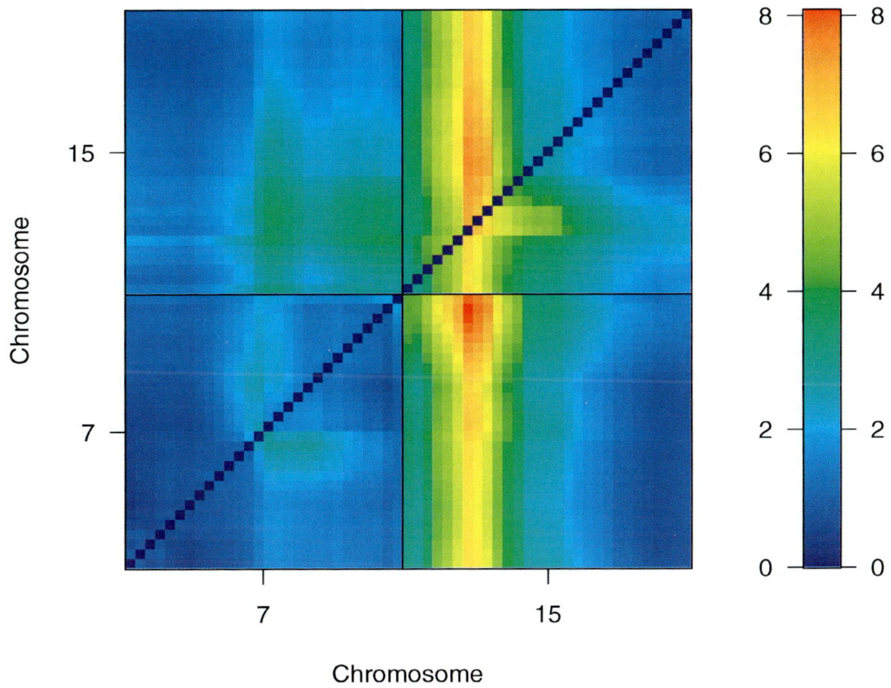

Figure 9.9. Results of a two-dimensional, two-QTL scan on chr 7 and 15, in the context of a model with additional QTL on chr 1, 4, and 6, with the `hyper` data. The two QTL being scanned were allowed to interact, and the first of them interacts with the chr 6 locus. The LOD scores displayed are for the five-QTL model relative to the three-QTL model. The x-axis corresponds to the first of the new QTL (which interacts with the chr 6 locus); the y-axis corresponds to the second of the new QTL.

```
Q4 15@17.5  15 17.5     2
Q5  1@43.3   1 43.3     2
```

The syntax of `addtoqtl` is much like that of `makeqtl`, though one also provides the QTL object to which additional QTL are to be added.

If we want to move the first QTL on chr 1 to a different position (say to 77.3 cM rather than 67.8 cM), we may use `replaceqtl`. We refer to the QTL that are to be replaced by their numeric index, with the argument `index`. (That is the first `1` below; the second `1` indicates the chromosome.)

```
> print( rqtl3 <- replaceqtl(hyper, rqtl2, 1, 1, 77.3) )

      name chr  pos n.gen
Q1  1@77.3   1 77.3     2
Q2  4@30.0   4 30.0     2
Q3  6@66.7   6 66.7     2
```

```
Q4 15@17.5  15 17.5     2
Q5  1@43.3   1 43.3     2
```

If we wish to reorder the QTL (e.g., according to their map positions), we may use `reorderqtl`. We may provide a vector of numeric indices specifying the new order.

```
> print( rqtl4 <- reorderqtl(rqtl3, c(4,3,2,1,5)) )

       name chr  pos n.gen
Q1 15@17.5  15 17.5     2
Q2  6@66.7   6 66.7     2
Q3  4@30.0   4 30.0     2
Q4  1@77.3   1 77.3     2
Q5  1@43.3   1 43.3     2
```

Alternatively, if `reorderqtl` is called with only the QTL object, the QTL are ordered according to their genomic positions.

```
> print( rqtl5 <- reorderqtl(rqtl4) )

       name chr  pos n.gen
Q1  1@43.3   1 43.3     2
Q2  1@77.3   1 77.3     2
Q3  4@30.0   4 30.0     2
Q4  6@66.7   6 66.7     2
Q5 15@17.5  15 17.5     2
```

Finally, `dropfromqtl` is used to drop a locus from a QTL object. To drop the proximal locus on chr 1 (now the first QTL in the object), we would do the following.

```
> print( rqtl6 <- dropfromqtl(rqtl5, 1) )

       name chr  pos n.gen
Q1  1@77.3   1 77.3     2
Q2  4@30.0   4 30.0     2
Q3  6@66.7   6 66.7     2
Q4 15@17.5  15 17.5     2
```

In `dropfromqtl`, we may refer to the QTL to be dropped either by their numeric index (through the argument `index`), by their chromosome and position (through the arguments `chr` and `pos`), or by their name (through the argument `qtl.name`).

9.3.7 `stepwiseqtl`

With the function `stepwiseqtl`, one may use the forward/backward stepwise search algorithm described in Sec. 9.1.3 to optimize the penalized LOD score criterion described in Sec. 9.1.4. In this section, we illustrate the use of this function through application to the `hyper` data.

We must first derive the appropriate penalties for the penalized LOD score. This requires a permutation test with a two-dimensional, two-QTL genome scan. While we had performed such permutations in Sec. 8.1, there we had used maximum likelihood via the EM algorithm to deal with missing genotype data, and here we are using multiple imputation. As the results may be somewhat different with the two approaches, we must rerun the permutation analysis. Extremely hefty computations are required, on the order of 100 hours, in total. Thus it is best to split the permutations into batches to be performed in parallel using multiple processors. Such parallel computations were described in Sec. 8.1; see page 223.

Recall that, due to the selective genotyping, it is best to do a stratified permutation test (permuting separately within the two strata defined by the extent of genotype data available). And so we first define these strata, and then perform the permutation test with `scantwo`.

```
> strat <- (nmissing(hyper) > 50)
> operm2 <- scantwo(hyper, method="imp", n.perm=1000,
+                   perm.strat=strat)
```

The `summary.scantwoperm` function may be used to obtain estimated thresholds.

```
> summary(operm2)

bp (1000 permutations)
    full  fv1  int  add  av1  one
5%  5.41 4.19 3.79 4.34 2.29 2.55
10% 5.09 3.91 3.51 3.97 2.05 2.28
```

The function `calc.penalties` is used to derive the penalties from the permutation results. By default, we use a significance level of 5%. We use `print` in the following so that we may simultaneously assign the penalties to an object and print the results.

```
> print(pen <- calc.penalties(operm2))

 main heavy light
2.553 3.793 1.641
```

Note that these are quite different from the penalties presented in Sec. 9.1.4 (Table 9.1 on page 252), derived by computer simulation (in an admittedly artificial situation). In particular, the heavy interaction penalty is quite a bit larger (3.8 vs. 2.6).

The penalties corresponding to multiple significance levels can be derived at the same time.

```
> calc.penalties(operm2, alpha=c(0.05, 0.20))
```

```
     main heavy light
5%  2.553 3.793 1.641
20% 1.983 3.197 1.610
```

With these penalties in hand, we are now prepared to apply the fully automated model search algorithm with `stepwiseqtl`. The search algorithm uses forward selection to a model with a fixed number of QTL, at each step searching for an additional additive QTL, or an additional QTL interacting with one of the QTL in the current model. The forward selection process is followed by backward elimination to the null model. The final chosen model is that with the maximal penalized LOD score, among all models visited.

```
> outsw1 <- stepwiseqtl(hyper, max.qtl=8, penalties=pen,
+                       verbose=FALSE)
```

The output is a QTL object (of class `"qtl"`, as created by the function `makeqtl`) defining the chosen model.

```
> outsw1

       name chr  pos n.gen
Q1   1@67.8   1 67.8     2
Q2   4@30.0   4 30.0     2
Q3   6@66.7   6 66.7     2
Q4 15@17.5   15 17.5     2

  Formula: y ~ Q1 + Q2 + Q3 + Q4 + Q3:Q4
  pLOD: 10.18
```

This is identical to the model obtained in Sec. 9.3.2 (page 264).

If `stepwiseqtl` is run with the argument `keeplodprofile=TRUE`, one may obtain the LOD profiles for the four QTL in the model (as with the function `refineqtl`), which may then be plotted with `plotLodProfile` (see Sec. 9.3.2).

With the argument `keeptrace=TRUE`, the output will include the sequence of models visited in the forward/backward search algorithm. (That is, we retain information on the single best model at each step of forward selection and backward elimination.)

Let us rerun `stepwiseqtl` with these options, and visualize the results.

```
> outsw2 <- stepwiseqtl(hyper, max.qtl=8, penalties=pen,
+                       verbose=FALSE, keeplodprofile=TRUE,
+                       keeptrace=TRUE)
```

The LOD profiles may be displayed as follows. As the model is identical to that considered in Sec. 9.3.2, the LOD profiles are identical to those displayed in Fig. 9.4 on page 265, and so the plot will not be repeated.

```
> plotLodProfile(outsw2)
```

The sequence of models visited by `stepwiseqtl` are retained as an *attribute* (called `"trace"`) of the output, `outsw2`. Attributes are a way of hiding additional information within an object. The entire set of attributes for an object may be obtained with the `attributes` function. It is often useful to just look at the names of the attributes.

```
> names(attributes(outsw2))

[1] "names"      "class"      "map"        "lodprofile"
[5] "formula"    "pLOD"       "trace"
```

Individual attributes may be obtained with the `attr` function. So we can pull out the trace of models with the following. This is a long list, with each component being a compact representation of a QTL model, and so we will print just the first of them.

```
> thetrace <- attr(outsw2, "trace")
> thetrace[[1]]

   chr  pos
Q1   4 29.5

  Formula: y ~ Q1
  pLOD: 5.539
```

It is nicer to look at a sequence of pictures rather than a long list of models. The function `plotModel` may be used to plot a graphical representation of a model, with nodes (i.e., dots) representing QTL and edges (i.e., line segments connecting two nodes) representing pairwise interactions among QTL. The argument `chronly` is used to print just the chromosome ID for each QTL (rather than chromosome and position). The penalized LOD score for each model is saved as an attribute, `"pLOD"`; we include them in the title of each subplot, but this requires another call to `attr`.

```
> par(mfrow=c(6,3))
> for(i in seq(along=thetrace))
+   plotModel(thetrace[[i]], chronly=TRUE,
+             main=paste(i, ": pLOD =",
+               round(attr(thetrace[[i]], "pLOD"), 2)))
```

As seen in Fig. 9.10, our chosen model is identified immediately (at step 4). Note that the model at step 3 (with additive QTL on chr 1, 4 and 6) has a lower penalized LOD score than the model at step 2 (with just the chr 1 and 4 QTL), but then the inclusion of the chr 15 QTL and the 6 × 15 interaction gave the largest penalized LOD score, among all models visited. With the addition of a QTL on chr 5 (at step 5), the pLOD decreased somewhat; the LOD score for the model increased, but not as much as the main effect penalty.

In the above, we used the penalized LOD score criterion that balances the use of the heavy and light interaction penalties (see Sec. 9.1.4). If we call

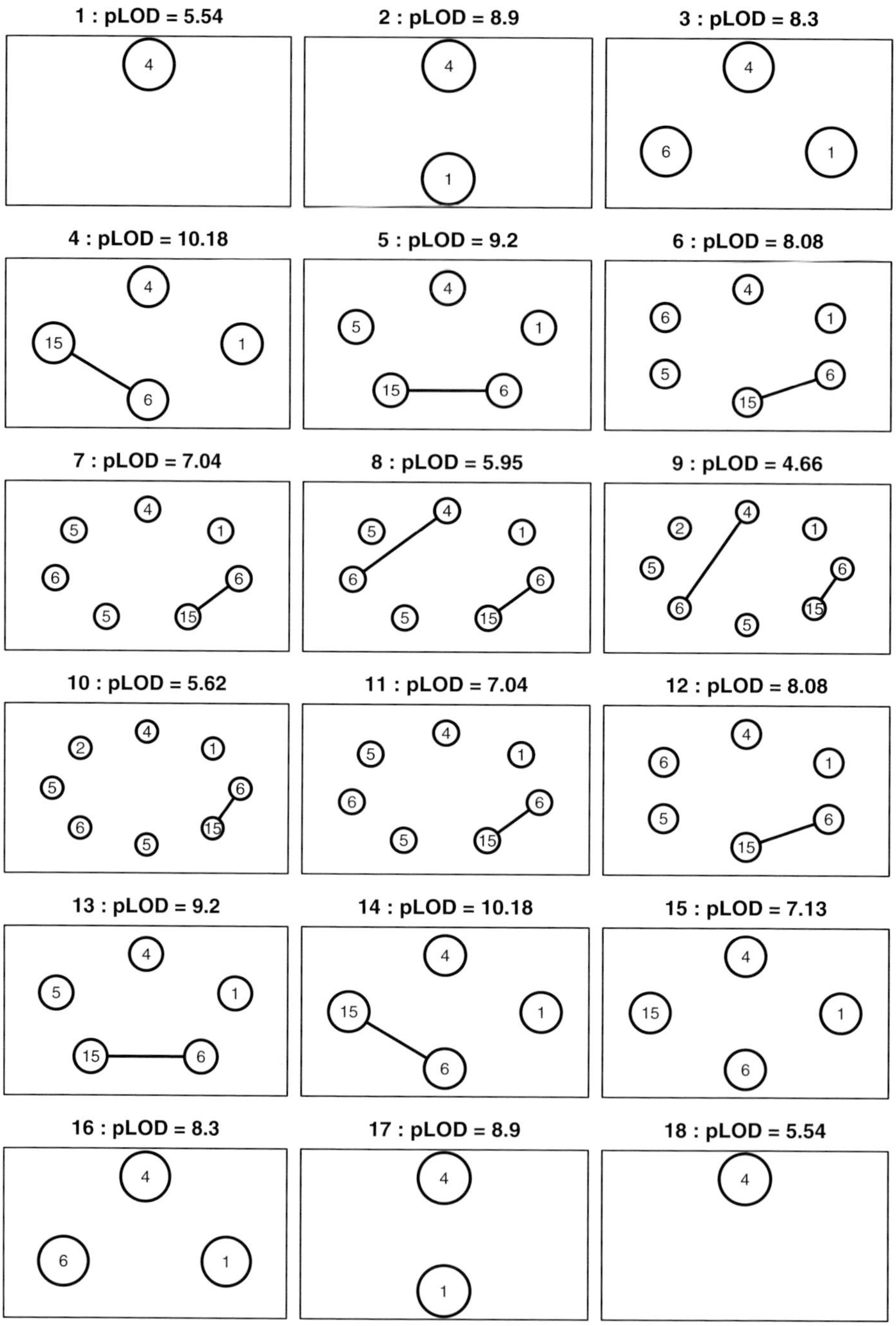

Figure 9.10. The sequence of models visited by the forward/backward search of `stepwiseqtl`, with the `hyper` data.

`stepwiseqtl` with just the first two penalties (the main effect penalties and the heavy interaction penalty), the light interaction penalty will be taken to be the same as the heavy interaction penalty, and so all pairwise interactions will be assigned the same penalty.

```
> outsw3 <- stepwiseqtl(hyper, max.qtl=8, penalties=pen[1:2],
+                         verbose=FALSE)
```

With only heavy interaction penalties, we choose the model with just the QTL on chr 1 and 4, and with no interactions.

```
> outsw3

     name chr  pos n.gen
Q1 1@67.8   1 67.8     2
Q2 4@29.5   4 29.5     2

  Formula: y ~ Q1 + Q2
  pLOD: 8.897
```

We could also consider more liberal penalties, such as those corresponding to a significance level of 20%.

```
> liberalpen <- calc.penalties(operm2, alpha=0.2)
> outsw4 <- stepwiseqtl(hyper, max.qtl=8, penalties=liberalpen,
+                         verbose=FALSE)
```

No additional QTL are obtained with these more liberal penalties. The penalties based on a significance level of 5% are conservative; the more liberal penalties can lead to a larger model (though, as seen here, not necessarily), but the false positive rate (the chance of extraneous loci being included in the chosen model) will be higher.

```
> outsw4

      name chr  pos n.gen
Q1  1@67.8   1 67.8     2
Q2  4@30.0   4 30.0     2
Q3  6@66.7   6 66.7     2
Q4 15@17.5  15 17.5     2

  Formula: y ~ Q1 + Q2 + Q3 + Q4 + Q3:Q4
  pLOD: 12.48
```

The `stepwiseqtl` function has a number of additional arguments. With `additive.only=TRUE`, one may consider only additive QTL models (allowing no interactions among QTL). With `scan.pairs=TRUE`, one may perform a two-dimensional, two-QTL scan at each step of forward selection. This could enhance our ability to identify interacting loci with limited marginal effects and pairs of loci that are linked in repulsion (having effects of opposite signs). However, the computational effort is great, and our limited experience suggests that it may not be necessary.

We won't explore the use of `scan.pairs=TRUE`, but let us see what happens when we focus solely on additive QTL models.

```
> outsw5 <- stepwiseqtl(hyper, max.qtl=8, penalties=pen,
+                       additive.only=TRUE, verbose=FALSE)
```

The chosen model then contains only the QTL on chr 1 and 4.

```
> outsw5

      name chr  pos n.gen
Q1 1@67.8    1 67.8     2
Q2 4@29.5    4 29.5     2

  Formula: y ~ Q1 + Q2
  pLOD: 8.897
```

Finally note that while, in the above, we started forward selection at the null QTL model, one may also begin the algorithm at any defined QTL model. For example, we could start at the first model that we considered in Sec. 9.3.1, determined from the results of `scanone` and `scantwo`. The starting model is indicated through the arguments `qtl` (a QTL object, created with `makeqtl`) and `formula` (a model formula).

```
> qtl <- makeqtl(hyper, chr=c(1, 4, 6, 15),
+                pos=c(68.3, 30, 60, 18))
> outsw6 <- stepwiseqtl(hyper, max.qtl=8, penalties=pen,
+                       qtl=qtl, formula=y~Q1+Q2+Q3*Q4,
+                       verbose=FALSE)
```

We get almost the same result, though we get to it slightly faster, as we skip the first few steps of forward selection.

```
> outsw6

       name chr  pos n.gen
Q1   1@67.8   1 67.8     2
Q2   4@29.5   4 29.5     2
Q3   6@60.0   6 60.0     2
Q4 15@17.5   15 17.5     2

  Formula: y ~ Q1 + Q2 + Q3 + Q4 + Q3:Q4
  pLOD: 10.13
```

9.4 Summary

The QTL mapping problem is best viewed as one of model selection: we seek to identify the set of QTL and epistatic interactions that are best supported by the data. While the sequence of hypothesis tests used for single- and two-QTL genome scans are a useful technique and often are sufficient, multiple-QTL mapping methods have the advantages of providing potentially increased power to detect QTL, of better separating linked QTL, and of more clearly defining epistatic interactions, and the hypothesis testing approach falls apart when one is faced with multiple-QTL models.

The most important aspect of the model selection approach to QTL mapping is the criterion for comparing models. We have described a penalized LOD score approach, with the aim to identify as many true QTL as possible while controlling the rate of inclusion of extraneous loci.

Bayesian methods for QTL mapping have a number of advantages over our more classical approach to the problem. They unify all aspects of the model selection problem, they provide a more natural expression of uncertainty in the results, and they more fully capture the uncertainty. However, the specification of a prior distribution on QTL models is difficult; specifying a target false positive rate, as is done in the classical approach, is arguably more natural. Moreover, the construction of the MCMC algorithms needed for the Bayesian methods requires great care, and their use in practice may require considerable training.

R/qtl includes a number of functions for the fit and exploration of multiple-QTL models. In addition to a fully automated method (`stepwiseqtl`), there are a number of functions for exploratory analyses.

9.5 Further reading

For general discussions of model selection, see Miller (2002) and Hastie *et al.* (2009). For a review of the model selection aspects of QTL mapping, see Sillanpää and Corander (2002).

The original papers describing multiple interval mapping (Kao and Zeng, 1997; Kao *et al.*, 1999; Zeng *et al.*, 1999) used a CEM algorithm (Meng and Rubin, 1993), in which each model parameter is updated one at a time. The implementation of a full EM algorithm (Dempster *et al.*, 1977) for this problem is straightforward (Ljungberg *et al.*, 2002; Chen, 2005), but it has not been widely used. Zeng *et al.* (1999) described the technique for trimming multi-locus QTL genotypes; they also described a useful model search algorithm, on which the method described in Sec. 9.1.3 was based.

Doerge and Churchill (1996) described the use of sequential permutation tests for mapping multiple QTL in the context of forward selection. The penalized LOD score for additive QTL models is equivalent to the BIC_δ criterion proposed by Broman and Speed (2002). Bogdan *et al.* (2004) and Baierl *et al.*

(2006) proposed alternative modifications to the BIC criterion for the consideration of epistatic interactions. The penalized LOD score for multiple QTL mapping with epistasis was described in Manichaikul *et al.* (2009).

For a discussion of the most recent approaches for Bayesian QTL mapping using Markov chain Monte Carlo, see Yi (2004), Yi *et al.* (2005, 2007), and the recent review of Yi and Shriner (2008). The key software package to consider is R/qtlbim (Yandell *et al.*, 2007), which works in conjunction with R/qtl. See `http://www.qtlbim.org`.

10

Case study I

In this chapter and the next, we present two case studies to illustrate the QTL mapping process in its entirety. We bring together tools from previous chapters and demonstrate their combined use to solve two moderately difficult problems. Both case studies have features that require special handling. In this sense they are not typical. On the other hand, almost every dataset has quirks that require an alert analyst to recognize them and respond accordingly. Our case studies illustrate the investigative process of QTL data analysis and improvisation using R/qtl.

In this chapter, we will consider the data of Orgogozo *et al.* (2006), included in the R/qtlbook package as the data set `ovar`. This is from a cross between two Drosophila species: *D. simulans* was crossed to *D. sechellia*, and the F_1 hybrid was crossed back to *D. simulans*. The phenotype of interest was ovariole number in females, a measure of fitness.

An initial cross produced 402 progeny; 383 had complete phenotype data. Initial genotyping focused on 94 individuals with extreme phenotype, though all individuals were genotyped for five morphological markers.

To increase the resolution of a major QTL identified on chromosome 3, an additional set of approximately 7000 flies were generated, and the 1050 individuals showing a recombination event between two morphological markers, *st* (bright red eyes) and *e* (dark brown body), were phenotyped and genotyped; 1038 individuals had complete phenotype data. The aim was to oversample recombinants in the QTL region, thereby increasing the fine mapping resolution.

There arc genotype data for 24 markers on 3 chromosomes. (The fourth chromosome had one marker, but it showed no effect and was excluded from further consideration.)

We begin with data diagnostics that show us the special features of these data: the selective genotyping and phenotyping strategies that were employed. We will then analyze the initial cross, followed by a combined analysis of both crosses. Finally, we discuss the strengths and limitations of the conclusions.

K.W. Broman, Ś. Sen, *A Guide to QTL Mapping with R/qtl*,
Statistics for Biology and Health, DOI 10.1007/978-0-387-92125-9_10,

10.1 Diagnostics

Before jumping into QTL mapping analyses, it is useful to perform exploratory analyses of the phenotype and genotype data. This will familiarize us with the data and help diagnose potential artifacts. It is not uncommon to devote half of the intellectual effort of a project on data diagnostics. The use of the various diagnostic tools described in Chap. 3 should become routine.

We must first load the R/qtl and R/qtlbook packages, and then the data set, `ovar`.

```
> library(qtl)
> library(qtlbook)
> data(ovar)
```

We first get a quick overview of the data.

```
> summary(ovar)

    Backcross

    No. individuals:    1452

    No. phenotypes:     4
    Percent phenotyped: 97.9 99 98.1 100

    No. chromosomes:    3
        Autosomes:      1 2 3

    Total markers:      24
    No. markers:        2 9 13
    Percent genotyped:  29.5
    Genotypes (%):      II:50.1  IE:49.9
```

There are a total of 1452 backcross individuals, with four phenotypes and genotypes at 24 markers. More than two-thirds of the genotype data are missing. The alleles I and E correspond to *D. simulans* and *D. sechellia*, respectively.

We make a summary plot as follows; see Fig. 10.1.

```
> plot(ovar)
```

The genetic map has some large gaps, with markers separated by as much as 40 cM. The phenotypes `on1` and `on2` are the ovariole counts for the two gonads. The phenotype `onm` is the average of the two counts. For several individuals, the ovariole count for one of the two gonads was missing, and so `onm` is missing. The fourth phenotype, `cross`, indicates which individuals were from the initial cross and which were from the second, selectively phenotyped cross.

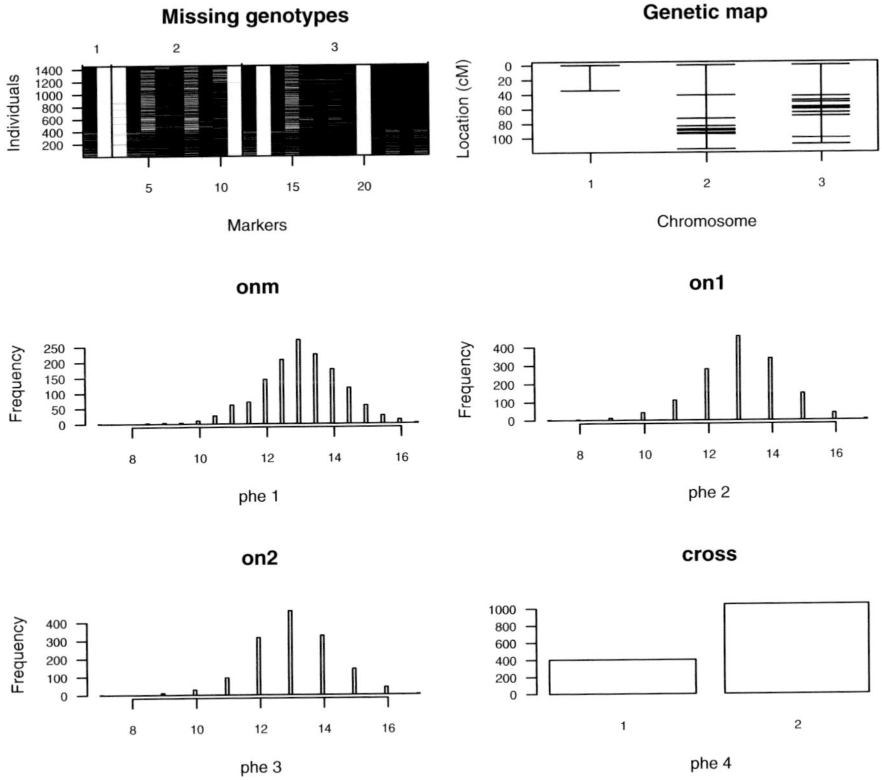

Figure 10.1. The summary plot of the ovar data provided by the plot.cross function, including the pattern of missing genotype data (upper left; black pixels indicate missing data), the genetic map of the typed markers (upper right), histograms of the three phenotypes, onm (average ovariole count), and on1 and on2 (gonad-specific ovariole counts), and a bar plot of the cross phenotype, indicating the numbers of individuals from the initial cross and from the second, selectively phenotyped cross.

Let us plot the two gonad-specific ovariole counts against each other. We use the function jitter to add a small amount of random noise to the counts so that we can distinguish multiple individuals with the same ovariole counts.

```
> plot(jitter(on2) ~ jitter(on1), data=ovar$pheno,
+       xlab="on1", ylab="on2", cex=0.6,
+       xlim=c(6.66, 17.53), ylim=c(6.66, 17.53))
```

We use cex to make the points smaller, and xlim and ylim to force the x- and y-axis limits to be the same.

Figure 10.2 indicates clear but not overly strong association between the two counts. We may use cor to calculate the sample correlation between the two counts. The argument use="complete" is required due to the missing

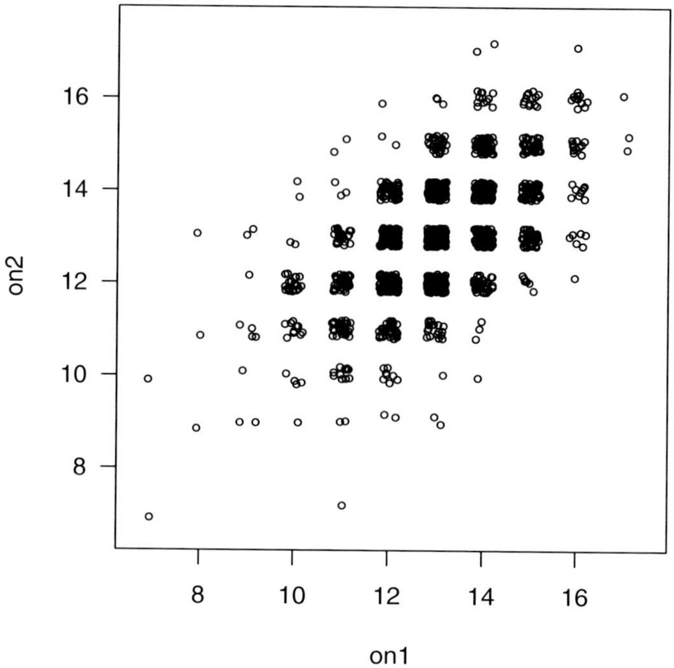

Figure 10.2. Scatterplot of `on1` and `on2`, the gonad-specific ovariole counts in the `ovar` data, with points randomly jittered so that overlapping points may be distinguished.

data; the correlation is then calculated using only those individuals with complete data.

```
> cor(pull.pheno(ovar,"on1"), pull.pheno(ovar,"on2"),
+       use="complete")
```

```
[1] 0.582
```

The phenotype `onm` should be the average of `on1` and `on2`. It is a good idea to check this.

```
> max(abs(pull.pheno(ovar,"onm") -
+            (pull.pheno(ovar,"on1")+pull.pheno(ovar,"on2"))/2),
+       na.rm=TRUE)
```

```
[1] 0
```

The phenotype `onm` should be missing if either `on1` or `on2` is missing. We should check this, too.

```
> table(apply(is.na(pull.pheno(ovar,1:3)), 1, paste,
+               collapse=":"))
```

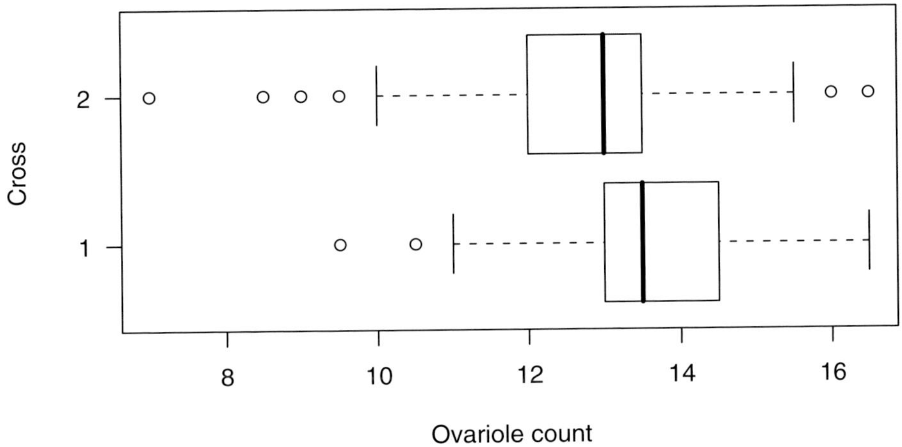

Figure 10.3. Box plot of onm for the two crosses forming the ovar data.

```
FALSE:FALSE:FALSE  FALSE:TRUE:FALSE   TRUE:FALSE:TRUE
             1419                 2                19
  TRUE:TRUE:FALSE    TRUE:TRUE:TRUE
                3                 9
```

The three TRUE/FALSE values indicate whether the onm, on1 and on2 phenotypes, respectively, are missing or not. There are 1419 individuals with no missing phenotype, 19 individuals missing onm and on2, 3 individuals missing onm and on1, and 9 individuals missing all three phenotypes. There are also two individuals for which onm is not missing but on1 is missing.

```
> ovar$pheno[!is.na(ovar$pheno$onm) & is.na(ovar$pheno$on1),]
```

```
     onm on1 on2 cross
1300  13  NA  13     2
1325  12  NA  12     2
```

We will fix these.

```
> ovar$pheno$onm[is.na(ovar$pheno$on1)] <- NA
```

A boxplot of the onm phenotypes in the two crosses will indicate whether there are systematic differences.

```
> boxplot(onm ~ cross, data=ovar$pheno, horizontal=TRUE,
+         xlab="Ovariole count", ylab="Cross")
```

As seen in Fig. 10.3, the ovariole counts were quite a bit smaller in the second cross. A t test will indicate whether this could reasonably ascribed to chance variation (though the figure is sufficiently convincing).

```
> t.test(onm ~ cross, data=ovar$pheno)
```

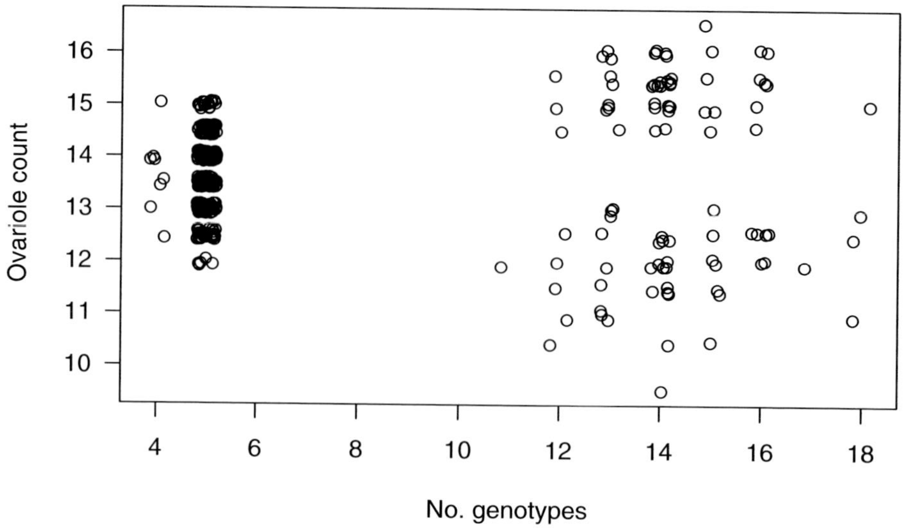

Figure 10.4. Plot of ovariole count against number of typed marker genotypes for the initial cross in the `ovar` data. On the right are points corresponding to the 94 genotyped individuals; on the left are the remaining individuals, for which only five morphological markers were scored.

```
        Welch Two Sample t-test

data:  onm by cross
t = 11.75, df = 710.2, p-value < 2.2e-16
alternative hypothesis: true difference in means is not equal to 0
95 percent confidence interval:
 0.6590 0.9235
sample estimates:
mean in group 1 mean in group 2
          13.64           12.85
```

Let us now study the genotype data in the first cross. We first use `subset.cross` to pull out the individuals from the first cross.

```
> ovar1 <- subset(ovar, ind=(pull.pheno(ovar, "cross")==1))
```

Note the selective genotyping.

```
> plot(jitter(ntyped(ovar1)),jitter(pull.pheno(ovar1,"onm")),
+      xlab="No. genotypes", ylab="Ovariole count")
```

As seen in Fig. 10.4, the individuals with the most genotype data have high ($\geq$ 14.5) or low ($\leq$ 13) phenotype, but there were no strict cutoffs: there is some overlap between the more and less highly genotyped groups.

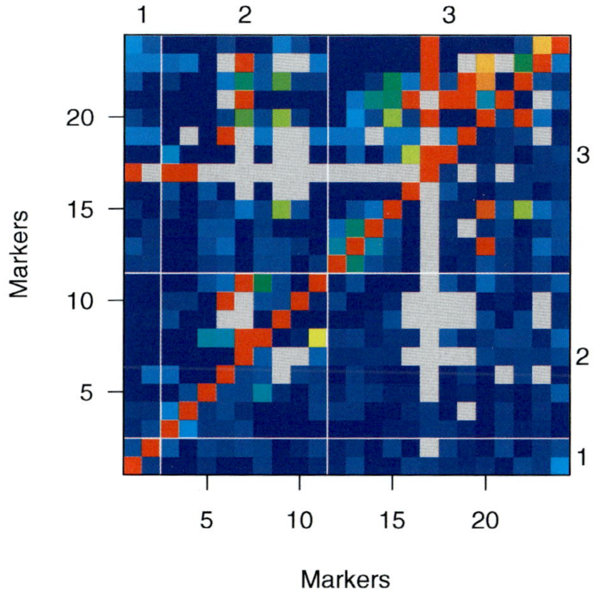

Figure 10.5. Plot of estimated recombination fractions (upper left) and LOD scores for a test of $r = 1/2$ (lower right) for all pairs of markers in the initial cross in the `ovar` data. Red indicates linkage, blue indicates no linkage, and gray indicates missing values (for marker pairs that were not both typed in any one individual).

The extensive missing genotype data cause some odd patterns in the estimated recombination fractions between markers, but it is nevertheless useful to plot these (see Sec. 3.4.1).

```
> ovar1 <- est.rf(ovar1)
> plot.rf(ovar1)
```

In Fig. 10.5, the LOD scores (in the lower right triangle) indicate no evidence for linkage between markers on different chromosomes.

The map associated with the data was established based on Macdonald and Goldstein (1999). It is a good idea to reestimate the intermarker distances, using the observed data, keeping marker order fixed, and plot the estimated map against that which came with the data.

```
> newmap <- est.map(ovar1, error.prob=0.001, verbose=FALSE)
> plot.map(ovar1, newmap)
```

There is some evidence for map expansion (see Fig. 10.6), though this may be partly due to the choice of map function. By default, `est.map` uses the Haldane map function to convert estimated recombination fractions to genetic distances. The choice of map function has a greater effect on larger marker intervals.

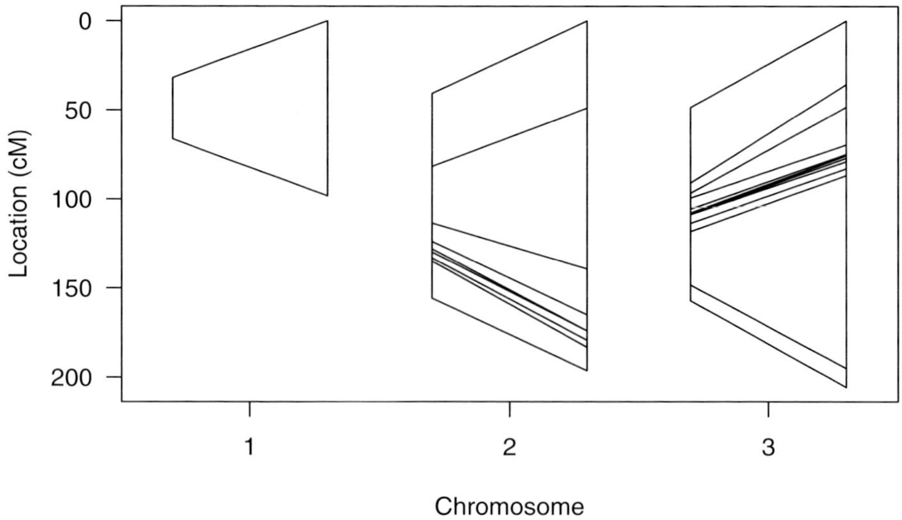

Figure 10.6. The genetic map in the `ovar` data plotted against the map estimated from the individuals in the initial cross. For each chromosome, the line on the left is the map provided with the data, and the line on the right is the map estimated using the Haldane mapping function; line segments connect the positions for each marker.

If we use the Kosambi map function (rather than the Haldane map function), the chromosomes are quite a bit smaller.

```
> newmap.k <- est.map(ovar1, err=0.001, map.function="kosambi")
> summary(newmap)

        n.mar length ave.spacing max.spacing
1           2   98.2        98.2        98.2
2           9  196.2        24.5        90.3
3          13  205.5        17.1       108.3
overall    24  500.0        23.8       108.3

> summary(newmap.k)

        n.mar length ave.spacing max.spacing
1           2   64.6        64.6        64.6
2           9  147.1        18.4        60.3
3          13  153.8        12.8        70.0
overall    24  365.5        17.4        70.0
```

We will stick with the map that came with the data, and we will use the Kosambi map function in our QTL analyses. There is not a compelling reason to replace the original map, and since crossover interference is known to exist in Drosophila, it seems prudent to use the Kosambi map function.

Finally, let us consider the chromosome 3 genotypes for the individuals in the second cross (which were selected to be recombinant between the markers *st* and *e*). The function `geno.crosstab` can be used to create a table of two-locus genotypes.

```
> ovar2 <- subset(ovar, ind=(pull.pheno(ovar,"cross")==2))
> geno.crosstab(ovar2, "st", "e")
```

```
    e
st    -  II  IE
   -  0   1   0
  II  0   0 566
  IE  0 481   2
```

There are two individuals that are not recombinant (these might be errors) and one individual that has missing genotype for the *st* marker.

Consider the same cross-tabulation for the first cross.

```
> geno.crosstab(ovar1, "st", "e")
```

```
    e
st    -  II  IE
   -  0   1   0
  II  0 146  72
  IE  0  47 136
```

There were 30% recombinants in the first cross.

Let us plot the chromosome 3 genotypes for a random set of 15 individuals from the second cross.

```
> toplot <- sort(sample(nind(ovar2), 15))
> plot.geno(ovar2, chr=3, ind=toplot)
```

As seen in Fig. 10.7, these individuals were genotyped just within the region of the selected recombination event.

10.2 Initial cross

We now turn to QTL mapping. We first focus on the initial cross of 402 individuals (of which 383 have complete phenotype data). In the last section, we created a cross object, `ovar1`, with just those individuals. To avoid repeated warning messages, we omit individuals for which the `onm` phenotype is missing.

```
> ovar1 <- subset(ovar1, ind=!is.na(pull.pheno(ovar1,"onm")))
```

We will use multiple imputation for the QTL mapping, because of the selective genotyping and our desire to fit multiple-QTL models (see Sec. 4.2.4). The extensive missing genotype data suggests that we should perform a large number of imputations; we will use 512. (We like powers of 2.) We begin with a call to `sim.geno`, to do the imputations.

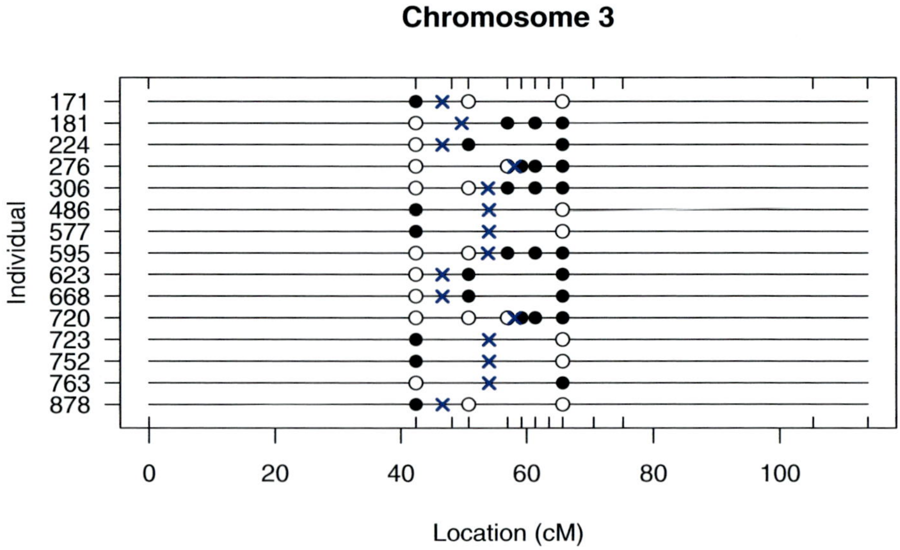

Figure 10.7. Chromosome 3 genotypes for a random set of individuals from the second `ovar` cross, showing the selective genotyping of recombinants. Blue ×'s indicate recombination events.

```
> ovar1 <- sim.geno(ovar1, n.draws=512, step=2, err=0.001,
+                   map.function="kosambi")
```

We now use `scanone` to perform a single-QTL genome scan by the multiple imputation approach.

```
> out1 <- scanone(ovar1, method="imp")
```

A quick summary and plot of the genome scan results (Fig. 10.8) indicates strong evidence for a QTL on chr 3, and some evidence for an additional QTL on chr 2.

```
> summary(out1)

     chr  pos    lod
per    1  4.5  0.213
SRPK   2 87.3  2.228
cpo    3 82.4 14.010

> plot(out1, ylab="LOD score")
```

The evidence for the QTL on chr 3 (with a LOD score of 14.01) is clear. It is worthwhile considering the estimated effect of this QTL. We can make a plot with `effectplot`, or we could get a numerical estimate of the effect with `fitqtl`. Let us do both.

First, we use `effectplot` to plot the estimated phenotype averages for each of the two QTL groups (see Sec. 4.6).

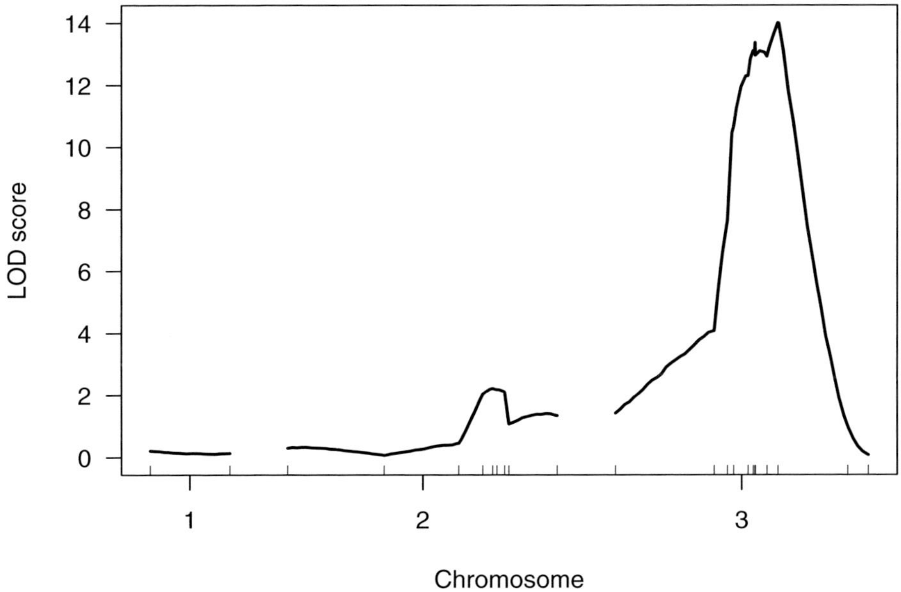

Figure 10.8. LOD curves from a genome scan by multiple imputation for the initial cross in the ovar data.

```
> effectplot(ovar1, mname1="cpo")
```

The plot, in Fig. 10.9, indicates that the *D. simulans* allele (I) is associated with one additional ovariole per gonad.

To use `fitqtl` to get a numerical estimate of the QTL effect, we first use `makeqtl` to create a QTL object and then call `fitqtl` with `dropone=FALSE`, as we can skip the drop-one-QTL analysis, and `get.ests=TRUE`, to get the estimates (see Sec. 9.3.1).

```
> qtl <- makeqtl(ovar1, chr=3, pos=82.4)
> summary(fitqtl(ovar1, qtl=qtl, dropone=FALSE, get.ests=TRUE))

Full model result
----------------------------------
Model formula: y ~ Q1

       df    SS     MS   LOD  %var Pvalue(Chi2) Pvalue(F)
Model   1  73.3 73.296 14.01 15.50    9.992e-16 1.110e-15
Error 381 399.5  1.049
Total 382 472.8

Estimated effects:
-----------------
```

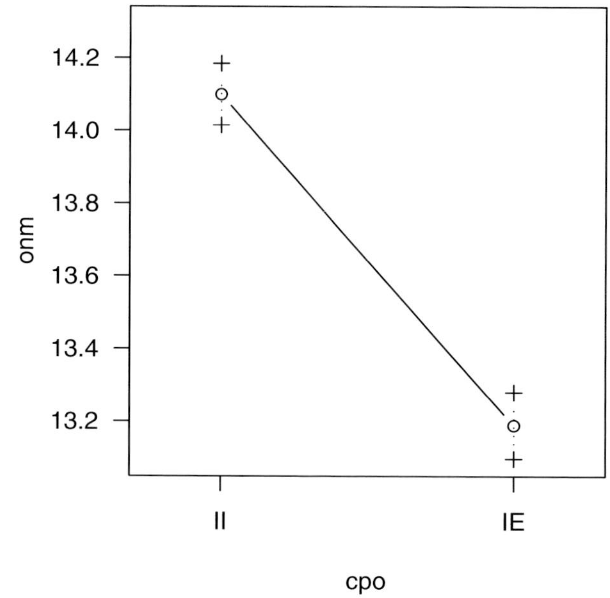

Figure 10.9. Estimated phenotype averages for the two groups defined by genotypes at marker *cpo*, for the initial cross in the `ovar1` data. Error bars are ± 1 SE.

```
               est        SE       t
Intercept 13.65411  0.05323 256.499
3@82.4    -0.89491  0.10602  -8.441
```

The estimated effect of the chr 3 QTL is -0.89 ± 0.11.

Given the large effect of the chr 3 QTL, let us repeat the genome scan, controlling for this locus. If we had complete genotype data at the inferred QTL, we could include it as an additive covariate in `scanone`. But in the current situation, with considerable missing data at the inferred QTL, we use `addqtl` to scan for an additional QTL using multiple imputation (see Sec. 9.3.4).

```
> out1.c3 <- addqtl(ovar1, qtl=qtl)
```

Again, we make a quick summary and plot (Fig. 10.10).

```
> summary(out1.c3)

         chr   pos   lod
c1.loc32   1  36.5 0.173
c2.loc90   2  90.0 3.531
slo        3 121.3 2.543

> plot(out1.c3, ylab="LOD score")
```

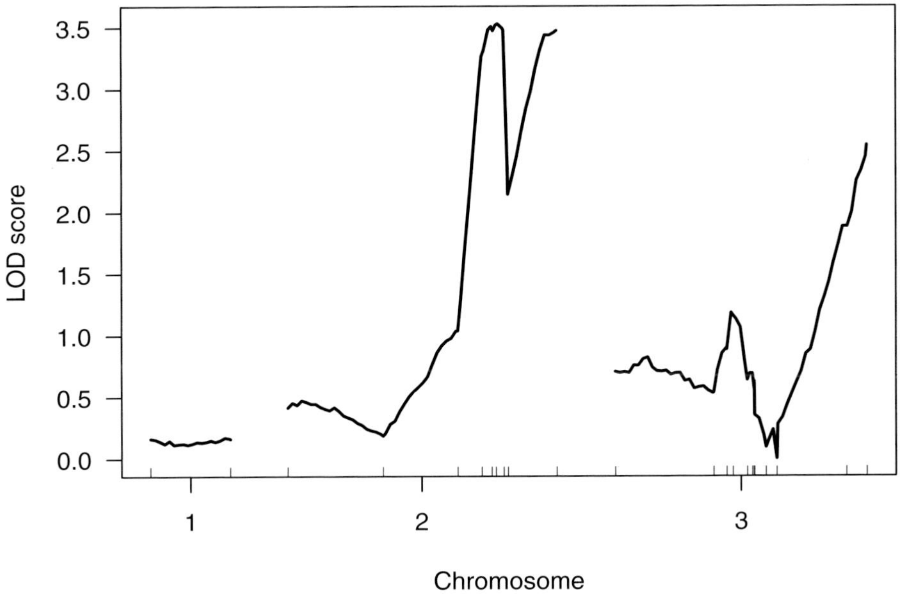

Figure 10.10. LOD curves from a genome scan by multiple imputation for the initial cross in the `ovar` data, adjusting for an inferred QTL at the marker *cpo*.

We have improved evidence for a QTL on chr 2 (the LOD increased from 2.23 to 3.53), and there is some evidence for an additional QTL at the distal end of chr 3.

Let us perform a two-dimensional, two-QTL scan on chr 3, to assess the evidence for a second QTL on the chromosome. We will include the inferred QTL on chr 2 as an additive covariate. This requires that we first create a new QTL object (containing just the chr 2 QTL) and run `addpair` (see Sec. 9.3.5).

```
> qtl2 <- makeqtl(ovar1, 2, 90)
> out1.ap <- addpair(ovar1, qtl=qtl2, chr=3, verbose=FALSE)
```

We can use `summary.scantwo` to pick off the largest peak for both a full model (with two interacting QTL) and an additive model (see Sec. 8.1).

```
> summary(out1.ap)

      pos1f pos2f lod.full lod.fv1 lod.int     pos1a pos2a
c3:c3  82.8   121     17.6    2.30   0.143      82.8   121
      lod.add lod.av1
c3:c3    17.5    2.16
```

The evidence for a second QTL on chr 3 remains, but it is not strong, and actually the evidence is weaker once the QTL on chr 2 is considered. (The LOD for two additive QTL on chr 3, versus a single QTL near the *cpo* marker, is 2.16 when the chr 2 locus is included in the model and is 2.54 when the chr 2

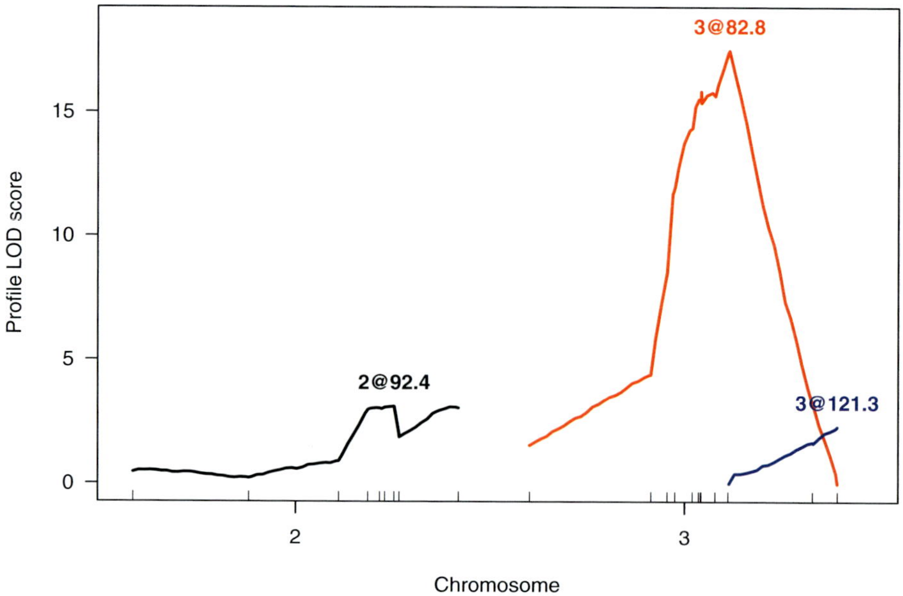

Figure 10.11. Profile LOD curves for a three-QTL model, for the initial cross in the `ovar` data.

locus is not included.) There is no evidence for an interaction between the loci on chr 3.

We now bring all three putative QTL into one model, and use `refineqtl` to improve our estimates of the QTL positions (see Sec. 9.3.2).

```
> qtl <- makeqtl(ovar1, c(2, 3, 3), c(90, 82.8, 121))
> rqtl <- refineqtl(ovar1, qtl=qtl, verbose=FALSE)
```

The QTL moved slightly:

```
> rqtl

       name chr   pos n.gen
Q1   2@92.4   2  92.4     2
Q2   3@82.8   3  82.8     2
Q3  3@121.3   3 121.3     2
```

We may use `plotLodProfile` to plot LOD profiles for the three QTL (see Sec. 9.3.2). The results appear in Fig. 10.11.

```
> plotLodProfile(rqtl, col=c("black","red","blue"),
+                ylab="Profile LOD score")
```

We are particularly interested in deriving a confidence interval for the location of the proximal QTL on chr 3, as the selective phenotyping in the second cross in these data was performed with the aim of more precisely mapping this locus. As described in Sec. 9.3.2, we may use `lodint` or `bayesint` to derive an approximate confidence interval for QTL location, based on the LOD profile calculated by `refineqtl`. These intervals need to be viewed with some caution, however, as they fail to account for the uncertainty in the location of the other QTL, and the performance of these intervals in the context of a multiple-QTL model is not well understood.

In the use of `lodint` and `bayesint`, we refer to the QTL by their numeric index within the QTL object. The proximal QTL on chr 3 is the second locus within `rqtl`, and so we use `qtl.index=2`.

```
> lodint(rqtl, qtl.index=2)

          chr  pos   lod
nos         3 77.8 15.66
c3.loc70    3 82.8 17.48
c3.loc74    3 86.8 15.62
```

```
> bayesint(rqtl, qtl.index=2)

          chr  pos   lod
c3.loc66    3 78.8 16.13
c3.loc70    3 82.8 17.48
c3.loc72    3 84.8 16.53
```

These intervals are reasonably small. (The 1.5-LOD support interval is 9.0 cM long, and the approximate 95% Bayesian credible interval is 6.0 cM long.)

It is surprising that the intervals start at 77.8 cM and so do not cover the region used to select recombinants in the second cross: selected individuals showed a recombination event between markers *st* (at 55.2 cM) and *e* (at 72.9 cM). If our confidence intervals are accurate, the selective phenotyping should not be expected to narrow the location of the QTL, as the selected recombination events will be flanking the QTL, rather than covering it. The region used to select recombinants was chosen based on an initial analysis with a simpler model and with the QTL Cartographer software, which indicated a QTL peak to the left of the *e* marker; see Fig. 2B in Orgogozo *et al.* (2006). Moreover, the *e* marker was the most distal morphological marker available in the *D. simulans* strain that was used for these crosses.

We have neglected to be precise about the strength of evidence for the three inferred QTL. The proximal QTL on chr 3 (with LOD score ~14) is clear, but in considering the other two QTL, we must take this large-effect QTL into account. Let us apply the multiple-QTL model comparison criterion described in Sec. 9.1.4.

We must first derive appropriate penalties for the penalized LOD score criterion, using a permutation test with a two-dimensional, two-QTL genome scan. The computational effort is forbidding, but it can be made feasible by the parallel use of multiple computers (see Sec. 8.1, page 223). The permutation test here took about four days of computer time, but we split it across 16 processors, so it took about six hours in real time.

Due to the selective genotyping in this initial cross (only 94 individuals chosen by their relatively extreme phenotypes were typed at most markers), it is best to perform a stratified permutation test (see Sec. 4.4.3).

```
> strat <- (nmissing(ovar1) < 15)
> operm <- scantwo(ovar1, method="imp", n.perm=1000,
+                   perm.strata=strat)
```

The significance thresholds can be calculated as follows.

```
> summary(operm)

onm (1000 permutations)
    full  fv1  int  add  av1  one
5%  3.83 2.87 2.09 3.01 1.77 1.70
10% 3.46 2.47 1.83 2.61 1.49 1.45
```

Most importantly, we may calculate the penalties as follows. Note the use of `print` to simultaneously print the results and assign them to the object `pen`.

```
> print(pen <- calc.penalties(operm))

 main heavy light
1.698 2.085 1.176
```

The significance thresholds and penalties are much smaller than others we have seen in this book, but keep in mind that the null distributions of the various LOD scores depend on not just the type of LOD score (e.g., the LOD score from a single-QTL genome scan versus the "full" LOD, comparing a model with two interacting loci to the null model, in a two-dimensional, two-QTL genome scan) and the type of cross (backcross versus intercross), but also on the genetic length of the genome. Drosophila has a much smaller genetic map than the mouse (which has been the focus of all of our previous examples), and so the significance thresholds and penalties are much smaller. Here, the penalty on main effects is just 1.70, and so all three of our QTL would be selected.

Let us complete our analysis of the initial cross in the `ovar` data by applying the stepwise model search algorithm described in Sec. 9.1.3, using the function `stepwiseqtl`. We can start forward selection at our current model, defined by `rqtl`.

```
> stepout1 <- stepwiseqtl(ovar1, penalties=pen, qtl=rqtl,
+                         max.qtl=8, verbose=FALSE)

> stepout1
```

```
        name chr   pos n.gen
Q1   2@92.4   2  92.4     2
Q2   2@94.2   2  94.2     2
Q3  2@112.0   2 112.0     2
Q4   3@82.8   3  82.8     2
Q5  3@121.3   3 121.3     2

  Formula: y ~ Q1 + Q2 + Q3 + Q4 + Q5 + Q1:Q2
  pLOD: 14.82
```

It is a bit of a surprise to see three QTL on chr 2, and the two tightly linked epistatic loci on chr 2 are rather suspicious. We often have difficulty with tightly linked QTL, particularly if they are allowed to interact. Let's look at a couple of plots of the phenotypes against the two-locus genotypes, using both `effectplot` and `plot.pxg`, to see what's going on. We use `find.marker` to find the marker closest to each QTL, for use in `plot.pxg`. In `effectplot`, we may refer directly to pseudomarkers (that is, the positions on the grid between markers), by their chromosome and cM position.

```
> mar <- find.marker(ovar1, 2, c(92.4, 94))
> par(mfrow=c(1,2))
> effectplot(ovar1, mname1="2@92.4", mname2="2@94")
> plot.pxg(ovar1, marker=mar)
```

The right panel in Fig. 10.12 suggests that the phenotypes for just a few individuals are leading to the large LOD score. (There are just three individuals with genotypes `IE` at marker *Amy-d* and `II` at marker *grh*. No individual has observed genotypes `II` at marker *Amy-d* and `IE` at marker *grh*; the points in the plot come from a single imputation, using data from surrounding markers.) Thus, we should discount the evidence for the two tightly linked QTL.

Let's omit the QTL at 94.2 cM (using `dropfromqtl`) and run `refineqtl` to refine the locations of the other QTL.

```
> qtl <- dropfromqtl(stepout1, 2)
> qtl <- refineqtl(ovar1, qtl=qtl, verbose=FALSE)
```

We can print the object to see if the QTL moved.

```
> qtl

        name chr   pos n.gen
Q1   2@92.4   2  92.4     2
Q2   2@94.0   2  94.0     2
Q3   3@82.8   3  82.8     2
Q4  3@121.3   3 121.3     2
```

The distal locus on chr 2 moved back to the 94 cM position, which we don't trust! We should probably stick with the three-QTL model, with just one QTL on chr 2.

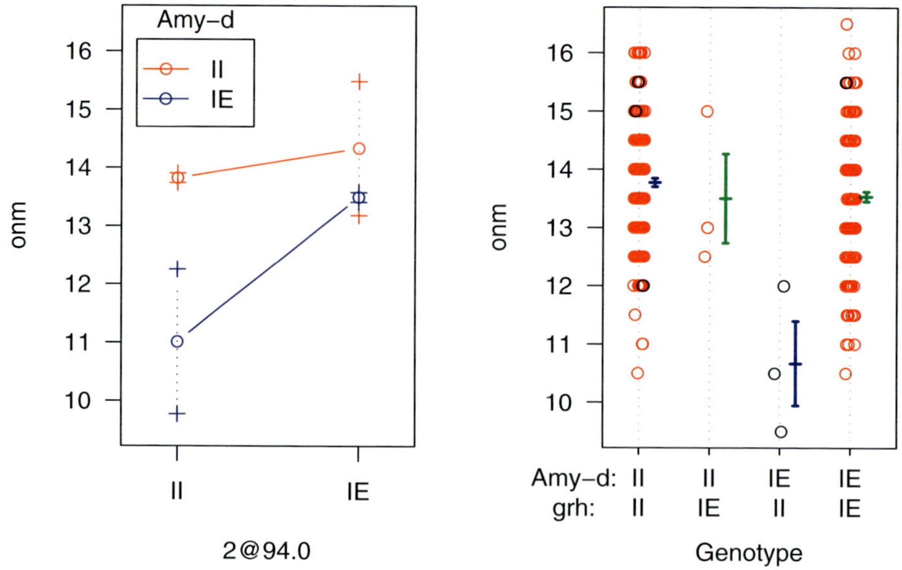

Figure 10.12. Plot of the `onm` phenotype against genotype at the two tightly linked loci on chr 2, for the initial cross in the `ovar` data. Red dots in the right panel are for imputed genotypes. Error bars are ± 1 SE.

10.3 Combined data

We now turn to an analysis of the combined data, including the second cross of 1050 individuals (of which 1038 have complete phenotype data) that were selected to be recombinant between the morphological markers, *st* (bright red eyes) and *e* (dark brown body).

Imputed genotypes (from `sim.geno`) take up a great deal of memory, and so it might be best to either remove them from the `ovar1` object (which we were working with in the last section) using `clean.cross`, or just remove the `ovar1` object from our workspace (using `rm`). One may use `object.size` to estimate the amount of memory (in bytes) that an object is taking up. Another useful tool is the function `gc`, which tells R to perform "garbage collection" to clean up memory and also gives information about R's current memory usage.

```
> object.size(ovar1)

[1] 237010992
```

Dividing by 1024^2 gives the result in megabytes (Mb).

```
> object.size(ovar1)/1024^2

[1] 226.0
```

We now use `clean` to remove all of the intermediate calculations from the object, and check the new size, in bytes.

```
> ovar1 <- clean(ovar1)
> object.size(ovar1)
```

```
[1] 103304
```

Dividing by 1024 gives the result in kilobytes (kb).

```
> object.size(ovar1)/1024
```

```
[1] 100.9
```

And so, by removing the intermediate calculations, the space used by `ovar1` went from 226 Mb to 101 kb.

We turn now to the full data, `ovar`. Let us first remove individuals missing the `onm` phenotype.

```
> ovar <- subset(ovar, ind=!is.na(pull.pheno(ovar,"onm")))
```

We use `sim.geno` to perform the multiple imputations. We will use the same settings as we had used for `ovar1` in the previous section.

```
> ovar <- sim.geno(ovar, n.draws=512, step=2, err=0.001,
+                  map.function="kosambi")
```

Because of the systematic difference in ovariole counts between the two crosses (see Fig. 10.3 on page 287), it would be best to include a cross indicator as an additive covariate in our analyses. In doing so, we assume that the effects of QTL are the same in the two crosses, but that there is an overall shift in the mean phenotype. The covariate must be numeric, but the coding of the two crosses (e.g., 0/1 or 1/2) has no influence on the LOD scores or estimated QTL effects that we calculate.

```
> cross <- pull.pheno(ovar, "cross")
> outc <- scanone(ovar, method="imp", addcovar=cross)
```

We will perform the same genome scan with just the individuals from the second cross, for comparison purposes. We do not need to rerun `sim.geno`, as the imputations will be retained in the output of `subset.cross`.

```
> out2 <- scanone(subset(ovar, ind=(cross==2)), method="imp")
```

We now plot the results for the combined data and for the two individual crosses.

```
> plot(outc, out1, out2, ylab="LOD score")
```

In the LOD curves in Fig. 10.13, we see that the results for the combined data (the black curves) are similar to those for the initial cross (the blue curves), though the LOD scores are much larger and the location of the QTL on chr 3 appears to be shifted to the left slightly. On chr 3, the LOD curve for

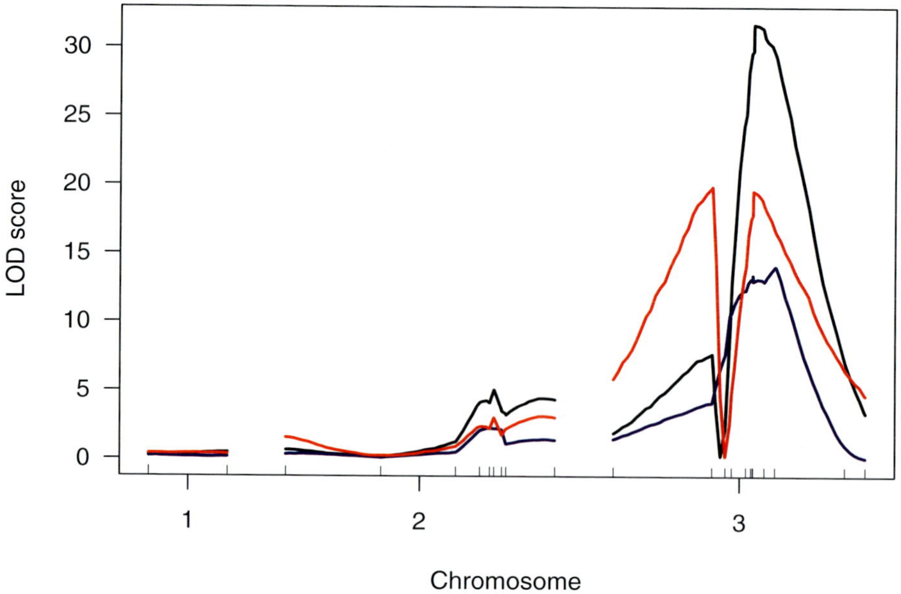

Figure 10.13. LOD curves from a genome scan by multiple imputation for the `ovar` data. The black curves are for the combined data, the blue curves are for the initial cross, and the red curves are for the second cross.

the second cross (the red curve) has an odd double peak. As the individuals were selected to be recombinant in the region between markers *st* and *e*, they have exactly opposite genotypes on either side of that region (and recall that they were not genotyped outside the region), and so a QTL to one side will show a mirror image on the other side (with the apparent QTL on the other side having an effect of the opposite sign).

We note the location of the maximum LOD score from the single-QTL genome scan, for the combined data.

```
> max(outc)

  chr  pos  lod
e   3 72.9 31.7
```

The QTL has moved to the left, from 82.4 cM (as inferred with the initial cross) to 72.9 cM (as inferred from the combined data).

Let us again adjust for this major locus, and scan for a second QTL. We will again consider the combined data as well as the second cross, on its own, and we will compare the results to those from the initial cross. Note that for `addqtl`, the covariates need to form a matrix or data frame, and so in the first line we convert our `cross` covariate to a data frame.

```
> cross <- data.frame(cross=cross)
> qtlc <- makeqtl(ovar, chr=3, pos=72.9)
```

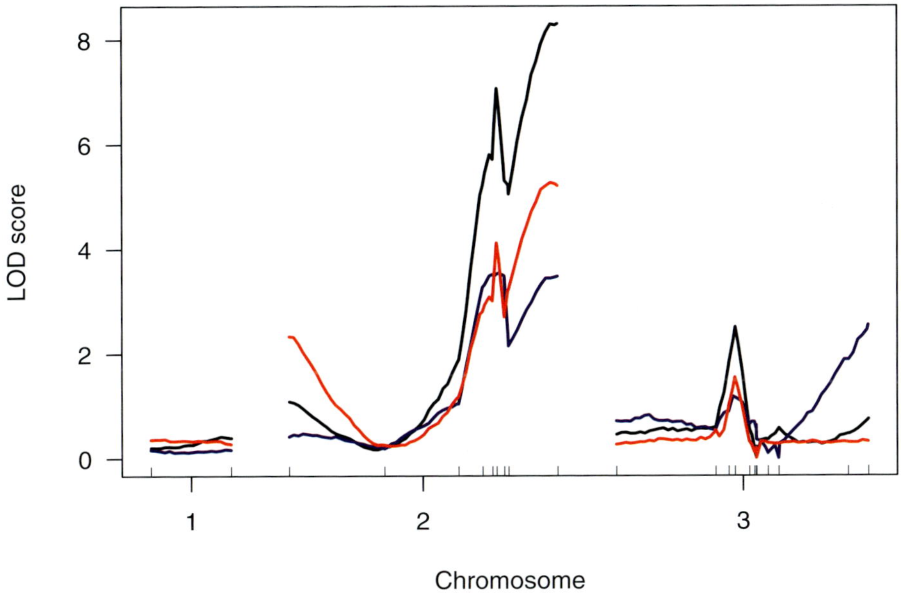

Figure 10.14. LOD curves from a genome scan by multiple imputation for the `ovar` data, adjusting for an inferred QTL on chr 3. The black curves are for the combined data, the blue curves are for the initial cross, and the red curves are for the second cross. Note that the position of the inferred QTL (on which we are conditioning) is different for the initial cross versus for the combined data and for the second cross.

```
> outc.c3 <- addqtl(ovar, qtl=qtlc, covar=cross)
> qtl2 <- makeqtl(subset(ovar, ind=(cross==2)), chr=3, pos=72.9)
> out2.c3 <- addqtl(subset(ovar, ind=(cross==2)), qtl=qtl2)
```

A plot of the LOD curves, controlling for the major locus on chr 3, is obtained as follows; see Fig. 10.14.

```
> plot(outc.c3, out1.c3, out2.c3, ylab="LOD score")
```

The results on chr 3 are quite different for the combined data; we now have clear evidence for a second QTL on chr 3. There remains strong evidence for a QTL on chr 2 (and possibly there are two QTL on chr 2).

Before proceeding further, let's run the necessary permutation tests that will help us to make sense of the statistical significance of our results. We will perform a permutation test with a two-dimensional, two QTL scan. It will be best to perform a stratified permutation test, as we had done for the permutation test with the data from initial cross in the last section. Here we will want three strata: the individuals from the initial cross that were selected for the initial genotyping, the other individuals from the initial cross, and the individuals from the second cross.

The `perm.strata` argument in `scantwo` should be a vector of indices that indicate which individuals are in which stratum. We can create such a vector as follows.

```
> strat <- pull.pheno(ovar, "cross")
> strat[strat==1 & nmissing(ovar)<15] <- 3
> table(strat)

strat
   1    2    3
 289 1036   94
```

The particular numeric codes that we assign to the three strata can be anything, and so we do whatever is most convenient.

We now are ready to perform the permutation test. We should again emphasize that the computation time is long, and so it is best to split the permutations into batches using multiple computers running in parallel. (These computations took 30 days of CPU time.)

```
> opermc <- scantwo(ovar, method="imp", addcovar=cross,
+                   n.perm=n.perm, perm.strata=strat)
```

We obtain the estimated significance thresholds and penalties for our multiple-QTL model comparison criterion as follows.

```
> summary(opermc)

onm (1000 permutations)
    full  fv1  int  add   av1  one
5%  3.01 1.99 1.53 2.46 1.090 1.63
10% 2.72 1.79 1.33 2.06 0.896 1.33

> print(penc <- calc.penalties(opermc))

  main  heavy  light
1.6285 1.5346 0.3603
```

The significance thresholds and penalties (particularly the interaction penalties) are quite a bit smaller than we had obtained based on the initial cross alone (see page 298).

Returning to the QTL analyses, we have reasonable evidence for at least one QTL on chr 2 and likely two QTL on chr 3. A simple approach to explore these models would be to perform two-dimensional, two-QTL scans on each chromosome, controlling for the major locus on the other chromosome. Let's begin with chromosome 3. We first create a QTL object containing the chr 2 locus, and then use `addpair` to scan for a pair of QTL on chr 3.

```
> qtl.c2 <- makeqtl(ovar, 2, 115)
> out.ap.c3 <- addpair(ovar, qtl=qtl.c2, chr=3, covar=cross,
+                      verbose=FALSE)
```

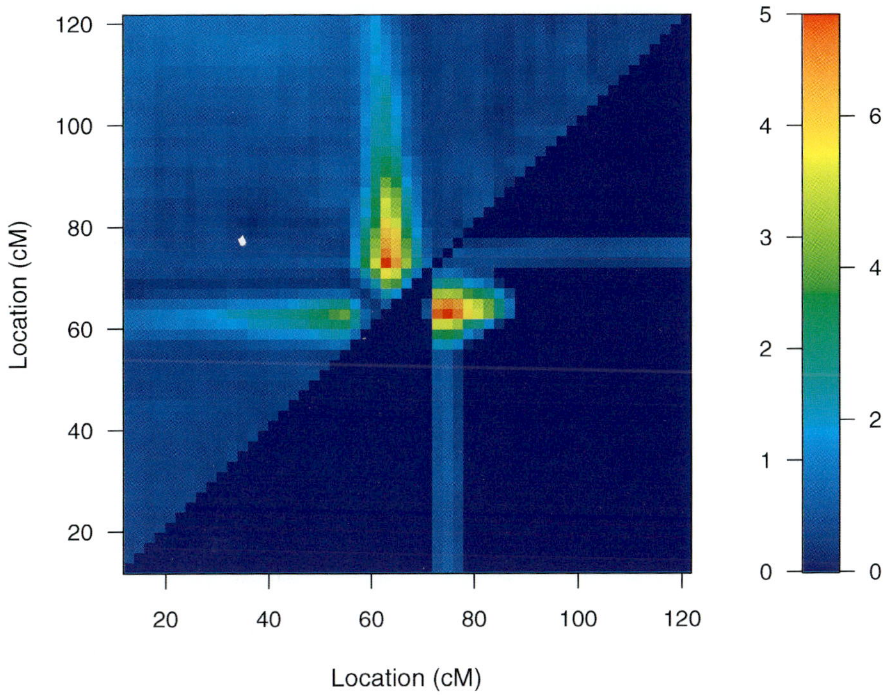

Figure 10.15. LOD scores from a two-dimensional, two-QTL scan on chr 3, controlling for a locus on chr 2, for the `ovar` data. LOD_i is displayed in the upper left triangle; LOD_{fv1} is displayed in the lower right triangle. In the color scale on the right, numbers to the left and right correspond to LOD_i and LOD_{fv1}, respectively.

The summary of the results indicates strong evidence for two interacting QTL on chr 3.

```
> summary(out.ap.c3)

      pos1f pos2f lod.full lod.fv1 lod.int     pos1a pos2a
c3:c3  62.8  74.8     42.9    7.33    4.68      62.8  74.8
      lod.add lod.av1
c3:c3    38.3    2.65
```

A plot of the LOD scores from the two-dimensional scan is useful for assessing the precision of localization of the two QTL. We will plot the interaction LOD scores in the upper left triangle (the default); in the lower right triangle, let us plot LOD scores comparing the full model (with two interacting loci) to the best single-QTL model.

```
> plot(out.ap.c3, lower="cond-int")
```

As seen in Fig. 10.15, the locations of the QTL are reasonably well defined.

We use a similar two-dimensional, two-QTL scan on chr 2, controlling for the major locus on chr 3.

```
> out.ap.c2 <- addpair(ovar, qtl=qtlc, chr=2, covar=cross,
+                      verbose=FALSE)
```

In summarizing the output of the two-dimensional, two-QTL scan on chr 2, we will include the permutation results to get approximate p-values. The permutation results do not formally apply to the present situation, as they concern the global null hypothesis (of no QTL), while the scan was conditional on the major locus on chr 3. However, they provide useful landmarks for evaluating the evidence.

```
> summary(out.ap.c2, perms=opermc, pval=TRUE)

        pos1f pos2f lod.full pval lod.fv1  pval lod.int pval
c2:c2      80   114     9.13    0    1.24 0.467   0.227    1
              pos1a pos2a lod.add pval lod.av1  pval
c2:c2             4   114     8.9    0    1.01 0.064
```

We see some evidence for a second QTL on chr 2, with p-value $= 6\%$; surprisingly, the two inferred QTL are on opposite ends of the chromosome, rather than at the two distal peaks seen in Fig. 10.14. There is no evidence for an interaction between the putative QTL on chr 2.

We plot the LOD scores to get a sense of the precision of localization of the two QTL. We will look at the LOD scores comparing two-locus models to the best single-locus model.

```
> plot(out.ap.c2, lower="cond-int", upper="cond-add")
```

The upper left triangle in Fig. 10.16, with LOD scores comparing models with two additive QTL to the best single-QTL model, are most interesting. While the location of the distal QTL (at around 114 cM) is quite well defined, the location of the proximal QTL is not at all well defined. It is inferred to be at the proximal tip of the chromosome, but large LOD scores are also seen at around 80–90 cM.

Let us bring the four QTL together into a single model and run `refineqtl` to refine the QTL positions.

```
> qtl <- makeqtl(ovar, c(2, 2, 3, 3), c(4, 114, 62.8, 74.8))
> rqtl <- refineqtl(ovar, qtl=qtl, covar=cross, verbose=FALSE,
+                   formula=y~cross+Q1+Q2+Q3+Q4+Q3:Q4)
```

The proximal QTL on chr 3 moved slightly.

```
> rqtl

        name chr   pos n.gen
Q1    2@4.0    2   4.0     2
Q2  2@114.0    2 114.0     2
```

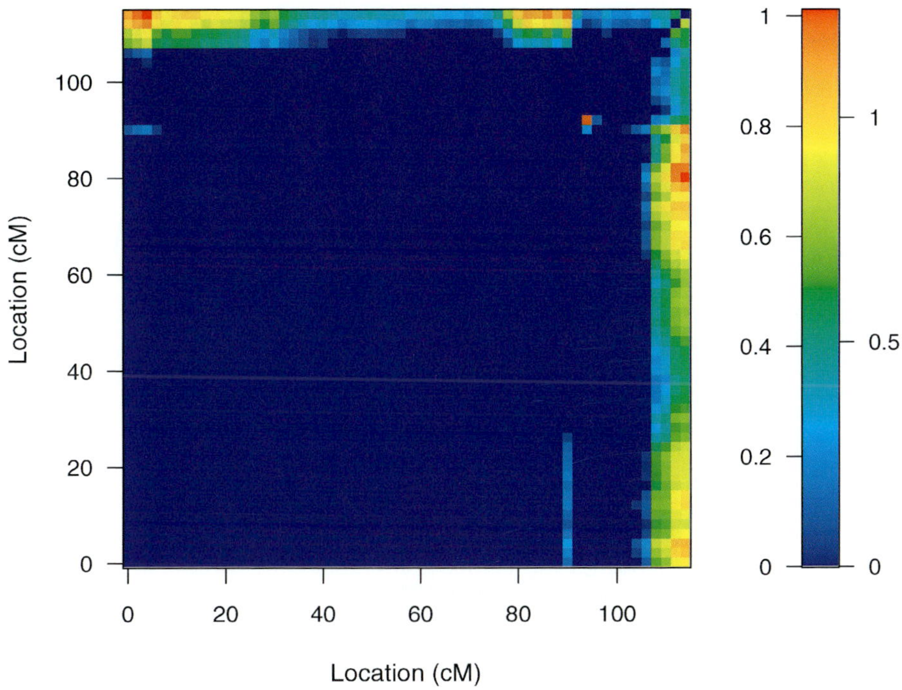

Figure 10.16. LOD scores from a two-dimensional, two-QTL scan on chr 2, controlling for a locus on chr 3, for the `ovar` data. LOD_{av1} is displayed in the upper left triangle; LOD_{fv1} is displayed in the lower right triangle. In the color scale on the right, numbers to the left and right correspond to LOD_{av1} and LOD_{fv1}, respectively.

```
Q3  3@63.6   3  63.6     2
Q4  3@74.8   3  74.8     2
```

We use `fitqtl` to perform a drop-one-QTL analysis with this four-QTL model. Note the use of `pvalues=FALSE` to omit the two columns of p-values.

```
> summary(fitqtl(ovar, qtl=rqtl, covar=cross,
+                  formula=y~cross+Q1+Q2+Q3+Q4+Q3:Q4),
+          pvalues=FALSE)

Full model result
----------------------------------
Model formula: y ~ cross + Q1 + Q2 + Q3 + Q4 + Q3:Q4

        df     SS     MS   LOD  %var
Model    6  450.8 75.139 76.63 22.02
Error 1412 1596.7  1.131
Total 1418 2047.5
```

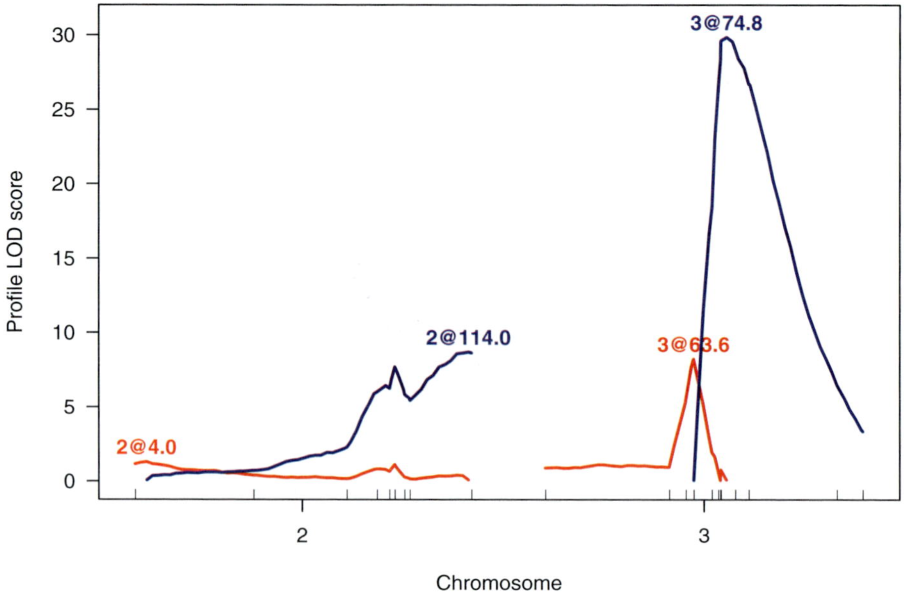

Figure 10.17. Profile LOD curves for a four-QTL model, for the full `ovar` data; the two QTL on chr 3 were allowed to interact.

```
Drop one QTL at a time ANOVA table:
----------------------------------
                df Type III SS     LOD    %var F value
cross            1    193.8113 35.3001  9.4656 171.391
2@4.0            1      6.3004  1.2135  0.3077   5.572
2@114.0          1     45.3005  8.6203  2.2124  40.060
3@63.6           2     42.7445  8.1403  2.0876  18.900
3@74.8           2    162.0437 29.7841  7.9141  71.649
3@63.6:3@74.8    1     27.0815  5.1823  1.3226  23.949
```

The evidence for the proximal QTL on chr 2 looks weak, but there is strong evidence for the others.

We may plot LOD profiles, to get another view of the localization of the QTL.

```
> plotLodProfile(rqtl, col=c("red","blue","red","blue"),
+                ylab="Profile LOD score")
```

In Fig. 10.17, we again see that the location of the proximal QTL on chr 2 is poorly defined. Perhaps we are being overly generous in calling this a QTL.

We might consider adding additional terms to our QTL model. First, let us use `addint` to explore the possibility of additional interactions (see Sec. 9.3.3). We use `pvalues=FALSE` to omit the two columns of p-values.

```
> addint(ovar, qtl=rqtl, covar=cross,
+         formula=y~cross+Q1+Q2+Q3+Q4+Q3:Q4,
+         pvalues=FALSE)

Model formula: y ~ cross + Q1 + Q2 + Q3 + Q4 + Q3:Q4

Add one pairwise interaction at a time table:
--------------------------------------------
                  df Type III SS     LOD    %var F value
cross:2@4.0        1     9.18153 1.77696 0.44842   8.161
cross:2@114.0      1     0.54432 0.10506 0.02658   0.481
cross:3@63.6       1     0.49906 0.09632 0.02437   0.441
cross:3@74.8       1     0.26691 0.05151 0.01304   0.236
2@4.0:2@114.0      1     0.31004 0.05984 0.01514   0.274
2@4.0:3@63.6       1     2.91815 0.56366 0.14252   2.583
2@4.0:3@74.8       1     0.34807 0.06718 0.01700   0.308
2@114.0:3@63.6     1     1.86901 0.36089 0.09128   1.654
2@114.0:3@74.8     1     3.31378 0.64016 0.16184   2.934
```

Note that QTL × covariate interactions are considered as well as QTL × QTL interactions. There appears to be some evidence for a difference in the effect of the proximal chr 2 locus in the two crosses; otherwise, we see no evidence for additional interactions.

Let us also use `addqtl` once more, to scan for an additional QTL.

```
> onemore <- addqtl(ovar, qtl=rqtl, covar=cross,
+                    formula=y~cross+Q1+Q2+Q3+Q4+Q3:Q4)
```

The summary of the results indicates the possibility of yet another QTL on chr 2; the conditional LOD score is the same as what we have for the proximal QTL on chr 2.

```
> summary(onemore)

          chr   pos   lod
c1.loc30    1  34.5 0.423
acc004516   2  89.1 1.164
slo         3 121.3 0.598
```

Let us add this additional QTL to the model, reorder the QTL according to their positions in the genome, and run `refineqtl` to refine the QTL positions.

```
> qtl2 <- addtoqtl(ovar, rqtl, 2, 89.1)
> qtl2 <- reorderqtl(qtl2)
> rqtl2 <- refineqtl(ovar, qtl=qtl2, covar=cross, verbose=FALSE,
+                     formula=y~cross+Q1+Q2+Q3+Q4+Q5+Q4:Q5)
```

The distal QTL on chr 2 shifted by 1 cM.

```
> rqtl2
```

```
      name chr   pos n.gen
Q1    2@4.0   2   4.0     2
Q2   2@89.1   2  89.1     2
Q3  2@115.0   2 115.0     2
Q4   3@63.6   3  63.6     2
Q5   3@74.8   3  74.8     2
```

We perform the drop-one-QTL analysis again, to assess the evidence for each QTL in the context of this five-QTL model.

```
> summary(fitqtl(ovar, qtl=rqtl2, covar=cross,
+                formula=y~cross+Q1+Q2+Q3+Q4+Q5+Q4:Q5),
+         pvalues=FALSE)

Full model result
----------------------------------
Model formula: y ~ cross + Q1 + Q2 + Q3 + Q4 + Q5 + Q4:Q5

        df     SS     MS   LOD  %var
Model    7  457.2 65.314 77.86 22.33
Error 1411 1590.3  1.127
Total 1418 2047.5

Drop one QTL at a time ANOVA table:
----------------------------------
              df Type III SS     LOD    %var F value
cross          1    194.6205 35.5733  9.5051 172.673
2@4.0          1      6.8879  1.3317  0.3364   6.111
2@89.1         1      6.6325  1.2824  0.3239   5.885
2@115.0        1     11.5318  2.2262  0.5632  10.231
3@63.6         2     41.8306  8.0000  2.0430  18.557
3@74.8         2    160.6383 29.6505  7.8454  71.261
3@63.6:3@74.8  1     26.5197  5.0959  1.2952  23.529
```

The evidence for each of the QTL on chr 2 is weak but non-negligible. With the QTL on chr 2 at 89 cM in the model, there is a big drop in the residual evidence for the most distal chr 2 locus.

Finally, let us apply our automated stepwise model selection procedures with `stepwiseqtl`.

```
> stepout1 <- stepwiseqtl(ovar, covar=cross, pen=pen,
+                         max.qtl=8, verbose=FALSE)

> stepout1

      name chr  pos n.gen
Q1  2@89.1   2 89.1     2
Q2  2@92.4   2 92.4     2
```

```
Q3  2@94.2   2  94.2      2
Q4 2@110.0   2 110.0      2
Q5  3@63.6   3  63.6      2
Q6  3@76.8   3  76.8      2

  Formula: y ~ cross + Q1 + Q2 + Q3 + Q4 + Q5 + Q6 + Q5:Q6 +
                 Q2:Q3
  pLOD: 43.96
```

The results are much like what we obtained above, though without the most proximal chr 2 locus, and with the addition of that untrustworthy pair of tightly linked epistatic loci on chr 2 seen in the analysis of the initial cross (see page 299).

Let us run stepwiseqtl again, this time using exclusively heavy interaction penalties.

```
> stepout2 <- stepwiseqtl(ovar, covar=cross, pen=pen[1:2],
+                         max.qtl=8, verbose=FALSE)
```

The results, with the exclusive use of heavy penalties, are identical to what we obtained above, with the mixture of heavy and light penalties.

```
> stepout2

       name chr   pos n.gen
Q1  2@89.1   2  89.1      2
Q2  2@92.4   2  92.4      2
Q3  2@94.2   2  94.2      2
Q4 2@110.0   2 110.0      2
Q5  3@63.6   3  63.6      2
Q6  3@76.8   3  76.8      2

  Formula: y ~ cross + Q1 + Q2 + Q3 + Q4 + Q5 + Q6 + Q5:Q6 +
                 Q2:Q3
  pLOD: 41.61
```

10.4 Discussion

Our principal aim in this case study was to illustrate many of the techniques for exploring multiple-QTL models. We have focused on the analysis techniques rather than the biological interpretation of the results; for the latter, see the original article describing these data (Orgogozo *et al.*, 2006).

Our primary tools include single- and two-dimensional scans for QTL. We repeat these, controlling for major loci, to identify additional QTL. A variety of paths may be taken. Hopefully, they all lead to similar conclusions.

The assessment of evidence for individual loci and epistatic interactions, in the context of a multiple-QTL model, can be difficult. We use the results of permutation tests as our guide, even though they formally apply only in a restricted context (a single- or two-QTL model). The independent segregation of chromosomes is extremely useful, in this regard, as it ensures that the null distribution of a maximized LOD score is approximately the same, whether or not one has controlled for the presence of additional QTL on other chromosomes.

There is extensive missing genotype information in the data of Orgogozo *et al.* (2006), due to selective genotyping in the initial cross and selective phenotyping (and incomplete genotyping) in the second cross. This is a nuisance when it comes to the QTL analysis; with the multiple imputation approach, a large number of imputations are required and so computation times can be great. The limited genotype information can make it difficult to separate multiple linked loci, and it exacerbates the common problem that is seen with tightly linked loci. One should be cautious about the fit of large QTL models in the presence of appreciable missing genotype information.

The selective phenotyping strategy used in the second cross can be extremely valuable for the fine-mapping of a QTL, but we were a bit disconcerted to see that our analysis of the initial cross placed the QTL outside of the interval that was used to select recombinants. With the combined data, we did infer the presence of a QTL within this selected region, and the estimated location of the major QTL moved closer to the region, but we must admit lingering concern that the extensively missing genotype information results in some bias in the estimated locations of the QTL.

QTL analysis is a complex model-building exercise. Our fully automated system for model selection may be useful to many scientists, but the exploratory tools are important supplements, and in either case, the final conclusions can be subject to considerable uncertainty.

11

Case study II

In this chapter, we present a second case study. This case study illustrates an investigation of interactions between QTL and a covariate. It also shows how we may deal with genotype data organized in linkage groups (as opposed to chromosomes). Data of that type are rare for established model organisms, but common for emerging ones without extensive genomic resources. As with the case study in Chap. 10, there are some unusual features here, but most QTL analyses will be unusual in one way or another.

We consider the data of Nichols *et al.* (2007), included in the R/qtlbook package as the data set `trout`. This is a set of doubled haploid individuals derived from a cross between the Oregon State University (OSU) and Clearwater (CW) River rainbow trout (*Oncorhynchus mykiss*) clonal lines. Doubled haploids are similar to a backcross, but with a single recombinant genome being doubled (rather than matched to a nonrecombinant genome), so that at any genomic position, individuals are homozygous for one of two possible genotypes.

Eggs from one of eight outbred females, two from Troutlodge (TL) and six from the Spokane (SP) hatchery, were irradiated to destroy maternal nuclear DNA and fertilized with sperm from a single F_1 male. The first embryonic cleavage was blocked by heat shock to restore diploidy. There are a total of 554 individuals, with between 8 and 168 individuals from each of the eight females.

The primary phenotype is time to hatch (`tth`). An additional "phenotype," `female`, indicates the maternal cytoplasmic environment (MCE; the source of the egg).

There are data on 171 markers on 28 linkage groups. The linkage groups are named as in Nichols *et al.* (2003), though a pair of markers are assigned to linkage group "un," as they do not connect to any of the linkage groups in Nichols *et al.* (2003). Some care is required in referring to the linkage groups in R/qtl code; in some cases we will need to refer to the linkage group names in double-quotes (e.g., `"1"`).

K.W. Broman, Ś. Sen, *A Guide to QTL Mapping with R/qtl*,
Statistics for Biology and Health, DOI 10.1007/978-0-387-92125-9_11,

As with all QTL data analyses, we begin with an exploration of the phenotype and genotype data. In this case, further light is shed on the linkage groups, and on MCE. Next, we analyze the data for QTL without regard to interactions with MCE. We follow by considering QTL × MCE interactions, and we conclude with a discussion of the analysis.

11.1 Diagnostics

We begin by exploring the phenotype and genotype data. We must first load the R/qtl and R/qtlbook packages, and then the data set, `trout`.

```
> library(qtl)
> library(qtlbook)
> data(trout)
```

Note that the cross type is `"dh"`, indicating doubled haploids. In R/qtl, they are treated just like a backcross, except in references to the names of the genotypes.

```
> class(trout)

[1] "dh"    "cross"
```

If one were to import such data into R using `read.cross`, it would initially be read as a backcross. One would use code something like the following.

```
> trout <- read.cross("csv", file="trout.csv",
+                     genotypes=c("C","O"), alleles=c("C","O"))
```

One would then need to change the cross type to `"dh"`, as follows.

```
> class(trout)[1] <- "dh"
```

Turning back to the data, let us first consider a quick summary.

```
> summary(trout)

    Doubled haploids

    No. individuals:    554

    No. phenotypes:     2
    Percent phenotyped: 100 100

    No. chromosomes:    28
        Autosomes:      1 2 3 5 6 7 8 9 10 11 12 13 14 15 16 17
                        18 19 20 21 22 23 24 25 27 29 31 un

    Total markers:      171
```

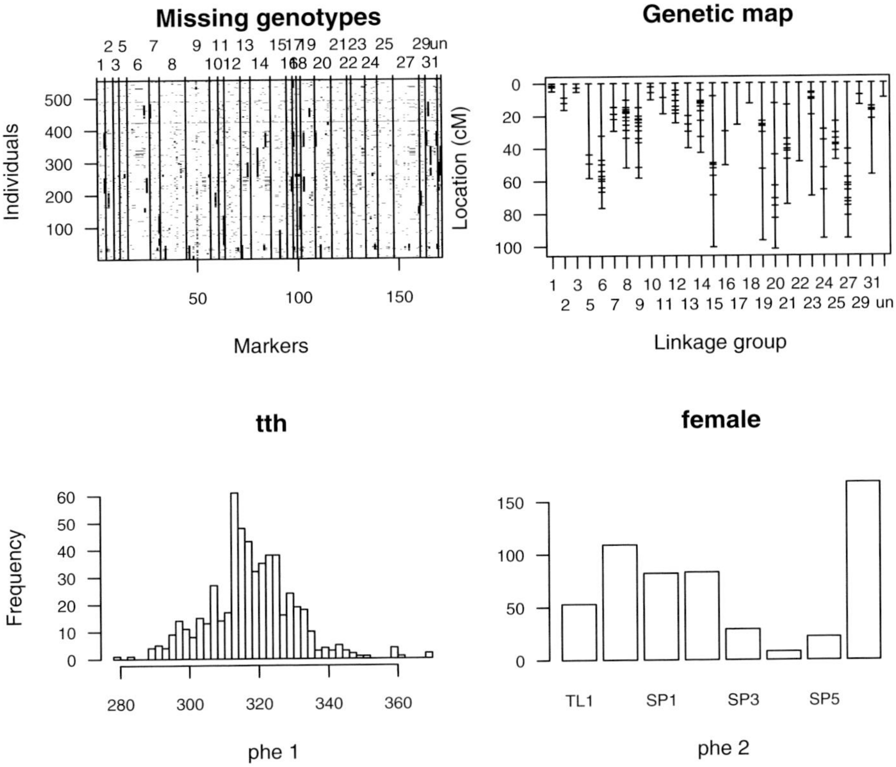

Figure 11.1. The summary plot of the `trout` data provided by the `plot.cross` function, including the pattern of missing genotype data (upper left; black pixels indicate missing data), the genetic map of the typed markers (upper right), a histogram of the `tth` phenotype (time to hatch), and a bar plot of the `female` phenotype, which indicates the maternal cytoplasmic environment (MCE; the source of the egg for each individual).

```
    No. markers:          4 4 3 4 11 5 13 12 4 3 8 5 10 8 3 2 2 7
                          8 8 2 7 6 8 13 3 6 2
    Percent genotyped:    95.2
    Genotypes (%):        CC:50  OO:50
```

This is just as described above. The C and O alleles refer to the CW and OSU clonal lines, respectively.

We make a summary plot as follows; see Fig. 11.1.

```
> plot(trout)
```

There are some very large gaps in the genetic map, but the marker genotype data are remarkably complete.

We first investigate whether there are differences in the `tth` phenotype, among the eight sources for eggs. We begin with a set of box plots.

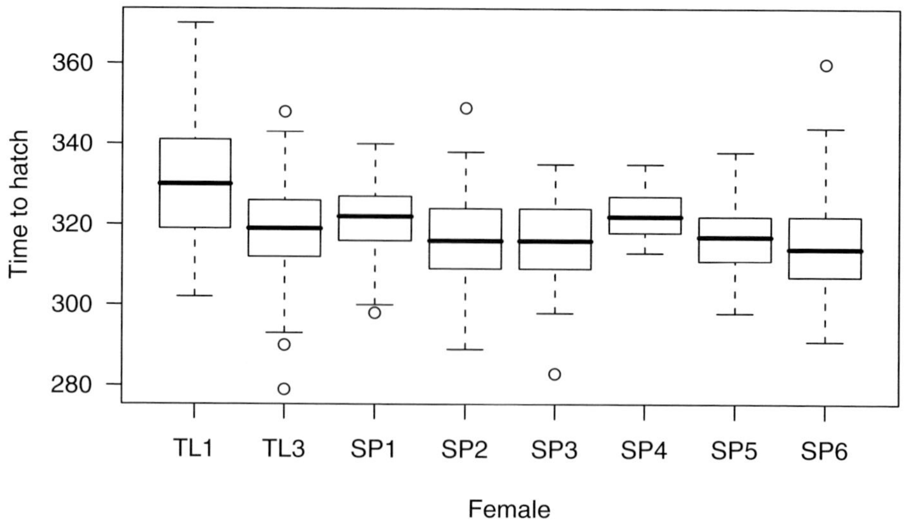

Figure 11.2. Box plots of the `tth` phenotype (time to hatch) according to the female source of the egg for the individuals, for the `trout` data.

```
> boxplot(tth ~ female, data=trout$pheno,
+         xlab="Female", ylab="Time to hatch")
```

There are clear differences among the females (see Fig. 11.2), particularly in that `tth` is larger for individuals whose egg came from the TL1 female.

An analysis of variance (ANOVA) will make clear that the observed differences cannot reasonably be ascribed to chance variation.

```
> anova(aov(tth ~ female, data=trout$pheno))

Analysis of Variance Table

Response: tth
           Df Sum Sq Mean Sq F value  Pr(>F)
female      7  12757    1822    13.5 3.6e-16
Residuals 546  73510     135
```

Turning now to the genotype data, we first look at the segregation patterns for each marker with `geno.table`. Since there are 171 markers, this will make quite a long table, and so we focus on those markers for which the two genotypes deviate significantly from the expected 1:1 ratio. Applying a Bonferroni correction for the number of markers, we pull out the unusual ones as follows.

```
> gt <- geno.table(trout)
> gt[gt$P.value < 0.05/totmar(trout),]
```

```
          chr missing  CC  OO   P.value
AGCCGT8     8       11 320 223 3.145e-05
agcagc11   13       49 187 318 5.562e-09
ACCAAG16   19       15 396 143 1.185e-27
```

There are three markers that are behaving oddly, with marker ACCAAG16 segregating closer to 3:1 than 1:1. We do not know whether this is due to segregation distortion or genotyping error (in which case we might omit these markers), and so we will leave them in, but we should pay attention to these regions in the later results.

We next turn to the estimated recombination fractions between all pairs of markers, estimated via `est.rf` and plotted via `plot.rf`. We use `alternate.chrid=TRUE` to make the names of the many linkage groups more easily distinguished.

```
> trout <- est.rf(trout)
> plot.rf(trout, alternate.chrid=TRUE)
```

The results (Fig. 11.3) are a bit of a surprise. There are five pairs of linkage groups that are quite tightly associated (with large LOD scores, in red), but not tightly linked (estimated recombination fractions > 0.5, in blue). Namely, the pairs of linkage groups 2 and 29, 10 and 18, 12 and 16, 14 and 20, and 27 and 31, all look to be tightly associated. We can focus on just these ten linkage groups, to make the observation more clear (see Fig. 11.4).

```
> plot.rf(trout, chr=c(2,29, 10,18, 12,16, 14,20, 27,31),
+         alternate.chrid=TRUE)
```

Let's look at a table of two-locus genotypes for a marker from a couple of these linkage groups, to figure out what is going on. We can use `find.marker` to pull out the first marker on each of linkage groups 10 and 18, and then use `geno.crosstab` to create a table of the two-locus genotypes.

```
> mar <- find.marker(trout, c(10,18), c(0,0))
> geno.crosstab(trout, mar[1], mar[2])
```

```
        AGCCAG12
agcagc9  -  CC  OO
      -  0   4   4
      CC 7  33 235
      OO 15 204  52
```

We see that most individuals have opposite genotypes at these two markers.

While we were surprised by these results, with a more complete understanding of the genome of this organism, they might have been anticipated. As described in Nichols *et al.* (2003), an ancestor of this species experienced an autotetraploidy event (a duplication of the genome). While diploidy has since been restored, a number of pairs of linkage groups are homeologous (meaning partially homologous), including the five pairs that we see to exhibit this odd

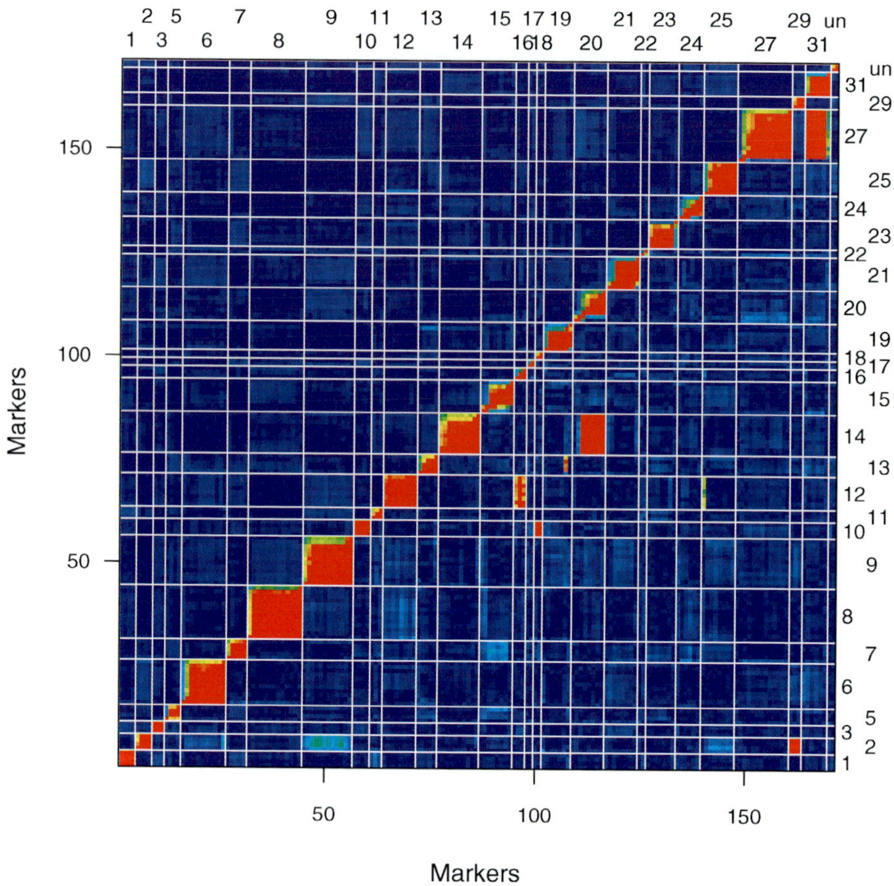

Figure 11.3. Plot of estimated recombination fractions (upper left) and LOD scores for a test of $r = 1/2$ (lower right) for all pairs of markers for the `trout` data. Red indicates linkage; blue indicates no linkage.

"reverse linkage." Johnson *et al.* (1987) observed this phenomemon in related species, of pseudolinkage between loci in separate linkage groups, with recombinants being more frequent than the parental types. It appears that these homeologous chromosomes are pairing at meiosis, though in a special way.

While we might leave things as they are, the tight negative association between these pairs of linkage groups would make the interpretation of QTL mapping results rather confusing, as a QTL on one linkage group could give a linkage signal on a second linkage group.

Alternatively, we could swap the genotypes in one of each of these pairs, and then merge the linkage groups. This would have the advantage of ensuring that a single QTL would show only one linkage signal, but the disadvantage that the parental origins of the alleles will not be clear. Moreover,

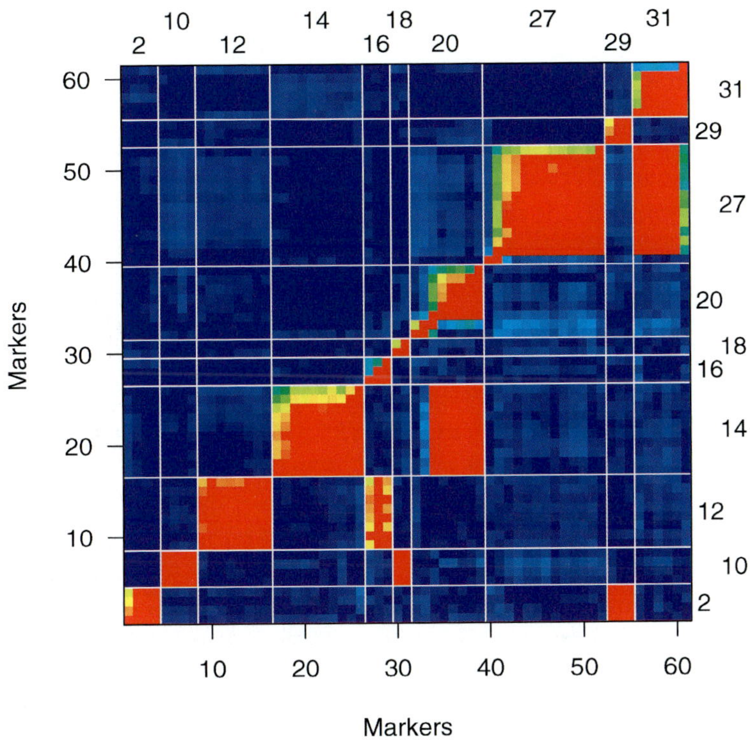

Figure 11.4. Plot of estimated recombination fractions (upper left) and LOD scores for a test of $r = 1/2$ (lower right) for all pairs of markers on selected linkage groups, for the `trout` data. Red indicates linkage; blue indicates no linkage.

our assumption, that the chromosomes are essentially joined end-to-end at meiosis, may be wrong. Nevertheless, we will follow this line of thinking. We will start by swapping the genotypes for the second of each of these pairs of linkage groups. (Note that the genotypes are coded 1 and 2, and so $3 - g$ will swap 1 for 2 and 2 for 1.)

```
> trout$geno[["29"]]$data <- 3 - trout$geno[["29"]]$data
> trout$geno[["18"]]$data <- 3 - trout$geno[["18"]]$data
> trout$geno[["16"]]$data <- 3 - trout$geno[["16"]]$data
> trout$geno[["20"]]$data <- 3 - trout$geno[["20"]]$data
> trout$geno[["31"]]$data <- 3 - trout$geno[["31"]]$data
```

The next step is to merge the pairs and then seek to establish marker order within the merged linkage groups.

We start with linkage groups 10 and 18, as they each have a small number of markers. First, we use `markernames` to obtain the names of the markers on linkage group 18 and then `movemarker` to move the markers from linkage group 18 to linkage group 10.

```
> lg18mar <- markernames(trout, 18)
> for(i in lg18mar)
+   trout <- movemarker(trout, i, 10)
```

We also change the name of the merged linkage group to "10.18."

```
> nam <- names(trout$geno)
> nam[nam=="10"] <- "10.18"
> names(trout$geno) <- nam
```

We now use `ripple` to consider all possible orders of the markers. Since there are only 6 markers (and so 360 possible marker orders), we will do a full likelihood analysis. We use the Kosambi map function and assume a 1% genotyping error rate.

```
> rip <- ripple(trout, chr="10.18", window=6, method="lik",
+                error.prob=0.01, map.function="kosambi",
+                verbose=FALSE)
```

Note that we referred to the chromosome ID in double-quotes. Chromosome identifiers are matched by name, with numbers being converted to character strings, and so one might use, for example `chr=5` in place of `chr="5"`. However, with the more complex chromosome names that we will be forming, it will be best to surround them in double-quotes, though we can actually mix numbers and character strings (and we will do so).

The summary of the `ripple` results indicates that we should switch the order of the markers. (We merged the groups at the wrong ends, and we might also switch the order of the third and fourth markers on linkage group 10.)

```
> summary(rip)

                         LOD chrlen
Initial   1 2 3 4 5 6    0.0   28.1
1         3 4 2 1 5 6    3.7   28.7
2         4 3 2 1 5 6    3.1   28.7
3         3 4 1 2 5 6    2.6   28.5
4         4 3 1 2 5 6    1.7   28.3
5         3 2 4 1 5 6    1.7   28.4
... [ 6  additional rows] ...
```

We use `switch.order` to switch the order of the markers to that with the maximum likelihood. The arguments `error.prob` and `map.function` are used in estimating the genetic map with the new order.

```
> trout <- switch.order(trout, "10.18", rip[2,],
+                       error.prob=0.01, map.function="kosambi")
```

The genetic map of the newly merged linkage groups is reasonably tight (with just a 15 cM gap between the two groups), which gives us some comfort that we are doing the right thing.

```
> pull.map(trout, chr="10.18")

ACGACA11 AGCAGT15  ACCAAG6  agcagc9 AGCCAG12 AGCCAG13
   0.000    1.992    2.754    3.445   18.031   28.692
```

We will omit most of the details for dealing with the other four pairs of linkage groups. The techniques are similar, though with many markers on a linkage group, an iterative process is required in establishing marker order (see Sec. 3.4.2). In the following, we merge the linkage groups and switch the orders to the best that we could find.

```
> lg29mar <- markernames(trout, 29)
> for(i in lg29mar)
+   trout <- movemarker(trout, i, 2)
> lg16mar <- markernames(trout, 16)
> for(i in lg16mar)
+   trout <- movemarker(trout, i, 12)
> trout <- switch.order(trout, chr=12, c(1:8,10,11,9),
+                       error.prob=0.01, map.function="kosambi")
> lg20mar <- markernames(trout, 20)
> for(i in lg20mar)
+   trout <- movemarker(trout, i, 14)
> trout <- switch.order(trout, chr=14, error.prob=0.01,
+                       c(10:1,14,16,15,17,18,13:11),
+                       map.function="kosambi")
> lg31mar <- markernames(trout, 31)
> for(i in lg31mar)
+   trout <- movemarker(trout, i, 27)
> trout <- switch.order(trout, chr=27, error.prob=0.01,
+                       c(1:6,8,7,9:13,15:18,14,19),
+                       map.function="kosambi")
```

Finally, we change the names of the linkage groups that have been merged.

```
> nam <- names(trout$geno)
> nam[nam=="2"] <- "2.29"
> nam[nam=="12"] <- "12.16"
> nam[nam=="14"] <- "14.20"
> nam[nam=="27"] <- "27.31"
> names(trout$geno) <- nam
```

There are a few remaining issues in the pairwise marker linkages: there is a marker on linkage group 19 that appears to be linked to linkage group 13, and there is a marker on linkage group 25 that appears to be linked to linkage group "12.16." Let us plot the estimated recombination fractions for those four linkage groups.

```
> plot.rf(trout, chr=c(13, 19, "12.16", 25))
```

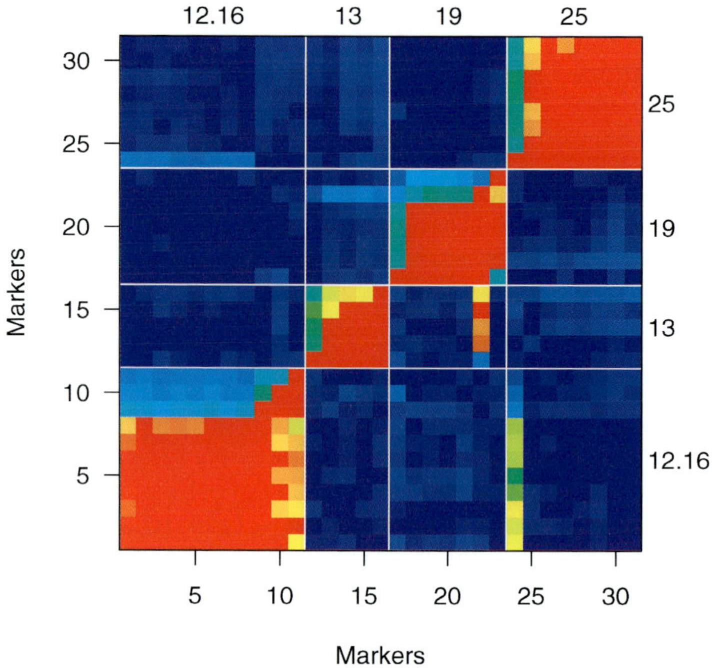

Figure 11.5. Plot of estimated recombination fractions (upper left) and LOD scores for a test of $r = 1/2$ (lower right) for all pairs of markers on linkage groups 13, 19, "12.16" and 25, for the `trout` data. Red indicates linkage; blue indicates no linkage.

The potentially problematic markers (see Fig. 11.5) are linked to another linkage group only weakly, but the large number of individuals results in reasonably large LOD scores. This may indicate that these linkage groups should also be merged, but it is likely best to keep these linkage groups separate, though we should keep in mind that a single QTL has the potential to result in LOD peaks on multiple linkage groups.

Let us reestimate the intermarker distances, using the observed data, keeping marker order fixed, and plot the estimated map against that which came with the data. We will use the Kosambi map function, as in Nichols *et al.* (2007).

```
> newmap <- est.map(trout, error.prob=0.01, verbose=FALSE,
+                   map.function="kosambi")
> plot.map(trout, newmap, alternate.chrid=TRUE)
```

As seen in Fig. 11.6, many linkage groups become slightly shorter, though overall there is good agreement. (The linkage groups that we had modified show no differences, but this is because, in switching the orders of markers, we have replaced the maps with those estimated from these data.) There are a few places (e.g., linkage group 8) where a number of markers appear to be

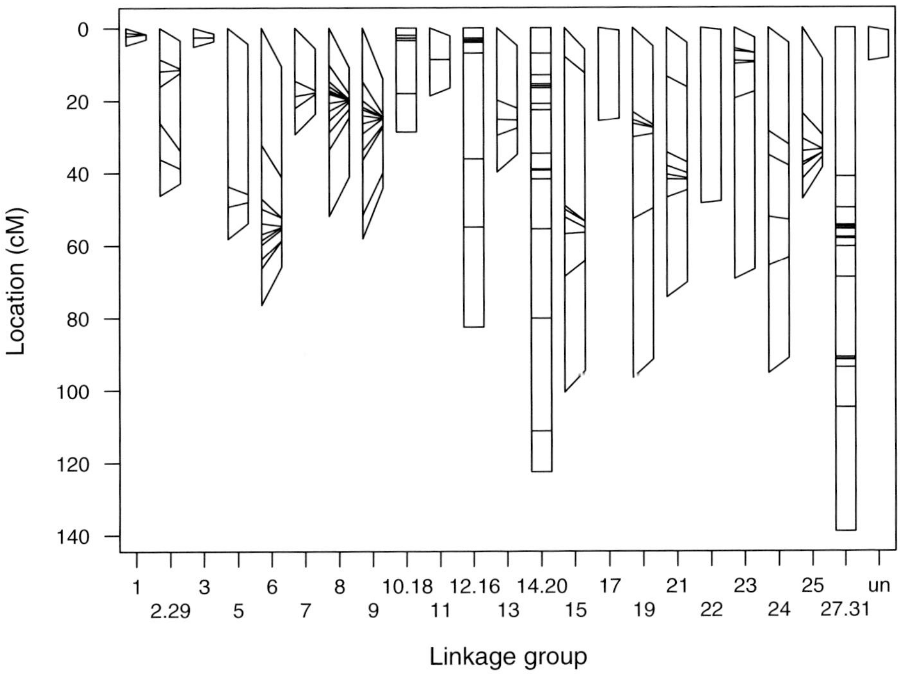

Figure 11.6. The genetic map in the `trout` data plotted against the map estimated from the genotype data. For each linkage group, the line on the left is that map provided with the data, and the line on the right is the estimated map; line segments connect the positions for each marker.

moved closer together. Overall, the estimated map is 1144 cM, while the initial map was 1328 cM (a difference of 14%).

We will replace the map in the data with our newly estimated one.

```
> trout <- replace.map(trout, newmap)
```

While more might be done to investigate the genotype data, let us trust the data and the genetic map and move on to the QTL mapping.

11.2 Initial QTL analyses

We will begin with the simpler aspects of the QTL analysis. In Sec. 11.3, we will investigate the possibility of QTL × MCE interactions. Since the marker genotype data are quite complete (though, admittedly, with a few large gaps between markers), we will use Haley–Knott regression (see Sec. 4.2.2). We must first calculate the conditional QTL genotype probabilities, given the available marker data.

```
> trout <- calc.genoprob(trout, step=1, err=0.01,
+                          map.function="kosambi")
```

While we will postpone the investigation of QTL × MCE interactions to the next section, the systematic differences in the phenotype among MCE groups (see Fig. 11.2 on page 316) suggests that we should include MCE as a set of additive covariates in the QTL mapping. In doing so, we assume that the effects of any QTL are the same in the eight groups, but we allow shifts in the average phenotype between the groups.

We form a matrix of indicators, with a column for each of the MCE groups except the first one.

```
> female <- pull.pheno(trout, "female")
> lev <- levels(female)
> nlev <- length(lev)
> femcov <- matrix(0, nrow=nind(trout), ncol=nlev-1)
> colnames(femcov) <- lev[-1]
> for(i in 2:nlev)
+   femcov[female==lev[i],i-1] <- 1
```

We now perform the genome scan, including `femcov` as a set of additive covariates.

```
> out <- scanone(trout, method="hk", addcovar=femcov)
```

We plot the LOD curves as follows.

```
> plot(out, ylab="LOD score", alternate.chrid=TRUE)
```

The results (see Fig. 11.7) indicate overwhelming support for a QTL on linkage group 8, with maximum LOD score = 43.2.

Let us perform a quick permutation test. With Haley–Knott regression, the permutations are quite fast, and so we will use 4000 permutations. As in the next section we will want to investigate the presence of QTL × MCE interactions, which will require matched permutation tests, with and without MCE as a set of interactive covariates, we use `set.seed` to define the seed for the random number generator, so that this can be repeated with the same permutations.

```
> set.seed(523938)
> operm <- scanone(trout, method="hk", addcovar=femcov,
+                  n.perm=4000)
```

We can now pull out the results from our initial scan that meet a 10% genome-wide significance threshold.

```
> summary(out, perms=operm, alpha=0.1, pvalues=TRUE)

             chr  pos   lod    pval
OmyFGT12       8 9.11 43.18 0.00000
```

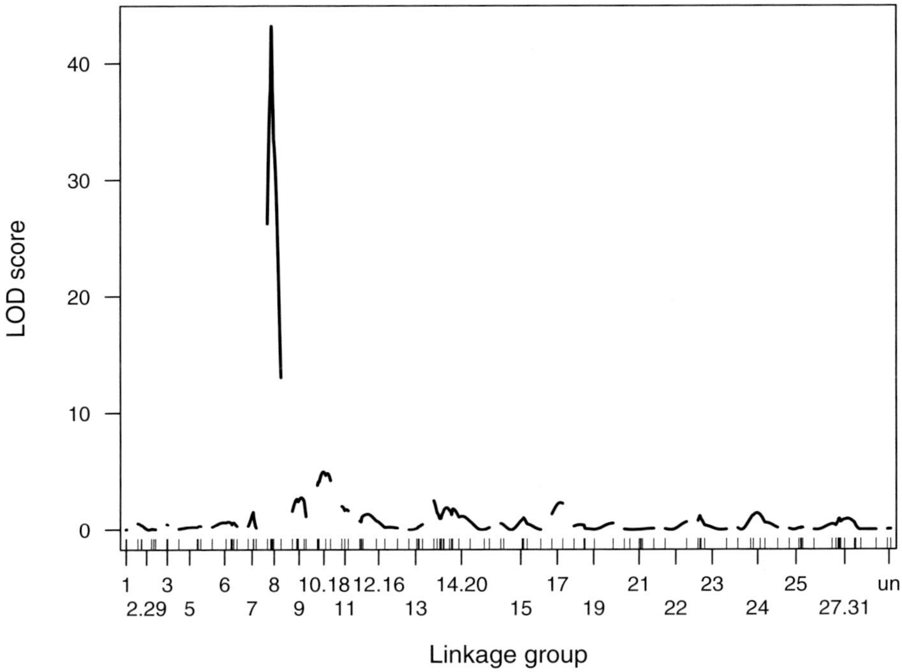

Figure 11.7. LOD curves from a genome scan by Haley–Knott regression for the `trout` data, with MCE groups included as additive covariates.

```
c9.loc20          9 20.00  2.76 0.03775
c10.18.loc13 10.18 13.00  4.97 0.00025
AGCAGT11     14.20  0.00  2.49 0.06950
```

In addition to the clear QTL on linkage group 8, we see evidence for QTL on linkage groups 9, 10.18, and 14.20.

Let us repeat the genome scan, controlling for the locus on linkage group 8. We use `makeqtl` to create a QTL object; as we will be performing Haley–Knott regression, we call `makeqtl` with `what="prob"` (rather than the default, `what="draws"`). We use `addqtl` to perform the scan.

```
> qtl <- makeqtl(trout, 8, 9.11, what="prob")
> out.c8 <- addqtl(trout, qtl=qtl, method="hk", covar=femcov)
```

While our permutation results do not formally apply in the present case, in which we are controlling for the locus on linkage group 8, they nevertheless provide a reasonable guide.

```
> summary(out.c8, perms=operm, alpha=0.1, pvalues=TRUE)

                chr pos  lod    pval
c9.loc21          9  21 2.59 0.05575
```

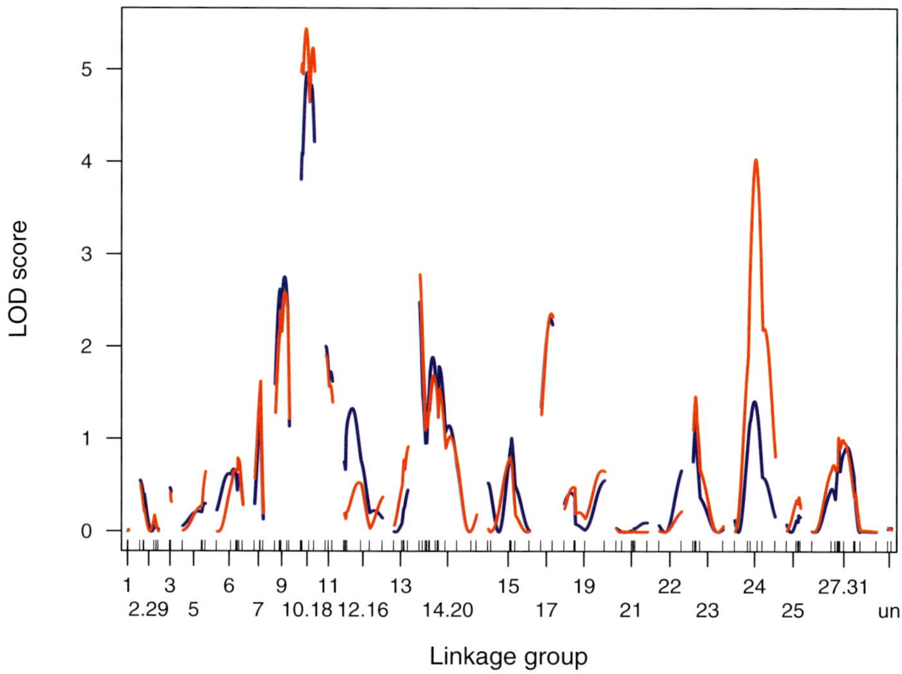

Figure 11.8. LOD curves from a genome scan by Haley–Knott regression for the `trout` data, with MCE groups included as additive covariates. The curves in blue are as in Fig. 11.7; those in red were calculated controlling for a QTL on linkage group 8.

```
c10.18.loc10 10.18  10 5.44 0.00025
AGCAGT11     14.20   0 2.78 0.03575
c17.loc20       17  20 2.36 0.09150
c24.loc44       24  44 4.03 0.00125
```

We see evidence for additional QTL on linkage groups 17 and 24. The three previously-identified QTL remain on the list, though their LOD scores have changed slightly.

We can plot the LOD curves, with and without controlling for the QTL on linkage group 8, as follows.

```
> plot(out, out.c8, chr = -8, col=c("blue","red"),
+       ylab="LOD score", alternate.chrid=TRUE)
```

Note the use of `chr = -8` to plot all but linkage group 8. The minus sign may also be used with character strings. For example, use of `chr="-un"` would result in a plot of all linkage groups except `"un"`.

As seen in Fig. 11.8, controlling for the QTL on linkage group 8 leads to a great increase in the evidence for a QTL on linkage group 24, and small increases and decreases in the LOD scores on many other linkage groups.

Let us bring all of the terms together into one model, and then refine our estimates of the locations of the QTL with `refineqtl`.

```
> qtl <- makeqtl(trout, c(8, 9, "10.18", "14.20", 17, 24),
+                c(9.11, 21, 10, 0, 20, 44), what="prob")
> rqtl <- refineqtl(trout, qtl=qtl, covar=femcov,
+                   method="hk", verbose=FALSE)
```

The location of the QTL changed just slightly.

```
> options(width=64)
> rqtl

          name   chr    pos n.gen
Q1       8@9.1     8  9.112     2
Q2      9@11.0     9 11.040     2
Q3 10.18@11.0 10.18 11.000     2
Q4  14.20@0.0 14.20  0.000     2
Q5     17@23.0    17 23.000     2
Q6     24@46.0    24 46.000     2
```

Let us fit the multiple-QTL model and perform a drop-one-QTL analysis with `fitqtl`. We do not find the two columns of p-values calculated by `fitqtl` to be particularly informative (as they fail to account for the scan across the genome), and they take up a lot of space, and so we omit them, using `pvalues=FALSE` in the `summary.fitqtl` function.

```
> summary(fitqtl(trout, qtl=rqtl, covar=femcov, method="hk"),
+         pvalues=FALSE)

Full model result
----------------------------------
Model formula: y ~ Q1 + Q2 + Q3 + Q4 + Q5 + Q6 + TL3 + SP1 + SP2
                    + SP3 + SP4 + SP5 + SP6


       df    SS     MS   LOD  %var
Model  13 41504 3192.6 78.92 48.11
Error 540 44762   82.9
Total 553 86266


Drop one QTL at a time ANOVA table:
----------------------------------
           df Type III SS       LOD       %var F value
8@9.1       1  21199.7196   46.6416    24.5748 255.747
9@11.0      1    911.2975    2.4245     1.0564  10.994
10.18@11.0  1   1972.8624    5.1886     2.2870  23.800
14.20@0.0   1    879.4502    2.3406     1.0195  10.609
17@23.0     1    840.3175    2.2374     0.9741  10.137
```

```
24@46.0      1    1437.5384     3.8027      1.6664  17.342
TL3          1    4876.5995    12.4400      5.6530  58.830
SP1          1    2592.7989     6.7738      3.0056  31.279
SP2          1    5167.0989    13.1420      5.9897  62.334
SP3          1    4184.0042    10.7497      4.8501  50.475
SP4          1     327.2468     0.8763      0.3793   3.948
SP5          1    2534.1374     6.6247      2.9376  30.571
SP6          1   10414.5254    25.1638     12.0726 125.638
```

Note that our major locus on linkage group 8 is responsible for an estimated 25% of the variation in the phenotype. The conditional LOD scores for the other five QTL are reduced a bit relative to what we had seen when they were considered individually, though still controlling for the QTL on linkage group 8. The evidence for QTL on linkage groups 10.18 and 24 remains strong. The evidence for the other three loci is much weaker, but they are still interesting.

We have not yet considered the possibility of epistatic interactions among QTL, and so we now use `addint` to look for epistatic interactions among the identified QTL; `qtl.only=TRUE` indicates that QTL × covariate interactions should not be considered.

```
> addint(trout, qtl=rqtl, qtl.only=TRUE, method="hk",
+          covar=femcov, pvalues=FALSE)
```

The output was extensive and not too interesting, and so we do not display it here. There is little evidence for interactions among any of these QTL; the largest interaction LOD score was 1.1, for the QTL on linkage groups 8 and 17.

Of course, we should also perform a two-dimensional, two-QTL genome scan, which will allow us to identify additional QTL with important interactions and also potential pairs of linked QTL (particularly those linked in repulsion, with effects of opposite sign, which would generally not show up in a single-QTL genome scan).

To make sense of the results of a two-dimensional genome scan, we will want results from a permutation test, which will be quite time consuming, and so let's get that going. [The computations took a total of about 7 days of CPU time; split across 16 processors (see Sec. 8.1, page 223), it took about 10 hours in real time.]

```
> operm2 <- scantwo(trout, method="hk", addcovar=femcov,
+                    n.perm=1000)
```

Interestingly, the significance thresholds are similar to what we had seen for the `hyper` data in Sec. 8.1.

```
> summary(operm2)

tth (1000 permutations)
    full  fv1  int  add  av1  one
5%  5.43 4.42 3.99 4.40 2.64 2.71
10% 5.12 4.05 3.72 4.04 2.39 2.32
```

Let us now perform the actual two-dimensional scan with the data. We use the argument `incl.markers=TRUE` so that calculations are performed at the markers as well as on the evenly-spaced grid.

```
> out2 <- scantwo(trout, method="hk", addcovar=femcov,
+                 incl.markers=TRUE)
```

The tabular summary of the two-dimensional scan output is simplest to interpret, so we will start with that. The p-values take up a lot of space, so we won't include them.

```
> summary(out2, perms=operm2, alpha=0.05)
```

```
              pos1f pos2f lod.full lod.fv1 lod.int     pos1a
c7 :c7        11.52 16.00     6.65    5.13  1.3915     11.52
c8 :c8         9.11 29.00    55.76   12.57 10.7777      9.11
c8 :c10.18     9.11  9.00    48.92    5.74  0.2927      9.11
c8 :c14.20     9.11  0.00    46.09    2.91  0.1275      9.11
c8 :c24        9.11 44.00    47.24    4.06  0.0232      9.11
c9 :c10.18    18.00 13.00     8.45    3.47  0.7140      9.77
c12.16:c12.16  2.00  6.95     8.10    6.76  6.1503      3.00
c13:c13        0.00 23.00     5.76    5.31  4.7070      8.00
              pos2a lod.add lod.av1
c7 :c7           15    5.26   3.741
c8 :c8           23   44.98   1.797
c8 :c10.18       10   48.63   5.444
c8 :c14.20        0   45.97   2.784
c8 :c24          44   47.22   4.035
c9 :c10.18       13    7.73   2.758
c12.16:c12.16     5    1.95   0.612
c13:c13          26    1.06   0.601
```

First note the pairs of linked QTL on linkage groups 7, 8, 12.16, and 13. The pair of QTL on linkage group 7 do not show strong evidence for an interaction, but the other three pairs show strong interaction effects. The pairs on linkage groups 7, 12.16 and 13 are relatively tightly linked, so we should be slightly skeptical. The other rows in the table indicate the multiple QTL that we had seen in our initial single-QTL scans; none show epistatic effects.

Let us consider a plot of the LOD scores for the pairs of linked QTL. We will plot the results for linkage group 8 separate from the others, as the extremely large LOD scores on linkage group 8 will make it difficult to study the others if they all are shown with the same color scale. We will focus on LOD_i, concerning epistatic interactions, and LOD_{fv1}, comparing the model with two interacting QTL to the best single-QTL model.

```
> plot(out2, lower="cond-int", chr=8)
> plot(out2, lower="cond-int", chr=c(7, "12.16", 13))
```

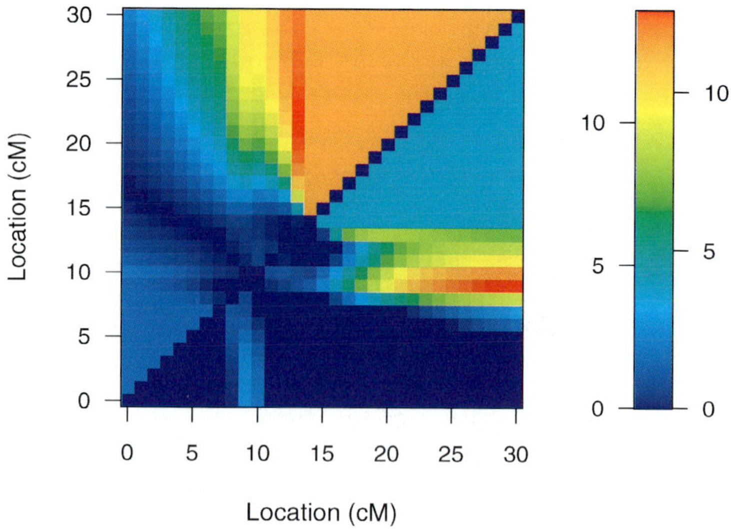

Figure 11.9. LOD scores, for linkage group 8, from a two-dimensional, two-QTL genome scan with the `trout` data. LOD_i is displayed in the upper left triangle; LOD_{fv1} is displayed in the lower right triangle. In the color scale on the right, numbers to the left and right correspond to LOD_i and LOD_{fv1}, respectively.

The LOD scores for linkage group 8 are shown in Fig. 11.9. The LOD scores for linkage groups 7, 12.16 and 13 are shown in Fig. 11.10. These figures do not provide much additional information, beyond what was obtained in the summary table above. We do get a sense of the precision of localization of the QTL, but not much else.

To alleviate our skepticism about these pairs of linked QTL, we plot the phenotypes against the two-locus genotypes at markers close to the inferred QTL positions. First, we use `find.marker` to identify the relevant markers; we also use `find.markerpos` to look at the positions of the selected markers, so that we are sure that the markers are near the QTL.

```
> mar7 <- find.marker(trout, 7, c(11.52, 16))
> find.markerpos(trout, mar7)

          chr   pos
agcagc3     7 11.52
agcatc13    7 18.05

> mar8 <- find.marker(trout, 8, c(9.11, 29))
> find.markerpos(trout, mar8)

          chr    pos
OmyFGT12    8  9.112
ACGACA8     8 30.215
```

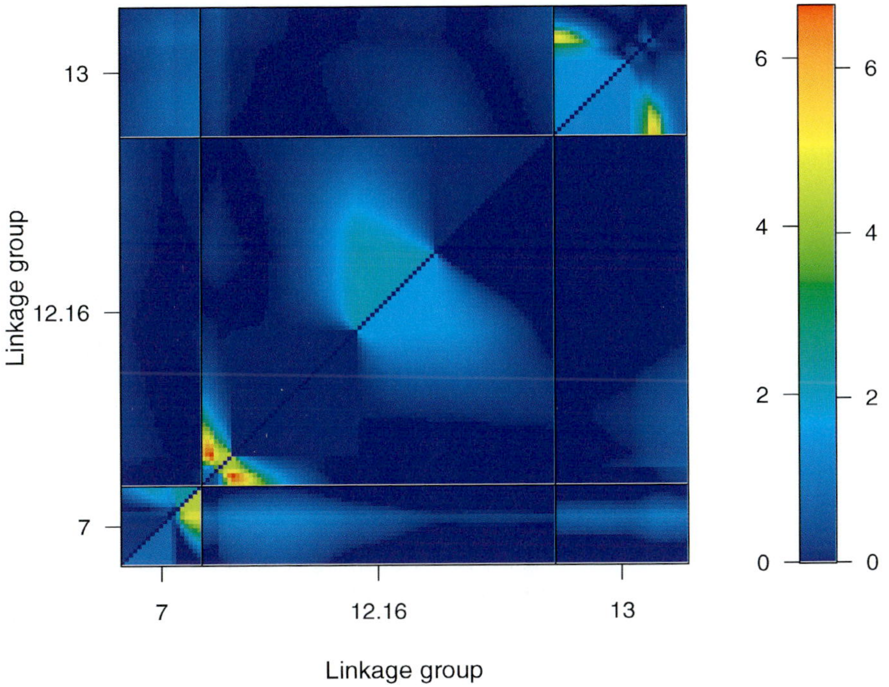

Figure 11.10. LOD scores, for selected linkage groups, from a two-dimensional, two-QTL genome scan with the `trout` data. LOD_i is displayed in the upper left triangle; LOD_{fv1} is displayed in the lower right triangle. In the color scale on the right, numbers to the left and right correspond to LOD_i and LOD_{fv1}, respectively.

```
> mar12.16 <- find.marker(trout, "12.16", c(2, 6.95))
> find.markerpos(trout, mar12.16)

           chr   pos
AGCATC6 12.16 2.794
ACGAGA5 12.16 6.952

> mar13 <- find.marker(trout, 13, c(0, 23))
> find.markerpos(trout, mar13)

          chr   pos
agcagc11   13  0.00
acgatg8    13 22.75
```

Now we call `plot.pxg` to create the plots of phenotypes against two-locus genotypes.

```
> par(mfrow=c(2,2))
> plot.pxg(trout, marker=mar7)
> plot.pxg(trout, marker=mar8)
```

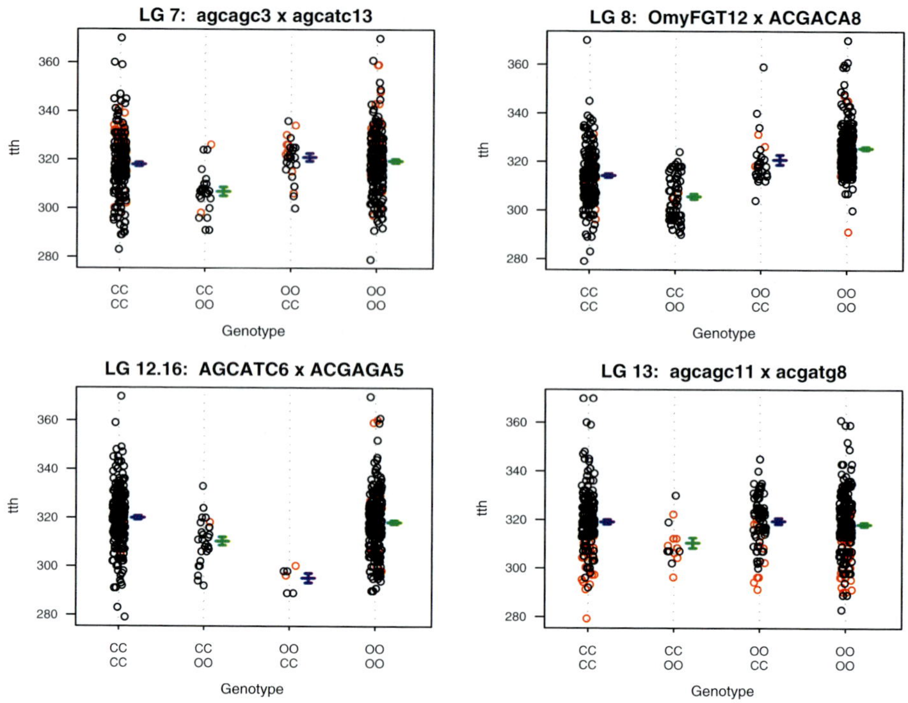

Figure 11.11. Plot of the `tth` phenotype against two-locus genotypes at four pairs of putative linked QTL, for the `trout` data. Points in red are imputed genotypes.

```
> plot.pxg(trout, marker=mar12.16)
> plot.pxg(trout, marker=mar13)
```

The plots of the phenotype against two-locus genotypes (Fig. 11.11) all look reasonable. For linkage group 12.16, the inference of a second epistatic locus depends on just a few individuals, which is worrisome, and to some extent this is also true for the loci on linkage groups 7 and 13. Nevertheless, these results are intriguing.

Let us bring all of our inferred QTL together into one model and look at the drop-one-QTL analysis from `fitqtl`. We will omit the putative loci on linkage groups 9, 14.20 and 17, which were weakly supported. We still have 10 QTL (and three pairs of interactions).

```
> qtl <- makeqtl(trout, c(7,7,8,8,"10.18","12.16","12.16",13,13,24),
+                c(11.52,16,9.11,29,11,2,6.95,0,23,46), what="prob")
> summary(fitqtl(trout, qtl=qtl, covar=femcov, method="hk",
+                formula=y~TL3+SP1+SP2+SP3+SP4+SP5+SP6+
+                         Q1+Q2+Q3*Q4+Q5+Q6*Q7+Q8*Q9+Q10),
+        pvalues=FALSE)

Full model result
----------------------------------
```

```
Model formula: y ~ TL3 + SP1 + SP2 + SP3 + SP4 + SP5 + SP6 + Q1 + Q2 +
                   Q3 + Q4 + Q5 + Q6 + Q7 + Q8 + Q9 + Q10 + Q3:Q4 +
                   Q6:Q7 + Q8:Q9

       df    SS     MS   LOD  %var
Model  20 47092 2354.6 94.97 54.59
Error 533 39175   73.5
Total 553 86266

Drop one QTL at a time ANOVA table:
----------------------------------
                       df Type III SS       LOD       %var F value
TL3                     1   6096.5117   17.4002     7.0671  82.948
SP1                     1   3164.2037    9.3443     3.6680  43.051
SP2                     1   5339.4284   15.3714     6.1895  72.647
SP3                     1   4476.5711   13.0166     5.1893  60.907
SP4                     1    481.7087    1.4702     0.5584   6.554
SP5                     1   1787.9517    5.3689     2.0726  24.326
SP6                     1  10119.5482   27.6421    11.7306 137.684
7@11.5                  1    793.7264    2.4131     0.9201  10.799
7@16.0                  1    500.5065    1.5272     0.5802   6.810
8@9.1                   2  15014.9633   39.0324    17.4054 102.145
8@29.0                  2   3569.1533   10.4895     4.1374  24.281
10.18@11.0              1   1448.3044    4.3673     1.6789  19.705
12.16@2.0               2    938.7691    2.8488     1.0882   6.386
12.16@7.0               2    913.7320    2.7737     1.0592   6.216
13@0.0                  2    971.0353    2.9456     1.1256   6.606
13@23.0                 2   1066.9214    3.2325     1.2368   7.258
24@46.0                 1   1476.6254    4.4511     1.7117  20.091
8@9.1:8@29.0            1   3075.7546    9.0928     3.5654  41.848
12.16@2.0:12.16@7.0     1    898.9660    2.7294     1.0421  12.231
13@0.0:13@23.0          1    881.1723    2.6760     1.0215  11.989
```

The support for the pairs of loci on linkage groups 7, 12.16 and 13 have dropped greatly, but there remains extremely strong support for the pair of QTL on linkage group 8, plus QTL on linkage groups 10.18 and 24.

Let us drop the loci on linkage groups 12.16 and 13, refine the QTL positions, and perform the drop-one analysis again.

```
> qtl2 <- dropfromqtl(qtl, 6:9)
> qtl2 <- refineqtl(trout, qtl=qtl2, covar=femcov, method="hk",
+                   formula=y~TL3+SP1+SP2+SP3+SP4+SP5+SP6+
+                   Q1+Q2+Q3*Q4+Q5+Q6, verbose=FALSE)
> summary(fitqtl(trout, qtl=qtl2, covar=femcov, method="hk",
+                formula=y~TL3+SP1+SP2+SP3+SP4+SP5+SP6+
+                          Q1+Q2+Q3*Q4+Q5+Q6),
+        pvalues=FALSE)

Full model result
----------------------------------
```

```
Model formula: y ~ TL3 + SP1 + SP2 + SP3 + SP4 + SP5 + SP6 + Q1
                     + Q2 + Q3 + Q4 + Q5 + Q6 + Q3:Q4

       df    SS      MS   LOD %var
Model  14 44943 3210.19 88.54 52.1
Error 539 41323   76.67
Total 553 86266

Drop one QTL at a time ANOVA table:
----------------------------------
            df Type III SS     LOD    %var F value
TL3          1   6194.3529 16.8028  7.1805  80.796
SP1          1   2919.7332  8.2130  3.3846  38.083
SP2          1   6663.2666 17.9841  7.7241  86.912
SP3          1   4902.4536 13.4868  5.6829  63.945
SP4          1    464.3894  1.3444  0.5383   6.057
SP5          1   1890.9159  5.3825  2.1920  24.664
SP6          1   9968.8767 25.9981 11.5560 130.028
7@11.5       1   1220.1786  3.5007  1.4144  15.915
7@17.0       1    700.8668  2.0232  0.8124   9.142
8@9.1        2  17976.8740 43.4503 20.8389 117.240
8@30.2       2   4266.6131 11.8206  4.9459  27.826
10.18@10.0   1   1686.1471  4.8112  1.9546  21.993
24@47.0      1   1521.7448  4.3504  1.7640  19.849
8@9.1:8@30.2 1   3790.5184 10.5577  4.3940  49.441
```

We now have good support for the proximal locus on linkage group 7, but not for the distal one. Let's drop the distal locus, refine the QTL positions, and repeat the drop-one-QTL analysis.

```
> qtl3 <- dropfromqtl(qtl2, 2)
> qtl3 <- refineqtl(trout, qtl=qtl3, covar=femcov, method="hk",
+                   formula=y~TL3+SP1+SP2+SP3+SP4+SP5+SP6+
+                   Q1+Q2*Q3+Q4+Q5, verbose=FALSE)
> summary(fitqtl(trout, qtl=qtl3, covar=femcov, method="hk",
+                formula=y~TL3+SP1+SP2+SP3+SP4+SP5+SP6+
+                          Q1+Q2*Q3+Q4+Q5),
+         pvalues=FALSE)

Full model result
----------------------------------
Model formula: y ~ TL3 + SP1 + SP2 + SP3 + SP4 + SP5 + SP6 + Q1
                     + Q2 + Q3 + Q4 + Q5 + Q2:Q3

       df    SS      MS   LOD  %var
Model  13 44278 3405.99 86.62 51.33
```

```
Error 540 41988   77.76
Total 553 86266

Drop one QTL at a time ANOVA table:
----------------------------------
             df Type III SS       LOD      %var F value
TL3           1   5995.4909   16.0567    6.9500  77.107
SP1           1   2854.5748    7.9126    3.3090  36.712
SP2           1   6543.2056   17.4221    7.5849  84.151
SP3           1   5059.5513   13.6870    5.8651  65.070
SP4           1    443.2697    1.2633    0.5138   5.701
SP5           1   1687.4758    4.7401    1.9561  21.702
SP6           1   9898.5955   25.4645   11.4745 127.303
7@11.5        1    865.3537    2.4541    1.0031  11.129
8@9.1         2  17614.6554   42.1427   20.4190 113.269
8@28.0        2   4822.5266   13.0794    5.5903  31.011
10.18@9.0     1   1787.9916    5.0167    2.0726  22.995
24@46.0       1   1482.9148    4.1754    1.7190  19.071
8@9.1:8@28.0  1   4179.5638   11.4156    4.8450  53.752
```

The support for remaining locus on linkage group 7 is no longer strong, so let's omit it and rerun `refineqtl` and `fitqtl`.

```
> qtl4 <- dropfromqtl(qtl3, 1)
> qtl4 <- refineqtl(trout, qtl=qtl4, covar=femcov, method="hk",
+                   formula=y~TL3+SP1+SP2+SP3+SP4+SP5+SP6+
+                   Q1*Q2+Q3+Q4, verbose=FALSE)
> summary(fitqtl(trout, qtl=qtl4, covar=femcov, method="hk",
+                formula=y~TL3+SP1+SP2+SP3+SP4+SP5+SP6+
+                          Q1*Q2+Q3+Q4),
+        pvalues=FALSE)
```

```
Full model result
----------------------------------
Model formula: y ~ TL3 + SP1 + SP2 + SP3 + SP4 + SP5 + SP6 + Q1
                    + Q2 + Q3 + Q4 + Q1:Q2

       df    SS     MS   LOD  %var
Model   12 43416 3618.0 84.18 50.33
Error  541 42851   79.2
Total  553 86266

Drop one QTL at a time ANOVA table:
----------------------------------
             df Type III SS       LOD      %var F value
TL3           1   5867.4051   15.4379    6.8015  74.078
```

```
SP1              1   2687.4846    7.3178     3.1153  33.930
SP2              1   6234.3561   16.3407     7.2269  78.710
SP3              1   4748.5963   12.6430     5.5046  59.952
SP4              1    442.3387    1.2355     0.5128   5.585
SP5              1   1518.2935    4.1887     1.7600  19.169
SP6              1   9591.8949   24.3002    11.1190 121.100
8@9.1            2  17486.1464   41.1692    20.2700 110.384
8@28.0           2   4623.4707   12.3264     5.3595  29.186
10.18@10.0       1   2027.7848    5.5623     2.3506  25.601
24@46.0          1   1312.6225    3.6298     1.5216  16.572
8@9.1:8@28.0     1   3972.6388   10.6658     4.6051  50.156
```

We have good support for the remaining four QTL, including the interaction between the two loci on linkage group 8.

Let us complete this initial analysis (ignoring the possibility of QTL × MCE interactions) with an application of the automated model selection approach accomplished with `stepwiseqtl`. We first calculate the penalties for our model comparison criterion, using the results of the permutation test with a two-dimensional, two-QTL genome scan.

```
> print(pen <- calc.penalties(operm2))
```

```
 main heavy light
2.711 3.990 1.711
```

Let us first consider strictly additive models, enforced with the argument `additive.only=TRUE`. Note that in this case, only the penalty on main effects is used.

```
> stepout1 <- stepwiseqtl(trout, covar=femcov, penalties=pen,
+                         method="hk", additive.only=TRUE,
+                         verbose=FALSE)
```

The chosen model includes loci on linkage groups 8, 9, 10.18 and 24.

```
> stepout1
```

```
        name   chr    pos n.gen
Q1     8@9.1     8  9.112     2
Q2      9@11     9 11.040     2
Q3 10.18@10 10.18 10.000     2
Q4     24@45    24 45.000     2

  Formula: y ~ TL3 + SP1 + SP2 + SP3 + SP4 + SP5 + SP6 + Q1 + Q2
                + Q3 + Q4
  pLOD: 44.51
```

We use `fitqtl` to perform a drop-one-QTL analysis, to look at the support for the individual terms in the model.

```
> summary(fitqtl(trout, qtl=stepout1, covar=femcov,
+                method="hk"), pvalues=FALSE)

Full model result
----------------------------------
Model formula: y ~ Q1 + Q2 + Q3 + Q4 + TL3 + SP1 + SP2 + SP3 +
                    SP4 + SP5 + SP6

       df    SS      MS  LOD  %var
Model   11 39866 3624.20 74.6 46.21
Error 542 46400   85.61
Total 553 86266

Drop one QTL at a time ANOVA table:
----------------------------------
         df Type III SS        LOD      %var F value
8@9.1     1   21858.0351   46.4352   25.3379 255.325
9@11      1    1071.7613    2.7471    1.2424  12.519
10.18@10  1    2213.2359    5.6055    2.5656  25.853
24@45     1    1624.4151    4.1395    1.8830  18.975
TL3       1    5275.8899   12.9553    6.1158  61.628
SP1       1    2312.3039    5.8504    2.6804  27.010
SP2       1    4676.5682   11.5520    5.4211  54.627
SP3       1    3697.8501    9.2244    4.2866  43.195
SP4       1     264.4609    0.6837    0.3066   3.089
SP5       1    2328.1130    5.8895    2.6988  27.195
SP6       1    9658.9287   22.7492   11.1967 112.827
```

Note that the locus on linkage group 9 just barely enters the model, as its conditional LOD score (that is, the $\log_{10}$ likelihood ratio comparing the four-QTL model to the model with the locus on linkage group 9 omitted) is 2.75 and the penalty on main effects was 2.71.

Let us repeat the stepwise analysis, allowing for epistatic interactions (and using the combination of heavy and light penalties on interactions). We use the default, of forward selection to a model with 10 QTL, followed by backward deletion.

```
> stepout2 <- stepwiseqtl(trout, covar=femcov, penalties=pen,
+                         method="hk", verbose=FALSE)
```

Allowing for interactions, we choose a model that includes the pair of interacting loci on linkage group 8, plus loci on linkage groups 10.18 and 24. This is identical to the model that we had chose through our exploratory search, above.

```
> stepout2
```

```
          name   chr    pos n.gen
Q1       8@9.1     8  9.112     2
Q2      8@28.0     8 28.000     2
Q3 10.18@10.0 10.18 10.000     2
Q4     24@46.0    24 46.000     2

  Formula: y ~ TL3 + SP1 + SP2 + SP3 + SP4 + SP5 + SP6 + Q1 + Q2
               + Q3 + Q4 + Q1:Q2
  pLOD: 52.37
```

Finally, let us study the estimated effects under this model. We use `fitqtl` with `get.ests=TRUE` (to obtain the estimated effects) and `dropone=FALSE` (to skip the drop-one-QTL analysis).

```
> summary(fitqtl(trout, qtl=stepout2, covar=femcov,
+                method="hk", dropone=FALSE, get.ests=TRUE,
+                formula=y~TL3+SP1+SP2+SP3+SP4+SP5+SP6+
+                          Q1*Q2+Q3+Q4))

Full model result
----------------------------------
Model formula: y ~ TL3 + SP1 + SP2 + SP3 + SP4 + SP5 + SP6 + Q1
                   + Q2 + Q3 + Q4 + Q1:Q2

       df    SS     MS   LOD  %var Pvalue(Chi2) Pvalue(F)
Model  12 43416 3618.0 84.18 50.33            0         0
Error 541 42851   79.2
Total 553 86266

Estimated effects:
-----------------
                  est     SE       t
Intercept    327.2417 1.2985 252.018
TL3          -12.8621 1.4944  -8.607
SP1           -9.1620 1.5729  -5.825
SP2          -14.1634 1.5964  -8.872
SP3          -16.0743 2.0760  -7.743
SP4           -8.0083 3.3888  -2.363
SP5          -10.1892 2.3273  -4.378
SP6          -15.5640 1.4143 -11.005
8@9.1         11.6583 1.2913   9.028
8@28.0        -0.3157 1.3856  -0.228
10.18@10.0     4.2497 0.8399   5.060
24@46.0        3.3667 0.8270   4.071
8@9.1:8@28.0  19.4213 2.7423   7.082
```

Most striking, of course, are the loci on linkage group 8. The distal locus on linkage group 8 has essentially no marginal effect, but it has a large influence on the effect of the proximal locus. Our estimated QTL model explains a substantial fraction of the phenotypic variance (50%).

11.3 QTL × covariate interactions

We now turn our attention to the search for potential QTL × MCE interactions in these data: loci that show varying effects across the eight MCE groups (defined by the female source of the eggs). As described in Sec. 7.2, there are three ways that we might go about identifying such interactions. First, we could focus on the QTL identified in Sec. 11.2, which showed clear marginal effects, and test for QTL × MCE interaction at those positions. Second, we may look for loci for which the combined effect of the QTL and its possible interactions with MCE are clear (after adjustment for the genome scan), and again test for the QTL × MCE interactions at those positions, with no further adjustment for multiple testing. Finally, we may look for positions for which the LOD score for the QTL × MCE interaction is large, adjusting for the genome scan.

We start with a genome scan including MCE as an interactive covariate.

```
> outi <- scanone(trout, method="hk", addcovar=femcov,
+                        intcovar=femcov)
```

The results are LOD scores for a test of the full model (including the QTL × MCE interaction) to the null model of no QTL, and so concern eight degrees of freedom (the effect of the QTL in each of the eight MCE groups). A large LOD score indicates that the QTL has an effect in at least one of the eight MCE groups.

It is best to compare these results side-by-side with the results, obtained in the previous section, in which MCE was included as an additive covariate but not an interactive covariate. For convenience, we combine the LOD scores together into one object, along with the difference, which concerns the test of QTL × MCE interaction.

```
> outi <- c(out, outi, outi-out, labels=c("a","f","i"))
```

We may then plot the three sets of LOD curves as follows.

```
> plot(outi, lod=1:3, ylab="LOD score", alternate.chrid=TRUE)
```

As seen in Fig. 11.12, we continue to have extremely strong evidence for the locus on linkage group eight, but note that there is little evidence for a QTL × MCE interaction at this locus.

To make sense of the statistical significance of the results, we perform a permutation test. It is critical that the permutations with MCE as an interactive covariate be precisely matched to those with MCE as a strictly additive

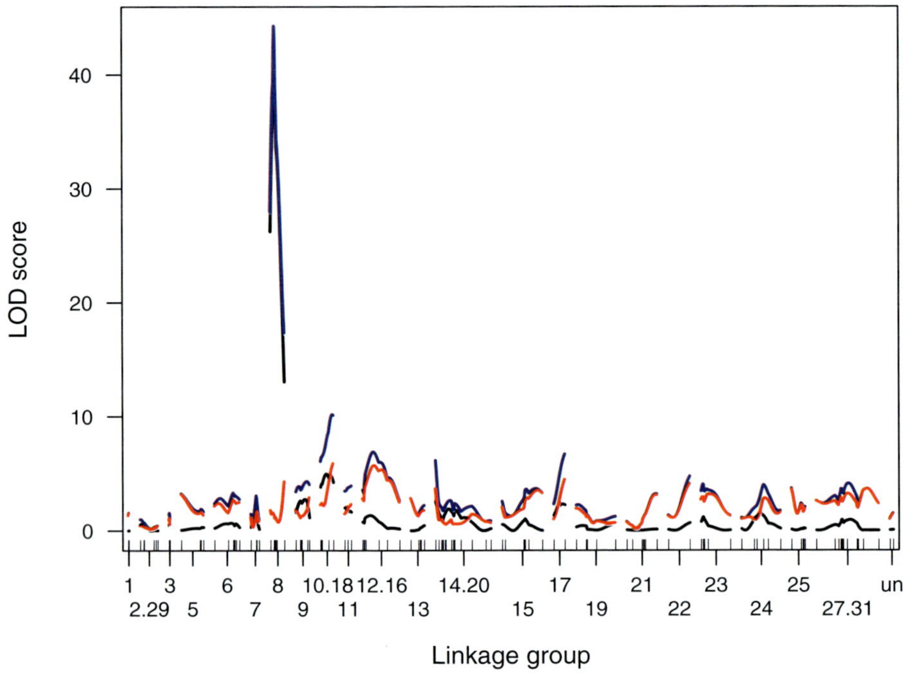

Figure 11.12. LOD curves from a genome scan by Haley–Knott regression for the `trout` data, with MCE groups included as additive covariates (in black), with MCE groups included as interactive covariates (in blue) and for the test of QTL × MCE interaction (in red).

covariate, so that the differences (which concern the test of QTL × MCE interaction) have meaning. And so we use `set.seed` to set the seed for the random number generator to be the same as was used in our permutations in Sec. 11.2.

```
> set.seed(523938)
> opermi <- scanone(trout, method="hk", addcovar=femcov,
+                   intcovar=femcov, n.perm=4000)
```

We again combine the results together into one object.

```
> opermi <- cbind(operm, opermi, opermi-operm,
+                 labels=c("a","f","i"))
```

With eight MCE groups and so seven degrees of freedom associated with the QTL × MCE interaction, the significance thresholds for the LOD scores with MCE as an interactive covariate are quite large.

```
> summary(opermi)

LOD thresholds (4000 permutations)
```

```
    lod.a lod.f lod.i
5%   2.66  6.93  5.36
10%  2.32  6.38  4.81
```

If we were to test for QTL × MCE pointwise (that is, not adjusting for the genome scan), we could make use of the approximation that LOD × $(2 \ln 10)$ follows a $\chi^2(\text{df} = 7)$ distribution under the null hypothesis (of no interaction). The pointwise 5% significance threshold is then

```
> qchisq(0.95, 7) / (2 * log(10))
```

```
[1] 3.055
```

The advantage of pasting the three sets of LOD curves together is that we can get a combined summary. If we use `format="allpheno"` in the call to `summary.scanone`, we get a row in the summary table for each position for which any one of the three LOD scores exceeds its chosen threshold.

```
> summary(outi, perms=opermi, alpha=0.05, pvalues=TRUE,
+         format="allpheno")
```

```
               chr   pos lod.a    pval lod.f   pval lod.i
OmyFGT12         8  9.11 43.18 0.00000 44.17 0.0000 0.988
c9.loc20         9 20.00  2.76 0.03775  4.09 0.8153 1.333
c10.18.loc13 10.18 13.00  4.97 0.00025  7.68 0.0177 2.705
c10.18.loc27 10.18 27.00  4.47 0.00050 10.06 0.0000 5.592
AGCCAG13     10.18 28.69  4.22 0.00125 10.00 0.0000 5.775
c12.16.loc26 12.16 26.00  1.17 0.78350  6.76 0.0638 5.593
               pval
OmyFGT12     0.9978
c9.loc20     0.9910
c10.18.loc13 0.7232
c10.18.loc27 0.0357
AGCCAG13     0.0283
c12.16.loc26 0.0357
```

With MCE as a strictly additive covariate (column `lod.a`), we see significant evidence for QTL on linkage groups 8, 9, and 10.18. The loci on linkage groups 8 and 9 show no evidence for QTL × MCE interaction ($\text{LOD}_i = \text{LOD}_f - \text{LOD}_a$ is small). However, the locus on linkage group 10.18 shows reasonably good evidence for an interaction. When allowing for QTL × MCE interaction, the QTL on linkage group 10.18 shifts a bit (from 13 cM to 27 cM), and the evidence for interaction becomes clear. In the analysis allowing QTL × MCE interaction, a locus on linkage group 12.16 nearly reaches significance, and shows a strong interaction effect.

Recall our three strategies for identifying QTL × MCE interactions. First, we could look at loci with clear marginal effects (adjusting for the genome scan), and test the interaction at these positions, pointwise. With this strategy, we identify loci on linkage groups 8, 9, and 10.18, but only the locus on linkage group 10.18 would show a QTL × MCE interaction. Second, we could look at significant loci in the scan allowing for QTL × MCE interaction, and again test for the interaction at these positions, pointwise. If we are strict in this approach, we identify just the loci on linkage groups 8 and 10.18, and again only the locus on linkage group 10.18 would show a significant QTL × MCE interaction. Finally, we could focus on the interaction LOD score alone, adjusting for the genome scan. This reveals the QTL × MCE interactions for loci on linkage groups 10.18 and 12.16.

In the above analysis, we consider just a single locus at a time. But our analysis in Sec. 11.2 revealed two interacting loci on linkage group 8 with very large effects, and so it would be best to repeat our scan for possible QTL × MCE interaction, accounting for these large-effect loci. This may be done with `addqtl`.

We first call `makeqtl` to create a QTL object containing our two QTL on linkage group 8.

```
> qtl <- makeqtl(trout, c("8", "8"), c(9.1, 28), what="prob")
```

We then call `addqtl` twice. First, we scan for a third QTL, with the MCE groups as strictly additive covariates. Second we scan for a third QTL, allowing the MCE groups to interact with the QTL being scanned but not with the first two QTL.

The model formulas are a bit cumbersome to create, as we must refer to the seven covariates by name. One can avoid some typing (and reduce the chance of errors) by using `paste` to create a character string representation of the model formula. We could then use `as.formula` to convert it to a formula, though actually `addqtl` and related functions accept the character representation, and so conversion with `as.formula` is not needed.

```
> addform <- paste("y~Q1*Q2+Q3+",
+                  paste(colnames(femcov), collapse="+"),
+                  sep="")
> addform

[1] "y~Q1*Q2+Q3+TL3+SP1+SP2+SP3+SP4+SP5+SP6"

> intform <- paste("y~Q1*Q2+Q3+",
+                  paste("Q3", colnames(femcov),
+                        sep="*", collapse="+"),
+                  sep="")
> intform

[1] "y~Q1*Q2+Q3+Q3*TL3+Q3*SP1+Q3*SP2+Q3*SP3+Q3*SP4+Q3*SP5+Q3*SP6"
```

Now we are ready to perform the scans with `addqtl`.

```
> out.aq <- addqtl(trout, qtl=qtl, method="hk", covar=femcov,
+                  formula=addform)
> outi.aq <- addqtl(trout, qtl=qtl, method="hk", covar=femcov,
+                   formula=intform)
```

We again paste the three sets of LOD scores into one object.

```
> outi.aq <- c(out.aq, outi.aq, outi.aq - out.aq,
+              labels=c("a","f","i"))
```

We use `summary.scanone` to pull out the interesting loci. We will assess significance using the results of our permutation tests not conditioning on linkage group 8, even though they are not strictly valid here.

```
> summary(outi.aq, perms=opermi, alpha=0.05, pvalues=TRUE,
+         format="allpheno")
```

```
              chr  pos lod.a    pval lod.f   pval lod.i   pval
c9.loc30        9 30.0  1.12 0.81800  7.32 0.0302 6.196 0.0155
AGCCAG15        9 30.2  1.07 0.85925  7.32 0.0302 6.251 0.0145
c10.18.loc9 10.18  9.0  5.57 0.00025  6.43 0.0922 0.866 0.9988
AGCCAG13    10.18 28.7  4.81 0.00025 10.56 0.0000 5.747 0.0305
c17.loc18      17 18.0  2.91 0.02675  5.26 0.3380 2.349 0.8413
c24.loc44      24 44.0  3.66 0.00450  5.45 0.2755 1.788 0.9573
```

The locus on linkage group 10.18 remains, and shows a clear QTL × MCE interaction. The locus on linkage group 12.16 has disappeared. Additional loci on linkage groups 17 and 24 are seen, but neither shows evidence for a QTL × MCE interaction. Most interesting is the locus on linkage group 9, which no longer shows a marginal effect, but does show a clear QTL × MCE interaction.

A plot of the LOD curves for selected linkage groups may be useful.

```
> plot(outi.aq, lod=1:3, ylab="LOD score", alternate.chrid=TRUE,
+      chr=c(8, 9, "10.18", "12.16", 17, 24))
```

The LOD curves in Fig. 11.13 are useful in giving a sense of the precision of localization of the QTL.

Of course, we should bring all of the QTL together into one model, as this gives the best assessment of the support for the individual loci. However, the drop-one-QTL analysis with `fitqtl` can be extremely cumbersome in the case that we have a multilevel factor as an interactive covariate: each term is dropped one at a time, and we really want to see what happens when the set of terms are omitted together.

We can, however, perform our own drop-one-QTL analysis, by repeatedly calling `fitqtl` with multiple model formulas. The formulas are long and cumbersome, but with careful use of `paste`, we can create them without too much difficulty. We illustrate the process here, though it is not for the faint-of-heart.

We first create a QTL object with all of the putative QTL; we will use `addtoqtl` to add additional terms to the object we had created, containing just the two loci on linkage group 8.

```
> qtl <- addtoqtl(trout, qtl, c(9, "10.18", 17, 24),
+                 c(30, 28.7, 18, 44))
```

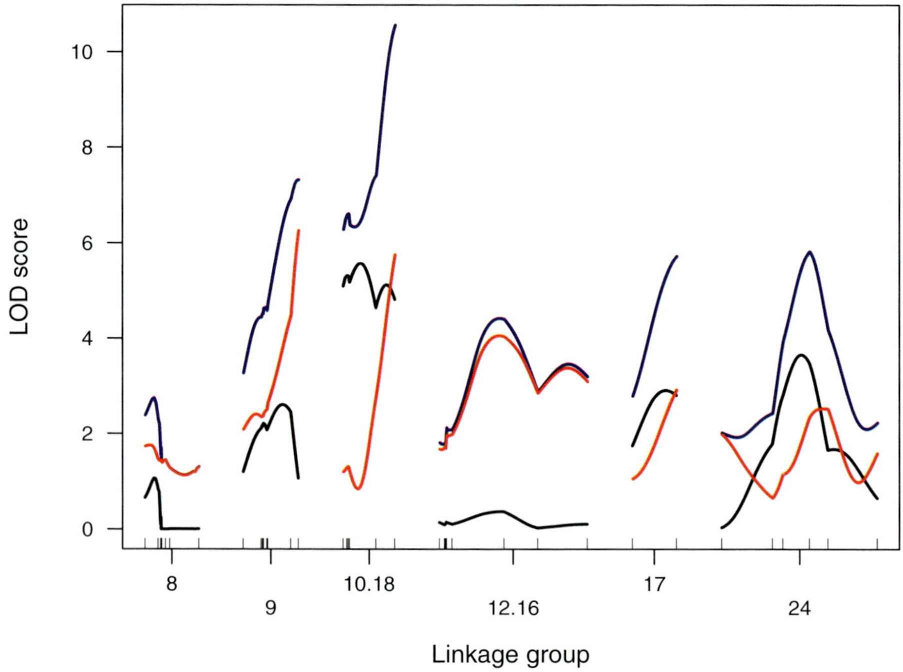

Figure 11.13. LOD curves for selected linkage groups from a genome scan by Haley–Knott regression for the `trout` data, controlling for two interacting loci on linkage group 8, with MCE groups included as additive covariates (in black), with MCE groups included as interactive covariates (in blue) and for the test of QTL × MCE interaction (in red).

Now we create our set of model formulas. We start with the full model, containing all of the QTL plus the QTL × MCE interactions for the loci on linkage groups 9 and 10.18.

```
> fullform <- paste("y~Q1*Q2+Q3+Q4+Q5+Q6",
+                   paste(colnames(femcov), collapse="+"),
+                   paste("Q3", colnames(femcov), sep=":",
+                         collapse="+"),
+                   paste("Q4", colnames(femcov), sep=":",
+                         collapse="+"), sep="+")
```

We will assume that evidence for the loci on linkage group 8 is so strong that we don't need to consider models that lack them, but we will want to fit the set of models with each of the other QTL missing.

```
> form.m9 <- paste("y~Q1*Q2+Q4+Q5+Q6",
+                  paste(colnames(femcov), collapse="+"),
+                  paste("Q4", colnames(femcov), sep=":",
```

```
+                          collapse="+"), sep="+")
> form.m1018 <- paste("y~Q1*Q2+Q3+Q5+Q6",
+                      paste(colnames(femcov), collapse="+"),
+                      paste("Q3", colnames(femcov), sep=":",
+                            collapse="+"), sep="+")
> form.m17 <- paste("y~Q1*Q2+Q3+Q4+Q6",
+                    paste(colnames(femcov), collapse="+"),
+                    paste("Q3", colnames(femcov), sep=":",
+                          collapse="+"),
+                    paste("Q4", colnames(femcov), sep=":",
+                          collapse="+"), sep="+")
> form.m24 <- paste("y~Q1*Q2+Q3+Q4+Q5",
+                    paste(colnames(femcov), collapse="+"),
+                    paste("Q3", colnames(femcov), sep=":",
+                          collapse="+"),
+                    paste("Q4", colnames(femcov), sep=":",
+                          collapse="+"), sep="+")
```

We also want formulas with just the QTL × MCE interactions for the loci on linkage groups 9 and 10.18 omitted.

```
> form.m9int <- paste("y~Q1*Q2+Q3+Q4+Q5+Q6",
+                      paste(colnames(femcov), collapse="+"),
+                      paste("Q4", colnames(femcov), sep=":",
+                            collapse="+"), sep="+")
> form.m1018int <- paste("y~Q1*Q2+Q3+Q4+Q5+Q6",
+                         paste(colnames(femcov), collapse="+"),
+                         paste("Q3", colnames(femcov), sep=":",
+                               collapse="+"), sep="+")
```

With our model formulas in hand, we can calculate the LOD scores for each of these seven models. Let us start with the full model. It would be best to first use `refineqtl` to get improved estimates of the locations of the QTL, in the context of this model.

```
> qtl <- refineqtl(trout, qtl=qtl, formula=fullform,
+                  method="hk", covar=femcov, verbose=FALSE)
```

We now use `fitqtl` to fit the model.

```
> full <- fitqtl(trout, qtl=qtl, formula=fullform, method="hk",
+                covar=femcov, dropone=FALSE)
```

In the summary of the result, we can see the LOD score comparing the full model to the null model (with none of the QTL or covariates).

```
> summary(full, pvalues=FALSE)

Full model result
----------------------------------
```

```
Model formula: y ~ Q1 + Q2 + Q3 + Q4 + Q5 + Q6 + TL3 + SP1 + SP2
                   + SP3 + SP4 + SP5 + SP6 + Q1:Q2 + Q3:TL3 +
                   Q3:SP1 + Q3:SP2 + Q3:SP3 + Q3:SP4 + Q3:SP5 +
                   Q3:SP6 + Q4:TL3 + Q4:SP1 + Q4:SP2 + Q4:SP3 +
                   Q4:SP4 + Q4:SP5 + Q4:SP6

       df    SS      MS   LOD  %var
Model  28 48552 1734.00 99.54 56.28
Error 525 37714   71.84
Total 553 86266
```

To pull out just the LOD score for the fit of the model (relative to the null model with no QTL), note that the output of `fitqtl` is a list, and one of the components, named `"lod"`, is simply this LOD score.

```
> print(fulllod <- full$lod)
[1] 99.54
```

We may now use `fitqtl` to fit our other six models.

```
> m9 <- fitqtl(trout, qtl=qtl, formula=form.m9, method="hk",
+             covar=femcov, dropone=FALSE)
> m1018 <- fitqtl(trout, qtl=qtl, formula=form.m1018,
+                method="hk", covar=femcov, dropone=FALSE)
> m17 <- fitqtl(trout, qtl=qtl, formula=form.m17, method="hk",
+              covar=femcov, dropone=FALSE)
> m24 <- fitqtl(trout, qtl=qtl, formula=form.m24, method="hk",
+              covar=femcov, dropone=FALSE)
> m9int <- fitqtl(trout, qtl=qtl, formula=form.m9int,
+                method="hk", covar=femcov, dropone=FALSE)
> m1018int <- fitqtl(trout, qtl=qtl, formula=form.m1018int,
+                   method="hk", covar=femcov, dropone=FALSE)
```

We are interested in the differences between the LOD score for the full model and the LOD scores for models with individual terms omitted. Let us start with the loci on linkage groups 17 and 24, for which we did not include QTL × MCE interactions.

```
> fulllod - m17$lod
[1] 2.854
> fulllod - m24$lod
[1] 2.749
```

The LOD score for each is above the main effect penalty (2.71).

For the loci on linkage groups 9 and 10.18, we first consider the difference between the full model and the models with both the QTL and the QTL × MCE interaction terms omitted.

```
> fulllod - m9$lod
```

```
[1] 7.933
```

```
> fulllod - m1018$lod
```

```
[1] 11.32
```

These are both well above the threshold from the permutation test including QTL × MCE interactions. We turn to the interaction LOD scores, comparing the full model to the models with just the QTL × MCE interaction terms omitted.

```
> fulllod - m9int$lod
```

```
[1] 6.867
```

```
> fulllod - m1018int$lod
```

```
[1] 6.046
```

Both are quite large, indicating that the loci on linkage groups 9 and 10.18 show clear QTL × MCE interactions.

In the LOD scores above, the loci other than the one under test are kept fixed at their estimated positions under the full model. This is what is done in the drop-one-QTL analysis in `fitqtl`, but it gives a somewhat rosy view of the support for the individual loci. Ideally, for each submodel, the locations for the remaining QTL would be reestimated, and each comparison would be between the full model with QTL in their best positions under the full model, to a submodel with QTL in their best positions under that submodel.

This is simple enough to accomplish in our "by hand" comparisons of the individual models. We simply run `refineqtl` for each submodel, followed by `fitqtl`.

```
> qtl.m9 <- refineqtl(trout, qtl=qtl, formula=form.m9,
+                     method="hk", covar=femcov,
+                     verbose=FALSE)
> qtl.m1018 <- refineqtl(trout, qtl=qtl, formula=form.m1018,
+                        method="hk", covar=femcov,
+                        verbose=FALSE)
> qtl.m17 <- refineqtl(trout, qtl=qtl, formula=form.m17,
+                      method="hk", covar=femcov,
+                      verbose=FALSE)
> qtl.m24 <- refineqtl(trout, qtl=qtl, formula=form.m24,
+                      method="hk", covar=femcov,
+                      verbose=FALSE)
> qtl.m9int <- refineqtl(trout, qtl=qtl, formula=form.m9int,
+                        method="hk", covar=femcov,
+                        verbose=FALSE)
```

```
> qtl.m1018int <- refineqtl(trout, qtl=qtl, method="hk",
+                           formula=form.m1018int,covar=femcov,
+                           verbose=FALSE)
```

We now call fitqtl for each of these reduced models, with the QTL in the positions estimated under the corresponding model.

```
> m9r <- fitqtl(trout, qtl=qtl.m9, formula=form.m9,
+               method="hk", covar=femcov, dropone=FALSE)
> m1018r <- fitqtl(trout, qtl=qtl.m1018, formula=form.m1018,
+                  method="hk", covar=femcov, dropone=FALSE)
> m17r <- fitqtl(trout, qtl=qtl.m17, formula=form.m17,
+                method="hk", covar=femcov, dropone=FALSE)
> m24r <- fitqtl(trout, qtl=qtl.m24, formula=form.m24,
+                method="hk", covar=femcov, dropone=FALSE)
> m9intr <- fitqtl(trout, qtl=qtl.m9int, formula=form.m9int,
+                  method="hk", covar=femcov, dropone=FALSE)
> m1018intr <- fitqtl(trout, qtl=qtl.m1018int, method="hk",
+                     formula=form.m1018int, covar=femcov,
+                     dropone=FALSE)
```

We recalculate the conditional LOD scores for each locus, first for the loci on linkage groups 17 and 24.

```
> fulllod - m17r$lod

[1] 2.849

> fulllod - m24r$lod

[1] 2.736
```

The LOD scores are slightly smaller, but both are still above main effect penalty (2.71).

Now we consider the loci on linkage groups 9 and 10.18.

```
> fulllod - m9r$lod

[1] 7.921

> fulllod - m1018r$lod

[1] 11.01

> fulllod - m9intr$lod

[1] 5.168

> fulllod - m1018intr$lod

[1] 5.798
```

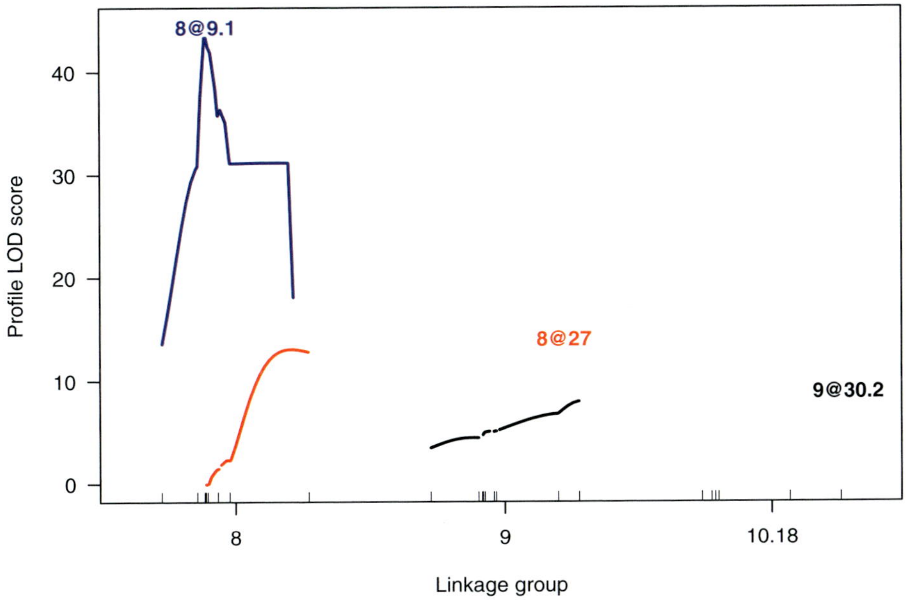

Figure 11.14. Profile LOD curves for a six-QTL model, including two epistatic QTL on linkage group 8 and QTL × MCE interactions for the QTL on linkage groups 9 and 10.18, for the `trout` data.

The biggest change is in the interaction LOD score for the locus on linkage group 9, which dropped from 6.87 to 5.17. But the evidence for both of these loci, and for their QTL × MCE interactions, remains clear.

Let us turn to interval estimates for the locations of the inferred QTL. We are particularly interested in the QTL on linkage group 10.18 (which is really the pasting together of linkage groups 10 and 18): is the QTL in the linkage group 10 part or the linkage group 18 part, or can we not tell?

First, let us plot the LOD profiles (see Sec. 9.3.2). Since we had used `refineqtl` on the object `qtl`, concerning the full model, we may immediately use `plotLodProfile`.

```
> plotLodProfile(qtl, col=c("blue","red",rep("black",4)),
+                ylab="Profile LOD score")
```

The LOD profiles appear in Fig. 11.14. Recall from Sec. 9.3.2 that in these curves, the location of one QTL is allowed to vary while other QTL are fixed at their best estimates. The LOD scores are for the comparison between the full model and the reduced model with the QTL of interest (and any of its interactions) omitted. For example, in the curve for the QTL on linkage group 10.18, that locus is allowed to vary while the locations of all other QTL are kept fixed, and the LOD scores compare the full model, with the locus on linkage group 10.18 in varying position, to the reduced model in which this QTL plus its QTL × MCE interactions are omitted.

The LOD profiles for the proximal locus on linkage group 8 looks a bit odd, but note that for linked QTL (and particularly for QTL linked as tightly as these two), the profile LOD curves give a poor representation of our uncertainty in the locations of the QTL. It would be best to allow the locations of the two QTL to vary jointly. We will return to that in a moment.

We were particularly interested in the QTL on linkage group 10.18, and of whether its location could be defined to linkage group 10 or linkage group 18, or whether this was uncertain. We may use `lodint` and `bayesint` to obtain approximate confidence intervals for the location of the QTL. The intervals should be considered with caution, as they fail to capture the uncertainty in the locations of the other QTL in the model, and also the performance of these intervals, in the context of multiple-QTL models, is not well understood. The 1.5-LOD support interval and 95% Bayesian credible interval are calculated as follows.

```
> lodint(qtl, qtl.index=4)

     chr   pos    lod
29 10.18 24.00  9.498
34 10.18 28.69 11.319
34 10.18 28.69 11.319
```

```
> bayesint(qtl, qtl.index=4)

     chr   pos   lod
31 10.18 26.00 10.38
34 10.18 28.69 11.32
34 10.18 28.69 11.32
```

The intervals are remarkably small: the LOD support interval spans 4.7 cM and the Bayesian interval spans 2.7 cM. It is hard to believe that the location of the QTL could be so precisely defined.

Now, consider the genetic map of the markers on linkage group 10.18.

```
> pull.map(trout, "10.18")

ACGACA11 AGCAGT15  ACCAAG6  agcagc9 AGCCAG12 AGCCAG13
   0.000    1.992    2.753    3.444   18.031   28.692
```

The first four markers were on linkage group 10, and the last two markers were on linkage group 18. Thus it appears that the QTL on the merged linkage group 10.18 is within the linkage group 18 part.

Returning to the case of the linked loci on linkage group 8, if we wished to better define the locations of these QTL in the context of our six-QTL model, we should perform a two-dimensional scan on linkage group 8, keeping the locations of the other four QTL at their best estimates. This can be accomplished with `addpair`. We first use `dropfromqtl` to drop the two QTL on linkage group 8 from our QTL object.

```
> qtl.m8 <- dropfromqtl(qtl, 1:2)
```

We need to create a model formula including our QTL × MCE interactions, and with the two additional QTL (to be scanned) allowed to interact. Note that, having dropped the QTL on linkage group 8, the numeric indices of the other four QTL have changed.

```
> theformula <- paste("y~Q1+Q2+Q3+Q4+Q5*Q6",
+                     paste(colnames(femcov), collapse="+"),
+                     paste("Q1", colnames(femcov), sep=":",
+                           collapse="+"),
+                     paste("Q2", colnames(femcov), sep=":",
+                           collapse="+"), sep="+")
```

Now we are ready to run `addpair`.

```
> out.ap <- addpair(trout, chr="8", qtl=qtl.m8, covar=femcov,
+                   formula=theformula, method="hk",
+                   incl.markers=TRUE, verbose=FALSE)
```

The results are the two-dimensional equivalent of the LOD profiles in Fig. 11.14. At each pair of positions for the two QTL, we compare the full model to the reduced model with these two QTL (and their interaction) omitted.

We may plot the two-dimensional LOD profile as follows. The argument `contour` is used to display an approximate 1.5-LOD support region. Note that the performance of such two-dimensonal regions is not well understood.

```
> plot(out.ap, contours=1.5)
```

As seen in Fig. 11.15, the 1.5-LOD support regions indicates that the location of the proximal QTL on linkage group 8 has been impressively well defined, to an interval spanning just over 1 cM. The location of the distal QTL is less well defined; the support region spans about 8 cM.

Finally, let us study the effects of the loci. We are particularly interested in the QTL on linkage groups 9 and 10.18, which exhibit QTL × MCE interactions. We will use `fitqtl` to get the estimated effects in the context of our six-QTL model.

```
> summary(fitqtl(trout, qtl=qtl, formula=fullform, covar=femcov,
+                method="hk", dropone=FALSE, get.ests=TRUE))

Full model result
----------------------------------

Model formula: y ~ Q1 + Q2 + Q3 + Q4 + Q5 + Q6 + TL3 + SP1 + SP2
                   + SP3 + SP4 + SP5 + SP6 + Q1:Q2 + Q3:TL3 +
                   Q3:SP1 + Q3:SP2 + Q3:SP3 + Q3:SP4 + Q3:SP5 +
```

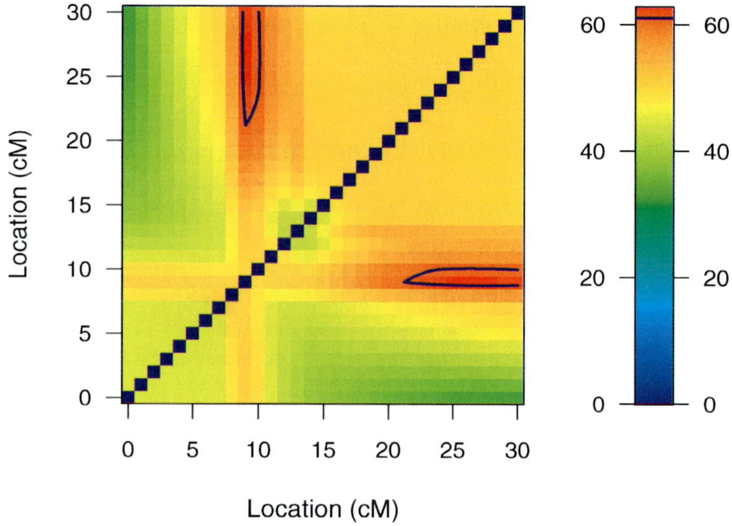

Figure 11.15. Two-dimensional profile LOD surface for the pair of interacting QTL on linkage group 8, in the context of a six-QTL model, including QTL × MCE interactions for the QTL on linkage groups 9 and 10.18 and additional loci on linkage groups 17 and 24, for the `trout` data.

```
                     Q3:SP6 + Q4:TL3 + Q4:SP1 + Q4:SP2 + Q4:SP3 +
                     Q4:SP4 + Q4:SP5 + Q4:SP6

       df    SS      MS   LOD  %var Pvalue(Chi2) Pvalue(F)
Model  28 48552 1734.00 99.54 56.28            0         0
Error 525 37714   71.84
Total 553 86266

Estimated effects:
-----------------
                    est     SE       t
Intercept      326.4836 1.3124 248.767
TL3            -12.1128 1.4839  -8.163
SP1             -8.5237 1.5711  -5.425
SP2            -14.0621 1.5777  -8.913
SP3            -15.2948 2.0203  -7.570
SP4             -6.0049 3.8896  -1.544
SP5            -11.1024 2.4090  -4.609
SP6            -15.2999 1.4161 -10.805
8@9.1           12.2976 1.3077   9.404
8@27.0          -1.1588 1.4225  -0.815
9@30.2           9.3167 2.5252   3.690
```

```
10.18@28.7          15.3082   3.0431   5.031
17@23.0              2.8183   0.7939   3.550
24@48.0              2.7082   0.7775   3.483
8@9.1:8@27.0        19.5517   2.8142   6.948
9@30.2:TL3         -12.9477   3.0492  -4.246
9@30.2:SP1          -3.0747   3.1703  -0.970
9@30.2:SP2          -5.8341   3.1489  -1.853
9@30.2:SP3         -12.1483   4.1183  -2.950
9@30.2:SP4          -3.3755   7.7862  -0.434
9@30.2:SP5         -14.3642   4.6790  -3.070
9@30.2:SP6          -8.3860   2.8493  -2.943
10.18@28.7:TL3      -7.5077   3.5224  -2.131
10.18@28.7:SP1 -12.7976   3.6071  -3.548
10.18@28.7:SP2 -11.4783   3.5959  -3.192
10.18@28.7:SP3 -10.2645   4.6146  -2.224
10.18@28.7:SP4 -11.6208   7.9593  -1.460
10.18@28.7:SP5 -13.8230   5.1415  -2.689
10.18@28.7:SP6 -15.0835   3.3221  -4.540
```

Consider the locus on linkage group 10.18, and recall that, for the linkage group 18 part of this merged "linkage group," we had swapped the two genotypes, and so the effects have the opposite sign than might be expected. The estimated coefficient in the row labeled `10.18@28.7` is the effect of the linkage group 10.18 locus in the first MCE group (TL1). That is, for the individuals whose egg came from the TL1 female, the difference in the average phenotype between the two genotypes is 15.3 ± 3.0. The other rows that begin `10.18@28.7` are for the differences in the QTL effect for the indicated MCE group and the QTL effect for the TL1 group. Relative to the TL1 group, the QTL on linkage group 10.18 has smaller effect in all other MCE groups; in some cases (e.g., SP6) the QTL appears to have no effect. Similar observations apply to the locus on linkage group 9.

The linked epistatic QTL on linkage group 8 are especially interesting. To make sense of the estimated effects, note that the marginal effects of the QTL are based on a coding of the two QTL genotypes as $\pm\ 1/2$, and the interaction term is the product of the two main effect terms. Thus the effect of the proximal QTL among individuals with the CC genotype at the distal locus is estimated to be $12.3 - 19.6/2 = 2.5$, while its effect among individuals with the OO genotype at the distal locus is estimated to be $12.3 + 19.6/2 = 22.1$. That is, the proximal locus on linkage group 8 has a large effect, but only among individuals with genotype OO at the distal locus.

11.4 Discussion

Our principal aim in this case study was to illustrate the exploration of possible QTL $\times$ covariate interactions. As with the first case study, we have focused on the analysis techniques rather than the biological interpretation of the results; for the latter, see the original article describing these data (Nichols *et al.*, 2007).

With these data, there was a single covariate to consider, maternal cytoplasmic environment (MCE): the source of the individuals' eggs in these doubled haploids. Nevertheless, the best approach for evaluating QTL × covariate interactions, with adjustment for the multiple hypothesis tests, is still not perfectly clear. We favor the use of a scan allowing for the possibility of QTL × covariate interactions, followed by pointwise tests of the significance of the interactions. Had we been presented with a set of covariates for which QTL × covariate interactions were to be explored, the multiple testing issue would be even more onerous.

Automated multiple-QTL analyses in R/qtl do not yet allow for the possibility of QTL × covariate interactions, and so we relied on a more exploratory approach based on single- and two-dimensional genome scans. Extension of the model comparison criteria of Sec. 9.1.4, to allow QTL × covariate interactions, deserves exploration. Likely one could use a significance threshold for the interaction LOD score as a penalty on a QTL × covariate interaction. If multiple covariates were to be considered, such a significance threshold should account for the search among the covariates as well as the scan across the genome.

Our treatment of the linked pairs of homeologous linkage groups in these data (swapping the genotypes on one linkage group and merging the two together) is unorthodox, and is not beyond criticism. An advantage of our approach is that we avoid the possibility of a single QTL giving linkage signals on multiple linkage groups. However, the gaps between the parts of the merged linkage groups may contain no DNA, and the swapping of genotypes on one part makes the interpretation of QTL effects more difficult. Clearly, we need a better understanding of the underlying mechanism giving rise to these pseudolinkages.

A

Installing R and R/qtl

Here we describe the installation of R and R/qtl on Windows, Mac OS X, and Unix. We also give a couple of tips for customizing the R environment.

The R statistical system is available at the Comprehensive R Archive Network (CRAN, `http://cran.r-project.org`). It's best to use a local mirror; you can find a list of links to these at `http://cran.r-project.org/mirrors.html`.

The main site for the R/qtl package is `http://www.rqtl.org`; it may also be obtained from CRAN. On CRAN, R/qtl is known as the `qtl` package.

A.1 Installing R

We include separate subsections on R installation for the three operating systems, as the details are quite different. For the definitive instructions on installing R, see the "R Installation and Administration" manual on the CRAN web site. (Click on "Manuals" on the left-side, under "Documentation," or go directly to `http://cran.r-project.org/doc/manuals/R-admin.pdf`).

A.1.1 Windows

Download the R program, which will be about 34 megabytes. It will be a file of the form `R-`*`version`*`-win32.exe`, where *`version`* is the version number, and so it will be something like `R-2.8.1-win32.exe`.

You can find the file by visiting CRAN and clicking on "Windows" and "base," or by going directly to `http://cran.r-project.org/bin/windows/base`.

Install R by executing this file and following the instructions; it is the usual sort of Windows setup program. We recommend installing R in `c:\R` rather than `c:\Program Files\R`, as the space in `Program Files` sometimes leads to difficulties. (Why didn't Microsoft use `Programs` rather than `Program Files`?)

K.W. Broman, Ś. Sen, *A Guide to QTL Mapping with R/qtl*,
Statistics for Biology and Health, DOI 10.1007/978-0-387-92125-9_BM2,

You can choose to have an R icon placed on the desktop or in the startup menu. Use one of these to invoke R.

A.1.2 Mac OS X

Download the R program, which will be about 63 megabytes. It will be a file of the form `R-`*`version`*`.dmg`, where *`version`* is the version number, and so it will be something like `R-2.8.1.dmg`.

You can find the file by visiting CRAN and clicking on "Mac OS X," or by going directly to `http://cran.r-project.org/bin/macosx`.

Double-click this file; it will create a drive on your desktop with the name `R-`*`version`*. Open that and then double-click `R.mpkg`. This will lead to the usual sort of installer; follow the instructions, placing the R program in your Applications folder. You will need an administrator password to install R.

To invoke R, double-click the R application that is now in your Applications folder.

You can also use the command-line version of R, as in Unix (below). To make this available, you need to put a soft-link to R in either `/usr/local/bin` or `/usr/bin`, using the following command from a terminal window. You will again need an administrator password. (If this paragraph makes no sense, you probably don't want to do this, and should stick with the graphical user interface.)

```
sudo ln -s /Library/Frameworks/R.framework/Resources/R/usr/bin/R
```

You may then invoke R by typing `R` at the prompt in an X Windows terminal.

A.1.3 Unix/Linux

Linux users may install an R binary using their distribution's package management system. This is convenient and adequate for most users. At the time of writing, binaries for Debian, Red Hat, SuSE, and Ubuntu are distributed on CRAN. For example, Debian and Ubuntu users may install R on their system using the command `sudo apt-get install r-base`. Further instructions are available on CRAN's download page. Linux users may also compile the source code; that is recommended if one wants to optimize R's performance.

R can be installed from source in the traditional way, using `configure` and `make`. We will give just limited details here, as we expect most users of Unix or Linux will be familiar with the procedure.

Download the R source file, available on the main page at CRAN. It will be a file of the form `R-`*`version`*`.tar.gz` where *`version`* is the version number, and so it will be something like `R-2.8.1.tar.gz`.

Uncompress the file somewhere (e.g., in `/tmp`) using `gunzip` and `tar`. Go into the directory and type `./configure`, and respond to the various queries (the defaults are generally okay). Then type `make` and finally `make install` (or `sudo make install`, if you are not the root user).

Invoke R by typing `R` at the Unix prompt.

A.2 Installing R/qtl

The simplest way to install R/qtl is to invoke R and then type the following command.

```
> install.packages("qtl")
```

You may later check for and install an updated version of this and other packages with the following command.

```
> update.packages()
```

An alternative solution for the installation of R/qtl is to download the relevant binary (for Windows or Mac OS X) or the source code for the package (for Unix). For Windows, this will be a file of the form `qtl_`*version*`.zip`; for Mac OS X, it will be of the form `qtl_`*version*`.tgz`; for Unix, it will be of the form `qtl_`*version*`.tar.gz`. Windows and Macintosh users may also wish to download the `.tar.gz` source file, as it contains all of the source code for the package. The files may be obtained from CRAN or from the R/qtl web site.

In Windows, unzip the file `qtl_`*version*`.zip`, placing the contents in the directory `$RHOME\library`, where `$RHOME` is something like `c:\R\R-2.8.1`. This should create a directory `$RHOME\library\qtl`. Then start R and type the following command, in order to get the help files of the QTL package added to the relevant indices.

```
> link.html.help()
```

In Mac OS X, uncompress the `qtl_`*version*`.tgz` file to the directory `/Library/Frameworks/R.framework/Resources/library/`. Alternatively, use some other directory (such as `~/Rlibs`). In the latter case, you will need to create a `~/.Renviron` file in your home directory containing a line like the following.

```
R_LIBS=/Users/auser/Rlibs
```

In Unix, you compile and install the source package using the following command (replacing `1.11-12` with the appropriate version number).

```
R CMD INSTALL qtl_1.11-12.tar.gz
```

Alternatively, you may install the package in a private directory; to install it in to `/home/auser/Rlibs`, type the following.

```
R CMD INSTALL --library=/home/auser/Rlibs qtl_1.11-12.tar.gz
```

With R and R/qtl both installed, start R and type the following to load the package.

```
> library(qtl)
```

A.3 Optimizing the R environment

We have a few suggestions to improve your use of R. First, if you prefer letter paper (8.5 × 11 in) to A4, we suggest creating a text file `.Renviron` containing the following line. (In Windows, place the file in `c:\`; in Mac OS X and Unix, place it in your home directory.)

```
R_PAPERSIZE=letter
```

Second, create a text file, `.Rprofile` (in the same place you put `.Renviron`) containing the following lines.

```
options(show.signif.starts=FALSE)
library(qtl)
```

This will prevent the display of the annoying asterices indicating "statistical significance" and will load the R/qtl package whenever you start R, so that you won't need to type `library(qtl)` every time.

For Windows users, we recommend turning off the "buffered output" by clicking Ctrl-W or by deselecting this item in "Misc" on the menu bar.

We recommend copying-and-pasting commands from a text editor, so that you may keep a record of what you have done in an analysis. In Unix and Mac OS X, we prefer to run R within Emacs; this is best done using Emacs Speaks Statistics (ESS), an Emacs extension that makes R easier to use. It is available at `http://ess.r-project.org`.

A.4 Working directories

It is best to have separate R working directories for separate projects, particularly because one's entire R workspace is read into memory, and so one will wish to keep the workspace as focused as possible.

In Windows, change the working directory by clicking on "Change Directory" in the File menu on the menu bar.

In the graphical user interface for Mac OS X, one may set the initial working directory in the preferences. (Click "Preferences" in the R menu on the menu bar, and then go to "Startup.") One may change the working directory in the "Misc" menu on the menu bar. One may load or save a workspace from the Workspace menu on the menu bar.

In Unix, one invokes R from a particular working directory; the `.RData` file in that directory, if it exists, is loaded.

One may also wish to save particularly large objects in separate files, to be loaded or removed from one's workspace as they are needed. An object or objects may be saved to a file with the `save` function, and subsequently read with `load`. With `save`, one must specify the `file` argument by name, as in the following examples.

```
> save(mycross, file = "mydata.RData")
> save(mycross, myoutput, file = "mydata.RData")
```

Note that these `.RData` files will work on all platforms (i.e., operating systems).

One's R workspace is saved in a single file, `.RData`, which is read when R is invoked and will be written upon exit (if one responds "y" to the query, "`Save workspace image?`"). It is valuable to use the function `save.image` to periodically save one's workspace, so that if the program crashes, one's results will not be lost. (By keeping code in a separate text file, though, it should be simple to rerun the analyses, if the results were not saved.) The `save.image` function is used as follows.

```
> save.image()
```

An `.RData` file, created by `save` or `save.image`, can also be attached to the R search tree, so that one may access the objects in the file without having them included in the workspace. The objects in the file cannot share the same name as an object in one's workspace, and if they are modified, the modified version will be placed in the workspace rather than back in the original file. Use `attach` as follows.

```
> attach("otherresults.RData")
```

Use `search` to see what files have been attached, and `detach` to detach them.

A.5 Documentation

R and R/qtl are distributed with extensive documentation. Each function has its own help file, with a complete description of the input and output, examples of its use, and references to related functions.

To view the help file for a function (e.g., `read.cross`), type one of the following.

```
> help(read.cross)
> ?(read.cross)
```

If you are not sure of the name of the relevant function, you can search the help files with a character string, as follows.

```
> help.search("read data")
```

We find it easiest to peruse the html version of the help files. In the R GUI on Windows and Mac OS X, you can get to these from "Help" on the menu bar. In Unix, type `help.start()`, and the help files will be loaded in your default browser. You can view all of the help files in a package by clicking "Packages" at the main help page, and then the name of the package. Within a help file, you can click on related functions (under "See Also") to view their help files.

The example code in a help file may be run by typing, for example:

```
> example(scanone)
```

All help files for R functions are available in a single PDF file at CRAN under "Manuals." There are a number of additional free tutorials on R, and a list of related books, at CRAN. [We should again emphasize the value of the books by Dalgaard (2002) and Venables and Ripley (2002).] The help files for R/qtl are available as a single PDF at its web page.

Also see the Frequently Asked Questions (FAQ) lists, available from `http://cran.us.r-project.org/faqs.html`. There is a general FAQ on R, as well as Windows- and Macintosh-specific FAQ lists. For a FAQ on R/qtl, see `http://www.rqtl.org/faq`.

A.6 Email lists

There are a number of mailing lists for discussion about R and R/qtl. Access to email lists about R is available at `http://www.r-project.org/mail.html`. The R-help list is the general R email list. The R-SIG-Mac list is a special interest group on R for the Macintosh. An archive of past posts to R-help is available at `https://www.stat.math.ethz.ch/pipermail/r-help`. One may search the R-help archive at `http://search.r-project.org`.

Two Google groups are available for R/qtl: Rqtl-announce for announcements of software updates and Rqtl-disc for general discussion. Announcements are also posted to Rqtl-disc, and so one need join only one of the groups. Posts may be received by email individually or in digest form, or may be read solely on the web. Go to `http://groups.google.com` or the R/qtl web page to find and join the groups.

B

List of functions in R/qtl

In this appendix, we list the major functions in R/qtl, organized by topic (rather than alphabetically, as they appear in the help files). Many of the functions listed are not discussed in the book. For those discussed, page numbers (in brackets) indicate the primary reference.

Sample data		
badorder		An intercross with misplaced markers
bristle3		Data on bristle number for Drosophila chromosome 3
bristleX		Data on bristle number for Drosophila X chromosome
fake.4way		Simulated data for a four-way cross
fake.bc		Simulated data for a backcross
fake.f2		Simulated data for an intercross
hyper	[33]	Backcross data on salt-induced hypertension
listeria	[33]	Intercross data on *Listeria monocytogenes* susceptibility
map10	[37]	A 10 cM genetic map modeled after the mouse genome
Input/output		
read.cross	[22]	Read data for a QTL experiment
write.cross	[33]	Write data for a QTL experiment to a file
Simulation		
sim.cross	[36]	Simulate a QTL experiment
sim.map	[37]	Generate a genetic map
Summaries		
qtlversion		Gives the version number of the installed R/qtl package
plot.cross	[35]	Plot various features of a cross object
plot.missing	[36]	Plot a grid of missing genotypes
geno.image		Plot a grid with colored pixels representing different genotypes
plot.pheno	[36]	Histogram or bar plot of a phenotype
plot.info	[70]	Plot the proportion of missing genotype information
summary.cross	[34]	Print a summary of a QTL experiment
summary.map	[38]	Print a summary of a genetic map
nchr, nind, nmar, nphe, totmar [36]		
nmissing	[71]	Number of missing genotypes by marker or individual

ntyped	[72]	Number of genotypes by marker or individual
find.pheno		Find the column number for a particular phenotype
find.marker	[57]	Find the marker closest to a specified position
find.flanking		Find the markers flanking a particular position
find.markerpos	[330]	Find the map positions of a marker
Data manipulation		
clean.cross	[45]	Remove intermediate calculations from a cross
drop.markers	[96]	Remove a set of markers
drop.nullmarkers	[200]	Remove markers without genotype data
fill.geno	[207]	Fill in holes in the genotype data by imputation or the Viterbi algorithm
strip.partials		Replace partially informative genotypes with missing values
markernames	[96]	Pull out the marker names from a cross
pull.map	[54]	Pull out the genetic map from a cross
pull.geno	[55]	Pull out the genotype data as a matrix
pull.pheno	[140]	Pull out a phenotype
replace.map	[65]	Replace the genetic map of a cross
jittermap	[84]	Jitter marker positions slightly so that no two coincide
subset.cross	[100]	Select a subset of chromosomes and/or individuals
c.cross		Combine multiple crosses
switch.order	[62]	Switch the order of markers on a chromosome
movemarker	[57]	Move a marker from one chromosome to another
HMM engine		
argmax.geno		Reconstruct underlying genotypes via the Viterbi algorithm
calc.genoprob	[84]	Calculate conditional genotype probabilities
sim.geno	[94]	Simulate genotypes given observed marker data
Diagnostics		
geno.table	[50]	Create a table of genotypes at each marker
geno.crosstab	[54]	Create a cross-tabulation of genotypes at two markers
checkAlleles	[54]	Identify markers with potentially switched alleles
calc.errorlod	[67]	Calculate genotyping error LOD scores
top.errorlod	[67]	List the genotypes with the highest error LOD values
plot.geno	[67]	Plot the observed genotypes, flagging likely errors
comparecrosses		Compare two cross objects, to see if they are the same
comparegeno	[52]	Calculate the proportion of matching genotypes for each pair of individuals
Genetic mapping		
est.rf	[53]	Estimate pairwise recombination fractions between markers
plot.rf	[55]	Plot recombination fractions
est.map	[64]	Estimate the genetic map
plot.map	[64]	Plot genetic map(s)
ripple	[60]	Assess marker order by permuting adjacent markers
summary.ripple	[61]	Print a summary of the ripple output
compareorder		Compare two orderings of markers on a chromosome
tryallpositions		Test all possible positions for a marker
formLinkageGroups		Partition markers into linkage groups

orderMarkers		Establish marker order, de novo
QTL mapping		
scanone	[84]	Genome scan with a single-QTL model
scantwo	[217]	Two-dimensional genome scan with a two-QTL model
lodint	[120]	Calculate a LOD support interval
bayesint	[120]	Calculate an approximate Bayes credible interval
scanoneboot	[121]	Nonparametric bootstrap to obtain a confidence interval for QTL location
plot.scanone	[79]	Plot output for a one-dimensional genome scan
add.threshold		Add a horizontal line at a LOD threshold to a genome scan plot
plot.scantwo	[217]	Plot output for a two-dimensional genome scan
summary.scanone	[79]	Print a summary of scanone output
summary.scantwo	[220]	Print a summary of scantwo output
max.scanone	[79]	Maximum peak in scanone output
max.scantwo	[238]	Maximum peak in scantwo output
−.scanone	[87]	Subtract LOD scores from multiple scanone results
+.scanone	[195]	Add LOD scores from multiple scanone results
−.scantwo	[87]	Subtract LOD scores from multiple scantwo results
+.scantwo		Add LOD scores from multiple scantwo results
c.scanone	[189]	Combine LOD score columns from multiple scanone results
c.scanoneperm	[223]	Combine multiple batches of permutation replicates from scanone
c.scantwoperm	[223]	Combine multiple batches of permutation replicates from scantwo
cbind.scanoneperm	[189]	Combine LOD score columns from multiple scanone permutation results
effectplot	[125]	Plot phenotype means of genotype groups defined by one or two markers or covariates
effectscan		Plot estimated QTL effects across the whole genome
plot.pxg	[126]	Like effectplot, but as a dot plot of the phenotypes
Multiple QTL models		
makeqtl	[259]	Make a qtl object for use by fitqtl
fitqtl	[260]	Fit a multiple QTL model
summary.fitqtl	[260]	Get a summary of the result of fitqtl
scanqtl	[258]	Perform a multidimensional genome scan
refineqtl	[263]	Refine the QTL locations in a multiple-QTL model
plotLodProfile	[264]	Plot LOD profiles for a multiple-QTL model
addqtl	[267]	Scan for an additional QTL, in a multiple-QTL model
addpair	[269]	Scan for an additional pair of QTL, in a multiple-QTL model
addint	[266]	Add pairwise interactions, one at a time, in a multiple-QTL model
plot.qtl	[260]	Plot the QTL locations on the genetic map
addtoqtl	[272]	Add to a QTL object
dropfromqtl	[274]	Drop a QTL from a QTL object
replaceqtl	[273]	Replace a QTL location in a QTL object with a different position

reorderqtl	[274]	Reorder the QTL in a QTL object
cim	[209]	A (relatively crude) implementation of composite interval mapping
stepwiseqtl	[276]	Stepwise selection for multiple QTL
calc.penalties	[275]	Calculate penalties for use with stepwiseqtl
plotModel	[277]	Plot a graphical representation of a multiple-QTL model

C

QTL mapping data sets

Here we give brief descriptions of the various QTL mapping data sets considered in this book. These data are included either in R/qtl or in the R/qtlbook package.

`ch3a` (R/qtlbook)

Reference:	Anonymous
Organism:	Mouse
Cross type:	Backcross
No. individuals:	234
No. markers:	166

These are anonymous data used to illustrate the process some of the data diagnostics discussed in Chap. 3.

`ch3b` (R/qtlbook)

Reference:	Anonymous
Organism:	Mouse
Cross type:	Intercross
No. individuals:	144
No. markers:	145

These are anonymous data used to illustrate the process some of the data diagnostics discussed in Chap. 3.

`ch3c` (R/qtlbook)

Reference:	Anonymous
Organism:	Mouse
Cross type:	Intercross
No. individuals:	100
No. markers:	101

These are anonymous data used to illustrate the process some of the data diagnostics discussed in Chap. 3.

`gutlength` (R/qtlbook)

Reference:	Owens *et al.* (2005); Broman *et al.* (2006)
Organism:	Mouse
Cross type:	Intercross
No. individuals:	1068
No. markers:	121

These data are from a mouse intercross between C3HeBFeJ and C57BL/6J, with one F_1 parent carrying the $Sox10^{Dom}$ mutation. Over 2000 mice were generated, but only those individuals heterozygous at $Sox10^{Dom}$ were genotyped and included in the data set. *Sox10* is on chromosome 15, and so that chromosome exhibits an unusual segregation pattern. Some mice received the mutation from their mother and some from their father. The primary phenotype is gut length (in cm). The phenotype "cross" indicates the cross used to generate an animal. A selective genotyping strategy was used with these data: 323 individuals with extreme aganglionosis phenotype (not included with this data set) were genotyped at more than 100 markers; the remaining 745 individuals were typed at fewer than 15 markers.

`hyper` (R/qtl)

Reference:	Sugiyama *et al.* (2001)
Organism:	Mouse
Cross type:	Backcross
No. individuals:	250
No. markers:	174

These data are from a mouse backcross using the C57BL/6J and A/J strains, with the F_1 mated back to C57BL/6J. All individuals are male. They were given water containing 1% NaCl for two weeks. The phenotype is blood pressure (actually the average of 20 blood pressure measurements from each of 5 days, obtained with a tail cuff.) A selective genotyping strategy was used, with the the upper 46 and lower 46 individuals, in terms of blood pressure, genotyped across the entire genome. The other individuals were genotyped at markers in regions showing initial evidence for a QTL. In some regions, additional markers were added within an interval, but only recombinant individuals were genotyped.

`iron` (R/qtlbook)

Reference:	Grant *et al.* (2006)
Organism:	Mouse
Cross type:	Intercross
No. individuals:	284
No. markers:	66

These data are from a mouse intercross with the C57BL/7J/Ola and SWR/Ola strains. There are two phenotypes: basal iron levels (in μg/g) in the liver and spleen. A selective genotyping strategy was used. The markers genotyped are quite sparse.

`listeria` (R/qtl)

Reference:	Boyartchuk *et al.* (2001)
Organism:	Mouse
Cross type:	Backcross
No. individuals:	120
No. markers:	133

These data are from a mouse intercross using the C57BL/6ByJ and BALB/cBYJ strains. There are 120 female intercross individuals (though only 116 were phenotyped). Mice were injected with *Listeria monocytogenes*; the phenotype is survival time (in hours). A large proportion of mice (35/116) survived past the 240-hour time point and were considered to have recovered from the infection; their phenotype was recorded as 264. The first 90 individuals were genotyped relatively completely; an additional 30 individuals were genotyped at around 90 of the 133 markers.

`nf1` (R/qtlbook)

Reference:	Reilly *et al.* (2006)
Organism:	Mouse
Cross type:	Backcross
No. individuals:	254
No. markers:	105

These data concern neurofibromatosis type 1, and are from a mouse backcross generated to identify modifiers to the *NPcis* mutation. The backcrosses used the F_1 hybrid of C57BL/6J and A/J, crossed back to C57BL/6J, and with individuals receiving the *NPcis* mutation from either their mother or father. The phenotype is dichotomous and indicates whether individuals were affected or unaffected with neurofibromatosis type 1. The genotype data are about 86% complete.

`ovar` (R/qtlbook)

Reference:	Orgogozo *et al.* (2006)
Organism:	Fruit fly
Cross type:	Backcross
No. individuals:	1452
No. markers:	24

These data are from a cross between two Drosophila species: *D. simulans* was crossed to *D. sechellia*, and the F_1 hybrid was crossed back to *D. simulans*. The phenotype of interest was ovariole number in females, a measure of fitness. In an initial cross of 402 individuals, 383 had complete phenotype data. Initial genotyping focused on 94 individuals with extreme phenotype. To increase the resolution of a major QTL identified on chromosome 3, a second cross of approximately 7000 flies was performed, though only 1050 individuals showing a recombination event between two morphological markers were phenotyped and genotyped; 1038 individuals had complete phenotype data.

`trout` (R/qtlbook)

Reference:	Nichols *et al.* (2007)
Organism:	Rainbow trout
Cross type:	Doubled haploids
No. individuals:	554
No. markers:	171

These data are doubled haploid individuals derived from a cross between the Oregon State University and Clearwater River rainbow trout (*Oncorhynchus mykiss*) clonal lines. Eggs from one of eight outbred females, two from Troutlodge and six from the Spokane hatchery, were irradiated to destroy maternal nuclear DNA and fertilized with sperm from a single F_1 male. The first embryonic cleavage was blocked by heat shock to restore diploidy. There are a total of 554 individuals, with between 8 and 168 individuals from each of the eight females. The primary phenotype is time to hatch. The genotype data are 95% complete.

D

Hidden Markov models for QTL mapping

An important aspect of the QTL mapping problem is the treatment of missing genotype data. If complete genotype data were available, QTL mapping would reduce to the problem of model selection in linear regression. However, in the consideration of loci in the intervals between the available genetic markers, genotype data is inherently missing. Even at the typed genetic markers, genotype data is seldom complete, as a result of failures in the genotyping assays or for the sake of economy (e.g., in the case of selective genotyping, where only individuals with extreme phenotypes are genotyped).

In standard interval mapping, one deals with the missing QTL genotype data by performing maximum likelihood under a mixture model, using a version of the EM algorithm. Central to this approach is the calculation of the distribution of QTL genotypes conditional on the observed multipoint marker data. In the multiple imputation approach to QTL mapping, one must be able to simulate from the joint distribution of the genotypes at positions on a grid along a chromosome, conditional on the observed marker data.

In this chapter, we discuss the use of algorithms developed for hidden Markov models (HMMs) to perform the tasks mentioned above and thus deal with the missing genotype data problem. Simpler approaches can and have been used. For example, in a backcross in the absence of genotyping errors, the QTL genotype probabilities, conditional on the marker data, are a simple function of the genotypes at the nearest flanking markers. The more refined algorithms described here have several advantages. First, we may allow for the presence of genotyping errors. Second, we may more easily deal with partially informative genotypes. (For example, in an intercross, at some markers the heterozygote may not be distinguishable from one of the homozygotes.) Third, the bookkeeping tasks in implementing these algorithms can be more simple. Fourth, the algorithms can be more easily extended to more complex experimental crosses (such as a four-way cross).

In the next section, we define hidden Markov models in the context of the analysis of experimental crosses. In the following sections, we describe the basic algorithms for calculating QTL genotype probabilities, simulating from the joint distribution of QTL genotypes, estimating genetic maps, and identifying genotyping errors. We conclude the chapter with a discussion of a practical issue in the implementation of these algorithms in computer programs.

D.1 Specification of the model

A Markov chain is a collection of random variables, $\{G_1, G_2, \ldots, G_n\}$, satisfying the Markov property $\Pr(G_{i+1}|G_i, \ldots, G_1) = \Pr(G_{i+1}|G_i)$ for all i. In a Markov chain, for any i, the "past," $\{G_1, \ldots, G_{i-1}\}$, and the "future," $\{G_{i+1}, \ldots, G_n\}$, are conditionally independent, given the "present," G_i. We focus on Markov chains for which the random variables $\{G_i\}$ take values in a common, finite set, $\mathcal{G}$.

A hidden Markov model (HMM) consists of an unobservable underlying Markov chain, $\{G_i\}$, and a set of observable random variables, $\{O_i\}$, where each O_i depends only on G_i. In other words, for each i, O_i, given G_i, is conditionally independent of everything else, $\{O_1, \ldots, O_{i-1}, O_{i+1}, \ldots, O_n, G_1, \ldots, G_{i-1}, G_{i+1}, \ldots, G_n\}$. It may be useful to keep in mind the illustration in Figure D.1.

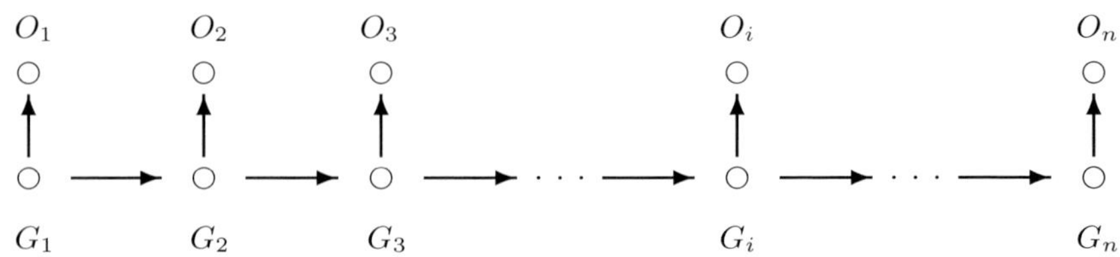

Figure D.1. Illustration of a hidden Markov model. G's indicate underlying genotypes; O's indicate observed marker phenotypes.

The hidden states, G_i, take values in a common, finite set, $\mathcal{G}$; the observed states, O_i, take values in another finite set, $\mathcal{O}$. The joint distribution of the O_i and G_i in the HMM is parameterized by three sets of probabilities, which we will call the initiation, transition and emission probabilities. The initiation probabilities define the distribution of the initial hidden state: $\pi(g) = \Pr(G_1 = g)$ for $g \in \mathcal{G}$. The transition probabilities complete the specification for the joint distribution of the underlying, hidden Markov chain: $t_i(g, g') = \Pr(G_{i+1} = g'|G_i = g)$ for $i = 1, \ldots, n-1$ and $g, g' \in \mathcal{G}$. The emission probabilities concern the conditional distribution of the observed states given the hidden states: $e_i(g, o) = \Pr(O_i = o|G_i = g)$ for $i = 1, \ldots, n$, $g \in \mathcal{G}$, and $o \in \mathcal{O}$. We will assume here that the emission probabilities are homogeneous, with $e_i(g, o) \equiv e(g, o)$ for all i, g, o.

We now begin to consider the application of HMMs to experimental crosses. Below, we will describe the backcross and intercross specifically, but first we define the relevant HMM in some generality.

One may focus on the genotypes for a single individual at a set of loci on a single chromosome. (We will focus on an autosome.) We let G_i, $i = 1, \ldots, n$, denote the true underlying genotypes for the individual at a set of n ordered loci, and let O_i denote the observed marker "phenotype" at locus i.

These loci may be genetic markers, or they may be "pseudomarkers," under consideration as putative QTL. The genotypes are often assumed to be phase-known genotypes, though for the intercross they need not be, as we will see below. Under the assumption of no crossover interference in meiosis, for many types of crosses, the G_i form a Markov chain. The set $\mathcal{G}$ corresponds to the possible values of these phase-known genotypes. The initiation probabilities correspond to a segregation model at a single locus; the transition probabilities are a function of the recombination fractions, r_i, between adjacent markers.

The set $\mathcal{O}$ corresponds to the set of possible observed marker phenotypes, which will include the possibility of missing values and partially informative phenotypes (such as in the case of a dominant or recessive marker). The emission probabilities involve a model for errors in genotyping, which we will assume to be common across markers, though in reality some markers are considerably more error-prone than others. It is important to point out, further, that one conditions on the observed pattern of missing data. This will become more clear below.

D.1.1 The backcross

Consider a backcross individual derived from two inbred strains, A and B, where the F_1 parent was crossed back to the A strain. We let $\mathcal{G} = \{AA, AB\}$, the possible genotypes at a locus. The set of possible marker phenotypes is $\mathcal{O} = \{A, H, -\}$, with the last symbol corresponding to a missing value. Note our attempt to use different symbols for the underlying genotypes and the observed marker phenotypes.

The initiation probabilities, assuming Mendel's rules, are simply $\pi(AA) = \pi(AB) = 1/2$. The transition probabilities are $t_i(AA, AB) = t_i(AB, AA) = r_i$, where r_i denotes the recombination fraction between loci i and $i+1$. Of course, $t_i(AA, AA) = t_i(AB, AB) = 1 - r_i$.

In forming the emission probabilities, we assume a constant error rate in genotyping, ϵ, so that $e(AA, A) = e(AB, H) = 1 - \epsilon$, and $e(AA, H) = e(AB, A) = \epsilon$. We condition on the observed pattern of missing data, and so $e(AA, -) = e(AB, -) = 1$. One may consider $- = \{A \text{ or } H\}$, so that $e(AA, -) = e(AA, A) + e(AA, H) = 1$.

One may consider, in forming the emission probabilities, more refined models for genotyping errors. For example, one may consider a locus-specific error rate, and one may allow the chance of a heterozygote being erroneously observed as a homozygote to be somewhat different than the converse.

Table D.1. The transition probabilities, $t_i(g, g') = \Pr(G_{i+1} = g'|G_i = g)$, for a phase-unknown intercross.

	g'		
g	AA	AB	BB
AA	$(1-r_i)^2$	$2r_i(1-r_i)$	r_i^2
AB	$r_i(1-r_i)$	$(1-r_i)^2 + r_i^2$	$r_i(1-r_i)$
BB	r_i^2	$2r_i(1-r_i)$	$(1-r_i)^2$

D.1.2 The intercross

Consider a single individual in the F_2 generation from an intercross between two inbred strains, A and B. One may consider the hidden states, G_i, to be either phase-known genotypes (with four possible states, $\{AA, AB, BA, BB\}$) or phase-unknown genotypes (with three possible states, $\{AA, AB, BB\}$). It is an interesting and useful fact that in either case the G_i form a Markov chain (under the assumption of no crossover interference).

We will focus on the phase-unknown case, with $\mathcal{G} = \{AA, AB, BB\}$. The initiation probabilities are again those implied by Mendel's rules: $\pi(AA) = \pi(BB) = 1/4$, $\pi(AB) = 1/2$. The transition probabilities are displayed in Table D.1, where r_i denotes the recombination fraction between markers i and $i+1$. Note that we assume that there are no sex differences in the recombination fractions.

As possible observed marker phenotypes, we let $\mathcal{O} = \{A, H, B, C, D, -\}$, with A, B, and H corresponding to the two homozygotes and the heterozygote, respectively, $-$ corresponding to a completely missing value, and with C and D allowing the treatment of dominant marker loci: we define C and D as in the popular computer software, MapMaker (Lander *et al.*, 1987), with $C = \{\text{not } A\} = \{B \text{ or } H\}$ and $D = \{\text{not } B\} = \{A \text{ or } H\}$.

The emission probabilities, for a simple genotyping error model, are shown in Table D.2, where we let ϵ denote the genotyping error rate. Note that we again condition on the pattern of missing genotype data, and so, for example, $\Pr(O_i = C|G_i) = \Pr(O_i = B|G_i) + \Pr(O_i = H|G_i)$.

D.2 QTL genotype probabilities

Having set up the hidden Markov model for experimental crosses, we now begin our discussion of the basic algorithms used in order to deal with missing genotype data in QTL mapping. We begin with the calculation of the conditional QTL genotype probabilities given multipoint marker data, which are critical for standard interval mapping with a single-QTL model. Using

Table D.2. The emission probabilities, $e(g, o) = \Pr(O_i = o|G_i = g)$, for a phase-unknown intercross.

	o					
g	A	H	B	C	D	$-$
AA	$1-\epsilon$	$\epsilon/2$	$\epsilon/2$	$\epsilon/2$	$1-\epsilon/2$	1
AB	$\epsilon/2$	$1-\epsilon$	$\epsilon/2$	$1-\epsilon/2$	$1-\epsilon/2$	1
BB	$\epsilon/2$	$\epsilon/2$	$1-\epsilon$	$1-\epsilon/2$	$\epsilon/2$	1

the notation developed in the previous section, we seek $\Pr(G_i = g|\boldsymbol{O})$, where $\boldsymbol{O} = (O_1, \ldots, O_n)$.

The brute-force approach for calculating this probability is to simply sum over all possible genotypes at the other loci.

$$\begin{aligned}\Pr(G_i = g_i|\boldsymbol{O}) &= \sum_{g_1} \cdots \sum_{g_{i-1}} \sum_{g_{i+1}} \cdots \sum_{g_n} \Pr(G_1 = g_1, \ldots, G_n = g_n|\boldsymbol{O}) \\ &\propto \sum_{g_1} \cdots \sum_{g_{i-1}} \sum_{g_{i+1}} \cdots \sum_{g_n} \pi(g_1) \prod_{j=1}^{n-1} t_j(g_j, g_{j+1}) \prod_{j=1}^{n} e(g_j, O_j)\end{aligned}$$

For the phase-known intercross, with three possible genotypes, this is a sum with 3^{n-1} terms; clearly this is unwieldy and unnecessary. That there are simple algorithms for this calculation, which make critical use of the conditional independence structure of the HMM, is the primary motivation for the use of HMMs in experimental crosses.

The approach we follow makes use of the following two sets of probabilities.

$$\begin{aligned}\alpha_i(g) &= \Pr(O_1, \ldots, O_i, G_i = g) \\ \beta_i(g) &= \Pr(O_{i+1}, \ldots, O_n|G_i = g)\end{aligned}$$

Note that, once the α's and β's have been calculated, the probability that is the focus of this section follows directly:

$$\begin{aligned}\Pr(G_i = g|\boldsymbol{O}) &= \Pr(G_i = g, \boldsymbol{O}) / \Pr(\boldsymbol{O}) \\ &= \alpha_i(g)\beta_i(g) / \textstyle\sum_{g'} \alpha_i(g')\beta_i(g').\end{aligned}$$

The α's and β's are calculated inductively, using what are called the forward and backward equations, respectively. We begin with the forward equations. First, note that

$$\alpha_1(g) = \Pr(O_1, G_1 = g) = \pi(g)\, e(g, O_1).$$

Now, assume that we have calculated $\alpha_i(g)$ for each $g \in \mathcal{G}$. Then we have

$$\begin{aligned}\alpha_{i+1}(g) &= \Pr(O_1,\ldots,O_i,O_{i+1},G_{i+1}=g)\\ &= \textstyle\sum_{g'} \Pr(O_1,\ldots,O_i,O_{i+1},G_i=g',G_{i+1}=g)\\ &= \textstyle\sum_{g'} \Pr(O_1,\ldots,O_i,G_i=g')\Pr(G_{i+1}=g|G_i=g')\Pr(O_{i+1}|G_{i+1}=g)\\ &= e(g,O_{i+1})\textstyle\sum_{g'} \alpha_i(g')\, t_i(g',g).\end{aligned}$$

In the third line above, we made use of the conditional independence structure of the HMM, noting that

$$\Pr(G_{i+1}=g|G_i=g',O_1,\ldots,O_i) = \Pr(G_{i+1}=g|G_i=g')$$

and

$$\Pr(O_{i+1}|G_{i+1}=g,G_i=g',O_1,\ldots,O_i) = \Pr(O_{i+1}|G_{i+1}=g).$$

Calculation of the β's proceeds similarly, though starting at the other end of the chain. We define $\beta_n(g) = 1$ for all $g \in \mathcal{G}$. Assuming that we have calculated $\beta_i(g)$ for all g, we have

$$\begin{aligned}\beta_{i-1}(g) &= \Pr(O_i,\ldots,O_n|G_{i-1}=g)\\ &= \textstyle\sum_{g'} \Pr(O_i,\ldots,O_n,G_i=g'|G_{i-1}=g)\\ &= \textstyle\sum_{g'} \Pr(O_{i+1},\ldots O_n|G_i=g')\Pr(G_i=g'|G_{i-1}=g)\Pr(O_i|G_i=g')\\ &= \textstyle\sum_{g'} \beta_i(g')\, t_{i-1}(g,g')\, e(g',O_i).\end{aligned}$$

Again, in the third line above, we made use of the conditional independence structure of the HMM.

In summary, in order to calculate the QTL genotype probabilities, conditional on multipoint marker data, $\Pr(G_i = g|\boldsymbol{O})$, we make use of the forward and backward equations to first calculate, for each i and g, $\alpha_i(g) = \Pr(O_1,\ldots,O_i,G_i = g)$ and $\beta_i(g) = \Pr(O_{i+1},\ldots,O_n|G_i = g)$. These algorithms are extremely efficient and can accommodate partially missing genotypes (such as are observed at dominant markers in an intercross) and a model for errors in genotyping.

D.3 Simulation of QTL genotypes

Central to the multiple imputation approach to QTL mapping is the simulation of QTL genotypes via their joint distribution conditional on the observed multipoint marker data. In this section, we describe how this is done. One considers a single chromosome and a single individual at a time. As will be seen, the simulation algorithm makes use of the β's defined in the previous section. Thus, one must first perform the backward equations described above.

The algorithm is quite simple. One first draws $g_1^\star$ from the distribution

$$\Pr(G_1 = g|\boldsymbol{O}) = \frac{\alpha_1(g)\beta_1(g)}{\sum_{g'} \alpha_1(g')\beta_1(g')}.$$

Genotypes for further loci are drawn iteratively: having drawn $g_1^\star, \dots, g_i^\star$, one draws $g_{i+1}^\star$ from the distribution

$$\begin{aligned}\Pr(G_{i+1} = g|\boldsymbol{O}, G_i = g_i^\star) &= \frac{\Pr(G_{i+1} = g, G_i = g_i^\star|\boldsymbol{O})}{\Pr(G_i = g_i^\star|\boldsymbol{O})} \\ &= \frac{\alpha_i(g_i^\star)\, t_i(g_i^\star, g)\, e(g, O_{i+1})\, \beta_{i+1}(g)}{\alpha_i(g_i^\star)\beta_i(g_i^\star)} \\ &= \frac{t_i(g_i^\star, g)\, e(g, O_{i+1})\, \beta_{i+1}(g)}{\beta_i(g_i^\star)}.\end{aligned}$$

We are again making critical use of the conditional independence structure of the HMM.

Note that the α's are not needed, except for $\alpha_1(g) = \pi(g)\, e(g, O_1)$. Thus the forward equations need not be performed. For each individual, one first uses the backward equations to calculate the β's and then simulates the chain from left to right, using the equations above. It should be no surprise that one may instead use the forward equations to calculate the α's, and then simulate the chain from right to left, using formulas analogous to those above.

D.4 Joint QTL genotype probabilities

In multiple interval mapping (MIM) with multiple linked QTL, it is important to calculate joint QTL genotype probabilities, conditional on the observed multipoint marker data.

We begin by describing the calculation of $\Pr(G_i = g, G_j = g'|\boldsymbol{O})$ for all i, j with $i < j$. As will be seen, one must first calculate the α's and β's defined above. One may start by calculating the case $j = i+1$ for each $i = 1, \dots, n-1$, as follows.

$$\begin{aligned}\Pr(G_i = g, G_{i+1} = g'|\boldsymbol{O}) &\propto \Pr(G_i = g, G_{i+1} = g', \boldsymbol{O}) \\ &= \Pr(O_1, \dots, O_i, G_i = g)\Pr(G_{i+1} = g'|G_i = g) \\ &\qquad \times \Pr(O_{i+1}|G_{i+1} = g')\Pr(O_{i+2}, \dots, O_n|G_{i+1} = g') \\ &= \alpha_i(g)\, t_i(g, g')\, e(g', O_{i+1})\, \beta_{i+1}(g')\end{aligned}$$

One uses the final line above and rescales the results so that they sum to 1.

The rest of the pairwise probabilities follow with the standard technique, using induction.

$$\begin{aligned}\Pr(G_i = g, G_j = g'|\boldsymbol{O}) &= \sum_{g''} \Pr(G_i = g, G_{j-1} = g'', G_j = g'|\boldsymbol{O}) \\ &= \sum_{g''} \Pr(G_i = g, G_{j-1} = g''|\boldsymbol{O})\Pr(G_j = g'|G_{j-1} = g'', \boldsymbol{O})\end{aligned}$$

Finally, one may wish to calculate the joint probabilities for multiple linked loci, conditional on the observed multipoint marker data. Again, the conditional independence structure of the HMM makes this a simple task: the joint distribution may be calculated based on pairwise probabilities whose calculation was described above. Consider $i_1 < i_2 < \cdots < i_k$, with each $i_j \in \{1, \ldots, n\}$; we have

$$\Pr(G_{i_1} = g_1, \ldots, G_{i_k} = g_k|\boldsymbol{O}) = \Pr(G_{i_1} = g_1, G_{i_2} = g_2|\boldsymbol{O}) \prod_{j=2}^{k-1} \Pr(G_{i_{j+1}} = g_{j+1}|G_{i_j} = g_j, \boldsymbol{O}).$$

The equations in this section do get a little bit complicated, but they are all formed of quite simple pieces. The central calculation is the use of the forward and backward equations to obtain the α's and β's.

D.5 The Viterbi algorithm

In some cases, it is useful to impute the underlying genotype data, calculating $\hat{\boldsymbol{G}} = \arg\max_{\boldsymbol{G}} \Pr(\boldsymbol{G}|\boldsymbol{O})$. The Viterbi algorithm solves this problem via dynamic programming.

First, define

$$\gamma_k(g) = \max_{g_1,\ldots,g_{k-1}} \Pr(G_1 = g_1, \ldots, G_{k-1} = g_{k-1}, G_k = g, O_1, \ldots, O_k).$$

These are calculated inductively, by an approach similar to that used in the forward equations (Sec. D.2). Let $\gamma_1(g) = \Pr(G_1 = g, O_1) = \pi(g)e(g, O_1)$. Given $\gamma_k(g)$ for all g, we have

$$\gamma_{k+1}(g) = e(g, O_{k+1}) \max_{g'} t_k(g', g)\gamma_k(g').$$

At the same time, we keep track of the values at which the maxima occurred: define $\delta_{k+1}(g) = \arg\max_{g'} t_k(g', g)\gamma_k(g')$. If the maximum is not unique, we can keep track of each of them or pick a random one. (We do the latter in R/qtl.)

To obtain the most probable sequence of underlying genotypes, we then take $\hat{G}_n = \arg\max_g \gamma_n(g)$ and, working backwards, $\hat{G}_{k-1} = \delta_k(\hat{G}_k)$.

The inferred genotypes obtained by the Viterbi algorithm should be used with great caution. If one treated the inferred genotypes as if they were the true values, an important source of uncertainty would be ignored.

Moreover, if intermarker positions are included and genotyping error is allowed, the results of the Viterbi algorithm can vary according to the density of intermarker positions that are used. The Viterbi algorithm identifies the most likely sequence of genotypes, but this sequence may have quite low probability and may exhibit features that are unlikely.

For example, consider three markers at a 10 cM spacing and a single backcross individual with observed marker genotypes AA–AB–AB at the three markers. If the Viterbi algorithm is applied with a genotyping error rate of

1%, and using just the three marker positions, the most likely sequence of underlying genotypes matches those observed. If, however, one considers positions at 1 cM steps across the region, the most likely sequence of underlying genotypes is such that the individual is heterozygous across the entire region. While it is probable that the individual is recombinant across the first interval and that the observed genotype at the first marker is not in error, if many intermarker positions are considered, this event is split across multiple sequences of genotypes (each corresponding to a different position for the recombination event), and so the sequence in which the initial genotype is in error and there is no recombination event ends up being most likely.

This issue leads us to recommend the use of simulation to impute genotypes (as described in Sec. D.3), rather than using the Viterbi algorithm to calculate the most probable sequence of underlying genotypes.

D.6 Estimation of intermarker distances

The calculations described above depend crucially on the order of the genetic markers and the recombination fractions between adjacent markers (i.e., the intermarker distances). In this section, we describe the derivation of joint maximum likelihood estimates (MLEs) of the recombination fractions between genetic markers, assuming that the order of the genetic markers is known. We omit from consideration the more difficult problem of determining marker order.

Taking the order of the genetic markers as fixed and known, the probability of the observed marker data for an individual, $\Pr(\boldsymbol{O})$, still depends on the recombination fractions between adjacent markers. For the sake of simplicity, this dependence has been neglected in our notation heretofore. Moreover, we have been considering a single individual at a time. In our discussion of the estimation of intermarker distances, however, it will be important to make this dependence clear. Let $\boldsymbol{r} = (r_1, \ldots, r_{n-1})$ denote the set of recombination fractions, and let $\boldsymbol{O}_k$ denote the observed marker data for individual k, for $k = 1, \ldots, N$.

We seek the MLE of $\boldsymbol{r}$, defined to be the value of $\boldsymbol{r}$ for which the likelihood is maximized, $\hat{\boldsymbol{r}} = \arg\max L(r)$, where $L(r) = \prod_{k=1}^{N} \Pr(\boldsymbol{O}_k|\boldsymbol{r})$. These estimates are obtained using a version of the EM algorithm (Dempster *et al.*, 1977).

We begin with initial estimates of the recombination fractions, $\hat{\boldsymbol{r}}^{(0)}$. The EM algorithm is an iterative algorithm: the estimated recombination fractions are successively improved, increasing the likelihood at each stage, until convergence. In each iteration, the updated estimates of the recombination fractions are the expected proportions of recombination events, across the N individuals, in each marker interval, given the current estimates of the recombination fractions.

At each iteration, we first perform the forward and backward equations for each individual, using the current estimates of the recombination fractions, $\hat{\boldsymbol{r}}^{(s)}$. We then calculate, for each interval i, $\gamma_{ki}(g, g'|\hat{\boldsymbol{r}}^{(s)}) = \Pr(G_{k,i} = g, G_{k,i+1} = g'|\boldsymbol{O}_k, \hat{\boldsymbol{r}}^{(s)})$. This is the probability that individual k has genotypes g and g' at markers i and $i+1$, given its multipoint marker data, and given the current estimates of the recombination fractions. The calculation of the γ's, based on the α's and β's for the corresponding individual, appears in Sec. D.4.

The updated estimate of the recombination fraction for interval i is then $\hat{r}_i^{(s+1)} = \sum_k \sum_{g,g'} \gamma_{ki}(g, g'|\hat{\boldsymbol{r}}^{(s)})\ p(g, g')/N$, where $p(g, g')$ is the proportion of recombination events across the interval (i.e., 0, 1/2, or 1) if the individual has genotypes g and g' at the markers defining the interval. Note that, in estimating the intermarker distances for an intercross, we use the phase-known (four-state) version of the HMM, so that the function $p(g, g')$ is well defined.

D.7 Detection of genotyping errors

Successful QTL mapping requires high-quality phenotype and genotype data. In this section, we describe an approach for identifying errors in the genotype data. For each marker and each individual, we calculate a LOD score, with large LOD scores indicating likely errors.

The presence of partially informative genotypes (e.g., at dominant markers in an intercross) makes this slightly tricky. Let us assume that the observed marker phenotypes, $o \in \mathcal{O}$, are subsets of the possible underlying marker genotypes, $\mathcal{G}$. For example, in the case of an intercross, where $\mathcal{G} = \{AA, AB, BB\}$, the set of possible marker phenotypes is $\mathcal{O} = \{A, H, B, C, D, -\}$, with, for example, $A = \{AA\}$ and $C = \{AB, BB\}$.

Let G_{ki} denote the true underlying genotype for individual k at marker i, and let O_{ki} denote the corresponding marker phenotype. We assume the simple model for genotyping errors that was described in Sec. D.1, and we assume the genotyping error rate, ϵ, is known. We seek to calculate

$$\begin{aligned}\mathrm{LOD}_{ki} &= \log_{10}\left\{\frac{\Pr(\boldsymbol{O}|G_{ki} \notin O_{ki}, \epsilon)}{\Pr(\boldsymbol{O}|G_{ki} \in O_{ki}, \epsilon)}\right\} \\ &= \log_{10}\left\{\frac{\Pr(G_{ki} \notin O_{ki}|\boldsymbol{O}, \epsilon)}{\Pr(G_{ki} \in O_{ki}|\boldsymbol{O}, \epsilon)} \times \frac{1-\epsilon}{\epsilon}\right\}\end{aligned}$$

The calculation of the probabilities in the above formula are by the forward and backward equations, described in Sec. D.2. While the calculations might be done allowing a constant genotyping error rate for all markers, we have found that, in the case of an apparent triple-recombination event, no individual genotype will be flagged as a likely error. We have found it best to instead perform the forward and backward equations separately for each individual and each marker, in each instance allowing only the genotype in question to be possibly in error; all other marker genotypes are assumed to be correct. While the computation time with this approach is greatly increased (so that it is probably not feasible for a very large number of markers and individuals), a broader set of possibly genotyping errors will be identified. Note that, while the genotyping error LOD scores depend on the specified error rate, ϵ, typical values, in the range 0.001 – 0.02, give indistinguisable results.

Genotyping error LOD scores below 3 or 4 are generally benign. Only when the LOD scores exceed 4 should they be given much consideration. It should be noted that genotyping errors can only be detected in the case of quite dense markers. At the same time, however, genotyping errors have little effect on the results of QTL mapping if the markers are not dense. Finally, if a particular marker gives many large error LOD scores, it may be that a problem with marker order is the cause (though, of course, the marker may also have a greater than typical frequency of errors).

D.8 A practical issue

In the case of many genetic markers (or of calculations on a dense grid), the direct calculation of α and β, as described above, will result in underflow: $\alpha_n(v) = \Pr(O_1, \ldots, O_n, G_n = v)$ can be extremely small. One method to deal with this is to calculate $\alpha' = \log \alpha$ and $\beta' = \log \beta$. In the forward equations, we must obtain $\alpha'_{i+1}(g) = \log e(g, O_{i+1}) + \log\{\sum_{g'} \alpha_i(g')t_i(g', g)\}$. This leads to the problem of calculating $\log(f_1 + f_2)$ on the basis of $g_i = \log f_i$, which may be facilitated by the following trick:

$$\begin{aligned}
\log(f_1 + f_2) &= \log(e^{g_1} + e^{g_2}) \\
&= \log\{e^{g_1}(1 + e^{g_2 - g_1})\} \\
&= g_1 + \log(1 + e^{g_2 - g_1})
\end{aligned}$$

A problem occurs when $g_2 \gg g_1$: the above formula will result in an overflow. In such a case one simply notes that $\log(f_1 + f_2) \approx g_2$.

D.9 Further reading

Baum *et al.* (1970) were the first to describe estimation for hidden Markov models, and derived the forward and backward equations. For other expositions of the use of HMMs, see Rabiner (1989) or Lange (1999, Sec. 23.3).

Churchill (1989) was the first to use HMMs explicitly in biology. HMMs have been used for a variety of biological applications, including the study of ion channels (Fredkin and Rice, 2001), multiple sequence alignment (Krogh *et al.*, 1994; Baldi *et al.*, 1994), gene finding (Henderson *et al.*, 1997), and protein structure prediction (Hubbard and Park, 1995).

Lander and Green (1987) described the multipoint estimation of genetic maps; their method was implemented for experimental crosses in the software MapMaker (Lander *et al.*, 1987). Jiang and Zeng (1997) described an alternative approach for dealing with missing and partially missing genotype data. Lincoln and Lander (1992) developed the LOD scores, defined above, for identifying genotyping errors in experimental crosses. Cartwright *et al.* (2007) described the estimation of a genetic map, allowing the genotyping error rate at markers to vary.

References

Ahmadiyeh N, Churchill GA, Shimomura K, Solberg LC, Takahashi JS, Redei EE (2003) X-linked and lineage-dependent inheritance of coping responses to stress. *Mamm. Genome*, **14**:748–757.

Baierl A, Bogdan M, Frommlet F, Futschik A (2006) On locating multiple interacting quantitative trait loci in intercross designs. *Genetics*, **173**:1693–1703.

Baldi P, Chauvin Y, Hunkapiller T, McClure MA (1994) Hidden Markov models of biological primary sequence information. *Proc. Natl. Acad. Sci. USA*, **91**:1059–1063.

Basten CJ, Weir BS, Zeng ZB (2002) QTL Cartographer: A reference manual and tutorial for QTL mapping. Program in Statistical Genetics, Bioinformatics Research Center, Department of Biostatistics, North Carolina State University.

Baum LE, Petrie T, Soules G, Weiss N (1970) A maximization technique occurring in the statistical analysis of probabilistic functions of Markov chains. *Ann. Math. Stat.*, **41**:164–171.

Beavis WD (1994) The power and deceit of QTL experiments: Lessons from comparative QTL studies. In Wilkinson DB, editor, *49th Annual Corn and Sorghum Research Conference*, pages 252–268, American Seed Trade Association, Washington, DC.

Belknap JK (1998) Effect of within-strain sample size on QTL detection and mapping using recombinant inbred mouse strains. *Behav. Genet.*, **28**:29–38.

Bogdan M, Ghosh JK, Doerge RW (2004) Modifying the Schwartz Bayesian Information Criterion to locate multiple interacting quantitative trait loci. *Genetics*, **167**:989–999.

Boyartchuk VL, Broman KW, Mosher RE, D'Orazio SEF, Starnbach MN, Dietrich WF (2001) Multigenic control of *Listeria monocytogenes* susceptibility in mice. *Nat. Genet.*, **27**:259–260.

Broman KW (1999) Cleaning genotype data. *Genet. Epidemiol.*, **17**(Suppl 1):S79–83.

Broman KW (2001) Review of statistical methods for QTL mapping in experimental crosses. *Lab Animal*, **30**(7):44–52.

Broman KW (2003) Mapping quantitative trait loci in the case of a spike in the phenotype distribution. *Genetics*, **163**:1169–1175.

Broman KW, Heath SC (2007) Managing and manipulating genetic data. In Barnes MR, Gray IC, editors, *Bioinformatics for Geneticists*, pages 17–31, Wiley, New York, 2nd edition.

Broman KW, Rowe LB, Churchill GA, Paigen K (2002) Crossover interference in the mouse. *Genetics*, **160**:1123–1131.

Broman KW, Sen S, Owens SE, Manichaikul A, Southard-Smith EM, Churchill GA (2006) The X chromosome in quantitative trait locus mapping. *Genetics*, **174**:2151–2158.

Broman KW, Speed TP (1999) A review of methods for identifying QTLs in experimental crosses. In Seillier-Moisenwitsch F, editor, *Statistics in Molecular Biology and Genetics*, volume 33 of *IMS Lecture Notes - Monograph Series*, pages 114–142, Institute of Mathematical Statistics, Hayward, CA.

Broman KW, Speed TP (2002) A model selection approach for the identification of quantitative trait loci in experimental crosses (with discussion). *J. R. Statist. Soc. B*, **64**:641–656, 737–775.

Broman KW, Wu H, Sen S, Churchill GA (2003) R/qtl: QTL mapping in experimental crosses. *Bioinformatics*, **19**:889–890.

Brown T (2006) *Genomes 3*. Wiley, New York.

Cartwright DA, Troggio M, Velasco R, Gutin A (2007) Genetic mapping in the presence of genotyping errors. *Genetics*, **176**:2521–2527.

Chen Z (2005) The full EM algorithm for the MLEs of QTL effects and positions and their estimated variances in multiple-interval mapping. *Biometrics*, **61**:474–480.

Christiansen T, Torkington N (2003) *Perl Cookbook*. O'Reilly Media, Sebastopol, CA, 2nd edition.

Churchill GA (1989) Stochastic models for heterogeneous DNA sequences. *B. Math. Biol.*, **51**:79–94.

Churchill GA, Doerge RW (1994) Empirical threshold values for quantitative trait mapping. *Genetics*, **138**:963–971.

Copenhaver GP, Housworth EA, Stahl FW (2002) Crossover interference in arabidopsis. *Genetics*, **160**:1631–1639.

Cox DR (1972) Regression models and life tables. *J. Roy. Stat. Soc. B*, **34**:187–220.

Cox DR, Hinkley DV (1974) *Theoretical Statistics*. Chapman and Hall, London.

Dalgaard P (2002) *Introductory Statistics with R*. Springer, New York.

Darvasi A (1998) Experimental strategies for the genetic dissection of complex traits in animal models. *Nat. Genet.*, **18**:19–24.

Darvasi A, Soller M (1992) Selective genotyping for determination of linkage between a marker locus and a quantitative trait locus. *Theor. Appl. Genet.*, **85**:353–359.

Dempster AP, Laird NM, Rubin DB (1977) Maximum likelihood from incomplete data via the EM algorithm (with discussion). *J. Roy. Stat. Soc. B*, **39**:1–38.

Diao G, Lin DY (2005) Semiparametric methods for mapping quantitative trait loci with censored data. *Biometrics*, **61**:789–798.

Diao G, Lin DY, Zou F (2004) Mapping quantitative trait loci with censored observations. *Genetics*, **168**:1689–1698.

Doerge RW, Churchill GA (1996) Permutation tests for multiple loci affecting a quantitative character. *Genetics*, **142**:285–294.

Doerge RW, Zeng ZB, Weir BS (1997) Statistical issues in the search for genes affecting quantitative traits in experimental populations. *Stat. Sci.*, **12**:195–219.

Draper NR, Smith H (1998) *Applied Regression Analysis*. Wiley, New York, 3rd edition.

Dupuis J, Siegmund D (1999) Statistical methods for mapping quantitative trait loci from a dense set of markers. *Genetics*, **151**:373–386.

Falconer DS, Mackay TFC (1996) *Introduction to Quantitative Genetics*. Prentice-Hall, Harlow, 4th edition.

Feenstra B, Skovgaard IM, Broman KW (2006) Mapping quantitative trait loci by an extension of the Haley–Knott regression method using estimating equations. *Genetics*, **173**:2269–2282.

Flint J, Valdar W, Shifman S, Mott R (2005) Strategies for mapping and cloning quantitative trait genes in rodents. *Nat. Rev. Genet.*, **6**:271–286.

Fredkin DR, Rice JA (2001) Fast evaluation of the likelihood of an HMM: Ion channel currents with filtering and colored noise. *IEEE Trans. Signal Process.*, **49**:625–633.

Gonick L, Smith W (1993) *The Cartoon Guide to Statistics*. HarperCollins, New York.

Gonick L, Wheelis M (1991) *The Cartoon Guide to Genetics*. HarperCollins, New York.

Grant GG, Robinson SW, Edwards RE, Clothier B, Davies RB, Judah DJ, Broman KW, Smith AC (2006) Multiple polymorphic genes determine "normal" hepatic and splenic iron status in mice. *Hepatology*, **44**:174–185.

Hackett CA, Weller JI (1995) Genetic mapping of quantitative trait loci for traits with ordinal distributions. *Biometrics*, **51**:1252–1263.

Haley CS, Knott SA (1992) A simple regression method for mapping quantitative trait loci in line crosses using flanking markers. *Heredity*, **69**:315–324.

Hastie T, Tibshirani R, Friedman J (2009) *The Elements of Statistical Learning: Data Mining, Inference, and Prediction*. Springer, New York, 2nd edition.

Henderson J, Salzberg SL, Fasman K (1997) Finding genes in human DNA with a hidden Markov model. *J. Comput. Biol.*, **4**:127–141.

Hubbard TJ, Park J (1995) Fold recognition and ab initio structure predictions using hidden Markov models and β-strand pair potentials. *Proteins*, **23**:398–402.

Ihaka R, Gentleman R (1996) R: A language for data analysis and graphics. *J. Comput. Graph. Stat.*, **5**:299–314.

Jansen RC (1992) A general mixture model for mapping quantitative trait loci by using molecular markers. *Theor. Appl. Genet.*, **85**:252–260.

Jansen RC (1993a) Interval mapping of multiple quantitative trait loci. *Genetics*, **135**:205–211.

Jansen RC (1993b) Maximum likelihood in a generalized linear finite mixture model by using the EM algorithm. *Biometrics*, **49**:227–231.

Jansen RC (2007) Quantitative trait loci in inbred lines. In Balding DJ, Bishop M, Cannings C, editors, *Handbook of Statistical Genetics*, volume 1, pages 589–622, Wiley, Chichester, 3rd edition.

Jansen RC, Stam P (1994) High resolution of quantitative traits into multiple loci via interval mapping. *Genetics*, **136**:1447–1455.

Jiang C, Zeng ZB (1995) Multiple trait analysis of genetic mapping for quantitative trait loci. *Genetics*, **140**:1111–1127.

Jiang C, Zeng ZB (1997) Mapping quantitative trait loci with dominant and missing markers in various crosses from two inbred lines. *Genetica*, **101**:47–58.

Jin C, Lan H, Attie AD, Churchill GA, Yandell BS (2004) Selective phenotyping for increased efficiency in genetic mapping studies. *Genetics*, **168**:2285–2293.

Johnson KR, Wright JE Jr, May B (1987) Linkage relationships reflecting ancestral tetraploidy in salmonid fish. *Genetics*, **116**:579–591.

Kao CH, Zeng ZB (1997) General formulas for obtaining the MLEs and the asymptotic variance-covariance matrix in mapping quantitative trait loci when using the EM algorithm. *Biometrics*, **53**:653–665.

Kao CH, Zeng ZB, Teasdale RD (1999) Multiple interval mapping for quantitative trait loci. *Genetics*, **152**:1203–1216.

Krogh A, Brown M, Mian IS, Sjölander K, Haussler D (1994) Hidden Markov models in computational biology: Applications to protein modeling. *J. Mol. Biol.*, **235**:1501–1531.

Kruglyak L, Lander ES (1995) A nonparametric approach for mapping quantitative trait loci. *Genetics*, **139**:1421–1428.

Lander ES, Botstein D (1989) Mapping Mendelian factors underlying quantitative traits using RFLP linkage maps. *Genetics*, **121**:185–199.

Lander ES, Green P (1987) Construction of multilocus genetic linkage maps in humans. *Proc. Natl. Acad. Sci. USA*, **84**:2363–2367.

Lander ES, Green P, Abrahamson J, Barlow A, Daly MJ, Lincoln SE, Newburg L (1987) MAPMAKER: an interactive computer package for constructing primary genetic linkage maps of experimental and natural populations. *Genomics*, **1**:174–181.

Lange K (1999) *Numerical Analysis for Statisticians*. Springer, New York.

Lincoln SE, Lander ES (1992) Systematic detection of errors in genetic linkage data. *Genomics*, **14**:604–610.

Liu BH (1998) *Statistical Genomics: Linkage, Mapping, and QTL Analysis*. CRC Press, Boca Raton, FL.

Ljungberg K, Holmgren S, Carlborg O (2002) Efficient algorithms for quantitative trait loci mapping problems. *J. Comput. Biol.*, **9**:793–804.

Ljungberg K, Holmgren S, Carlborg O (2004) Simultaneous search for multiple QTL using the global optimization algorithm DIRECT. *Bioinformatics*, **20**:1887–1895.

London SJ, Colditz GA, Stampfer MJ, Willett WC, Rosner B, Speizer FE (1989) Prospective-study of relative weight, height, and risk of breast-cancer. *J. Am. Med. Asso.*, **262**:2853–2858.

Lynch M, Walsh B (1998) *Genetics and Analysis of Quantitative Traits*. Sinauer, Sunderland, MA.

Macdonald SJ, Goldstein DB (1999) A quantitative genetic analysis of male sexual traits distinguishing the sibling species *Drosophila simulans* and *D. sechellia*. *Genetics*, **153**:1683–1699.

Manichaikul A, Dupuis J, Sen S, Broman KW (2006) Poor performance of bootstrap confidence intervals for the location of a quantitative trait locus. *Genetics*, **174**:481–489.

Manichaikul A, Moon JY, Sen S, Yandell BS, Broman KW (2009) A model selection approach for the identification of quantitative trait loci in experimental crosses, allowing epistasis. *Genetics*, **181**:1077–1086.

Manichaikul A, Palmer AA, Sen S, Broman KW (2007) Significance thresholds for quantitative trait locus mapping under selective genotyping. *Genetics*, **177**:1963–1966.

Manly KF, Cudmore RH Jr, Meer JM (2001) Map Manager QTX, cross-platform software for genetic mapping. *Mamm. Genome*, **12**:930–932.

Martínez O, Curnow RN (1992) Estimating the locations and the sizes of the effects of quantitative trait loci using flanking markers. *Theor. Appl. Genet.*, **85**:480–488.

McIntyre LM, Coffman CJ, Doerge RW (2001) Detection and localization of a single binary trait locus in experimental populations. *Genet. Res.*, **78**:79–92.

McPeek MS (1996) An introduction to recombination and linkage analysis. In Speed T, Waterman MS, editors, *Genetic Mapping and DNA sequencing*, volume 81 of *IMA Volumes in Mathematics and Its Applications*, pages 1–14, Springer, New York.

McPeek MS, Speed TP (1995) Modeling interference in genetic recombination. *Genetics*, **139**:1031–1044.

Meng XL, Rubin DB (1993) Maximum likelihood estimation via the ECM algorithm: A general framework. *Biometrika*, **80**:267–278.

Miller A (2002) *Subset Selection in Regression*. Chapman & Hall/CRC, Boca Raton, FL, 2nd edition.

Moreno CR, Elsen JM, Le Roy P, Ducrocq V (2005) Interval mapping methods for detecting QTL affecting survival and time-to-event phenotypes. *Genet. Res.*, **85**:139–149.

Nichols KM, Broman KW, Sundin K, Young JM, Wheeler PA, Thorgaard GH (2007) Quantitative trait loci × maternal cytoplasmic environment interaction for development rate in *Oncorhynchus mykiss*. *Genetics*, **175**:335–347.

Nichols KM, Young WP, Danzmann RG, Robison BD, Rexroad C, Noakes M, Phillips RB, Bentzen P, Spies I, Knudsen K, Allendorf FW, Cunningham BM, Brunelli J, Zhang H, Ristow S, Drew R, Brown KH, Wheeler PA, Thorgaard GH (2003) A consolidated linkage map for rainbow trout (*Oncorhynchus mykiss*). *Anim. Genet.*, **34**:102–115.

Orgogozo V, Broman KW, Stern DL (2006) High-resolution quantitative trait locus mapping reveals sign epistasis controlling ovariole number between two Drosophila species. *Genetics*, **173**:197–205.

Owens SE, Broman KW, Wiltshire T, Elmore JB, Bradley KM, Smith JR, Southard-Smith EM (2005) Genome-wide linkage analysis identifies novel modifier loci of aganglionosis in the $Sox10^{Dom}$ model of Hirschsprung disease. *Hum. Mol. Genet.*, **14**:1549–1558.

Purcell S, Cherny SS, Sham PC (2003) Genetic Power Calculator: design of linkage and association genetic mapping studies of complex traits. *Bioinformatics*, **19**:149–150.

Rabiner LR (1989) A tutorial on hidden Markov models and selected applications in speech recognition. *Proc. IEEE*, **77**:257–286.

Reilly KM, Broman KW, Bronson RT, Tsang S, Loisel DA, Christy ES, Sun Z, Diehl J, Munroe DJ, Tuskan RG (2006) An imprinted locus epistatically influences *Nstr1* and *Nstr2* to control resistance to nerve sheath tumors in a neurofibromatosis type 1 mouse model. *Cancer Res.*, **66**:62–68.

Rice JA (2006) *Mathematical Statistics with Data Analysis*. Duxbury Press, Belmont, CA, 3rd edition.

Schwartz RL, Phoenix T, Foy BD (2008) *Learning Perl*. O'Reilly Media, Sebastopol, CA, 5th edition.

Sen S, Churchill GA (2001) A statistical framework for quantitative trait mapping. *Genetics*, **159**:371–387.

Sen S, Johannes F, Broman KW (2009) Selective genotyping and phenotyping strategies in a complex trait context. *Genetics*, **181**:1613–1626.

Sen S, Satagopan JM, Broman KW, Churchill GA (2007) R/qtldesign: inbred line cross experimental design. *Mamm. Genome*, **18**:87–93.

Sen S, Satagopan JM, Churchill GA (2005) Quantitative trait loci study design from an information perspective. *Genetics*, **170**:447–464.

Sillanpää MJ, Corander J (2002) Model choice in gene mapping: what and why. *Trends Genet.*, **18**:301–307.

Silver L (1995) *Mouse Genetics: Concepts and Applications*. Oxford University Press, Oxford.

Solberg LC, Baum AE, Ahmadiyeh N, Shimomura K, Li R, Turek FW, Churchill GA, Takahashi JS, Redei EE (2004) Sex- and lineage-specific inheritance of depression-like behavior in the rat. *Mamm. Genome*, **15**:648–662.

Soller M, Brody T, Genizi A (1976) On the power of experimental designs for the detection of linkage between marker loci and quantitative loci in crosses between inbred lines. *Theor. Appl. Genet.*, **47**:35–39.

Speed TP (1996) What is a genetic map function? In Speed T, Waterman MS, editors, *Genetic Mapping and DNA Sequencing*, volume 81 of *IMA Volumes in Mathematics and Its Applications*, pages 65–88, Springer, New York.

Strickberger MW (1985) *Genetics*. Macmillan, New York, 3rd edition.

Sugiyama F, Churchill GA, Higgins DC, Johns C, Makaritsis KP, Gavras H, Paigen B (2001) Concordance of murine quantitative trait loci for salt-induced hypertension with rat and human loci. *Genomics*, **71**:70–77.

Symons RC, Daly MJ, Fridlyand J, Speed TP, Cook WD, Gerondakis S, Harris AW, Foote SJ (2002) Multiple genetic loci modify susceptibility to plasmacytoma-related morbidity in Eμ-v-*abl* transgenic mice. *Proc. Natl. Acad. Sci. USA*, **99**:11299–11304.

Venables WN, Ripley BD (2002) *Modern Applied Statistics with S*. Springer, New York, 4th edition.

Visscher PM, Haley CS, Knott SA (1996a) Mapping QTLs for binary traits in backcross and F2 populations. *Genet. Res.*, **68**:55–63.

Visscher PM, Thompson R, Haley CS (1996b) Confidence intervals in QTL mapping by bootstrapping. *Genetics*, **143**:1013–1020.

Wahlsten D, Metten P, Phillips TJ, Boehm SL, Burkhart-Kasch S, Dorow J, Doerksen S, Downing C, Fogarty J, Rodd-Henricks K, Hen R, McKinnon CS, Merrill CM, Nolte C, Schalomon M, Schlumbohm JP, Sibert JR, Wenger CD, Dudek BC, Crabbe JC (2003) Different data from different labs: lessons from studies of gene–environment interaction. *J. Neurobiol.*, **54**:283–311.

Wall L, Christiansen T, Orwant J (2000) *Programming Perl*. O'Reilly Media, Sebastopol, CA, 3rd edition.

Whittaker JC, Thompson R, Visscher PM (1996) On the mapping of QTL by regression of phenotype on marker-type. *Heredity*, **77**:23–32.

Wu R, Ma C, Casella G (2007) *Statistical Genetics of Quantitative Traits: Linkage, Maps and QTL*. Springer, New York.

Xu S (1998) Iteratively reweighted least squares mapping of quantitative trait loci. *Behav. Genet.*, **28**:341–355.

Xu S, Atchley WR (1996) Mapping quantitative trait loci for complex binary diseases using line crosses. *Genetics*, **143**:1417–1424.

Yandell BS, Mehta T, Banerjee S, Shriner D, Venkataraman R, Moon JY, Neely WW, Wu H, von Smith R, Yi N (2007) R/qtlbim: QTL with Bayesian Interval Mapping in experimental crosses. *Bioinformatics*, **23**:641–643.

Yi N (2004) A unified Markov chain Monte Carlo framework for mapping multiple quantitative trait loci. *Genetics*, **167**:967–975.

Yi N, Shriner D (2008) Advances in Bayesian multiple quantitative trait loci mapping in experimental crosses. *Heredity*, **100**:240–252.

Yi N, Shriner D, Banerjee S, Mehta T, Pomp D, Yandell BS (2007) An efficient Bayesian model selection approach for interacting quantitative trait loci models with many effects. *Genetics*, **176**:1865–1877.

Yi N, Yandell BS, Churchill GA, Allison DB, Eisen EJ, Pomp D (2005) Bayesian model selection for genome-wide epistatic QTL analysis. *Genetics*, **170**:1333–1344.

Zeng ZB (1993) Theoretical basis for separation of multiple linked gene effects in mapping quantitative trait loci. *Proc. Natl. Acad. Sci. USA*, **90**:10972–10976.

Zeng ZB (1994) Precision mapping of quantitative trait loci. *Genetics*, **136**:1457–1468.

Zeng ZB, Kao CH, Basten CJ (1999) Estimating the genetic architecture of quantitative traits. *Genet. Res.*, **74**:279–289.

Zhao H, Speed TP (1996) On genetic map functions. *Genetics*, **142**:1369–1377.

Zhao H, Speed TP, McPeek MS (1995) Statistical analysis of crossover interference using the chi-square model. *Genetics*, **139**:1045–1056.

Index

Springer
the language of science

springer.com

Applied Statistical Genetics with R
For Population-based Association Studies

Andrea S. Foulkes

Statistical genetics has become a core course in many graduate programs in public health and medicine. This book presents fundamental concepts and principles in this emerging field at a level that is accessible to students and researchers with a first course in biostatistics. Extensive examples are provided using publicly available data and the open source, statistical computing environment, R.

Content: Genetic association studies.- Elementary statistical principles.- Genetic data concepts and tests.- Multiple comparison procedures.- Methods for unobservable phase.- Classification and regression trees.- Additional topics in high-dimensional data analysis.

2009. Approx. 270 p. Softcover (Use R)
ISBN: 978-0-387-89553-6

Bioconductor Case Studies

Florian Hahne,
Wolfgang Huber, Robert Gentleman, and Seth Falcon

In this volume, the authors present a collection of cases to apply Bioconductor tools in the analysis of microarray gene expression data.

Content: The ALL data set.- R and Bioconductor introduction.- Processing affymetrix expression data.- Two color arrays.- Fold changes, log-ratios, background correction, shrinkage estimation and variance stabilization.- Easy differential expression.- Differential expression.- Annotation and metadata.- Supervised machine learning.- Unsupervised machine learning.- Using graphs for interactome data.- Graph layout.- Gene set enrichment analysis.- Hypergeometric testing used for gene set enrichment analysis.- Solutions to exercises.- References.- Index.

2008. XII, 284 p. Softcover (Use R)
ISBN: 978-0-387-77239-4

R Through Excel

Richard M. Heiberger
Erich Neuwirth

The book is designed as a computational supplement to introductory statistics texts and the authors provide RExcel examples covering the topics of the introductory course.

Content: Getting started.- Using RExcel and R Commander.- Getting data into R.- Normal and t Distributions.- Normal and t workbook.- t-tests.- One-way ANOVA.- Simple linear regression.- What is least squares?- Multiple regression-Two X-variables.- Polynomial regression.- Multiple regression-Three or more X-variables.- Contingency tables and the Chi-Square test.- Appendix: Installation of RExcel.

2009. XXIV, 338 p. Softcover (Use R)
ISBN: 978-1-4419-0051-7

Easy Ways to Order► Call: Toll-Free 1-800-SPRINGER • E-mail: orders-ny@springer.com • Write: Springer, Dept. S8113, PO Box 2485, Secaucus, NJ 07096-2485 • Visit: Your local scientific bookstore or urge your librarian to order.